# Dynamic Point Symbols in Earthquake Mapping

*Dynamic Point Symbols in Earthquake Mapping: Visualization Research in Space and Time* examines how animated symbols enhance digital map comprehension, particularly for earthquake data visualization. It provides much-needed empirical evidence for the understandability, usability, and utility of congruently using dynamic visualization to depict a particular dynamic environmental process. Through rigorous mixed-methods research combining online surveys and eye-tracking experiments, this book compares static symbols against various motion types across different cartographic tasks and provides evidence-based guidelines for implementing motion in maps, balancing innovation with usability. It serves as a practical resource with recommendations for cartographers and GIS (Geographic Information Systems) professionals.

Key Features:

- Combines large-scale data collection with detailed, lab-based visual behavior analysis, offering breadth and depth, and illustrates how people interact with animated map symbols in both natural and controlled conditions.
- Evaluates multiple motion forms like pulsation, vibration, and pictorial motion across tasks, helping readers understand what type of motion to use and when.
- Includes real-world applications in crisis mapping, emergency response, and public communication.
- Offers quantitative comparisons that allow practitioners to optimize both clarity and speed in their map designs.
- Provides recommendations on how to transform complex findings into ready-to-use guidance for cartographers, GIS specialists, and UX designers working with digital geovisualization.

This book is an excellent resource for GIS professionals, researchers, academics, and cartographers interested in web mapping, emergency response mapping, real-time data visualization, and graduate students taking advanced GIS and cartography courses, and those aiming to enhance their existing skillset in geovisualization mapping.

# Dynamic Point Symbols in Earthquake Mapping

## Visualization Research in Space and Time

Paweł Cybulski

CRC Press
Taylor & Francis Group
Boca Raton London New York

CRC Press is an imprint of the
Taylor & Francis Group, an **informa** business

Designed cover image: Getty Images

First edition published 2026
by CRC Press
2385 NW Executive Center Drive, Suite 320, Boca Raton FL 33431

and by CRC Press
4 Park Square, Milton Park, Abingdon, Oxon, OX14 4RN

*CRC Press is an imprint of Taylor & Francis Group, LLC*



Funding provided by Adam Mickiewicz University, Poznań, Poland.



ISBN: 978-1-041-21693-3 (hbk)
ISBN: 978-1-041-21694-0 (pbk)
ISBN: 978-1-003-72853-5 (ebk)

DOI: 10.1201/9781003728535

Typeset in Times
by codeMantra

Access the Support Material: www.Routledge.com/9781041216933

*To the tradition of Polish cartography,*
*whose intellectual heritage has shaped this work.*

*To my academic mentor, Professor Beata Medyńska-Gulij,*
*for her guidance, wisdom, and faith in my ideas.*

*And to my wife,*
*for her love, patience, and constant support.*

# Contents

Preface ... xi
About the Author ... xiii

**Chapter 1** Animated Symbols in Digital Cartography ... 1

1.1 Why Movement Matters in Modern Maps ... 1
1.2 Evaluating What Works—And Why ... 6
1.3 The Study at a Glance: Aims, Design, and Key Questions ... 8

**Chapter 2** How We See and Study Maps: From Earthquakes to Visual Perception ... 12

2.1 The Nature of Earthquakes: Mapping a Moving Earth ... 12
2.2 How Maps Tell Earthquake Stories: Visual Variables and Communication ... 15
2.3 When Maps Begin to Move: Dynamic Symbols and Animated Cartography ... 23
2.3.1 Interaction as Exploration: Making Animated Maps Responsive ... 26
2.3.2 From Vision to Design: Understanding Motion in Cartographic Animation ... 29
2.3.3 Designing Motion: Types of Dynamic Point Symbols ... 32
2.4 Seeing Through the Eyes of the User: Eye Tracking in Map Evaluation ... 36
2.4.1 What Eye Movements Reveal: Fixations, Saccades, and Scan Paths ... 39
2.4.2 From Gaze to Insight: Eye Tracking in Cartographic Research ... 41
2.5 What We Already Know and What We Still Don't: Research on Dynamic Symbols ... 44
2.5.1 Where the Gaps Remain: The Need for Research ... 51

**Chapter 3** How the Study Was Done: Combining Online Surveys and Eye-Tracking ... 53

3.1 Planning the Study ... 54
3.1.1 How Earthquake Features Were Visualized on Maps ... 56
3.1.2 Where the Earthquake Data Came From ... 60

3.1.3 Key Technical and Design Decisions ... 62
3.1.4 Designing and Analyzing the Map Tasks ... 64
3.2 Creating the Digital Maps ... 66
3.2.1 Preparing the Map Background and Earthquake Data ... 67
3.2.2 Setting the Time Frame for the Earthquake Data ... 69
3.2.3 Choosing and Designing the Static Map Symbols ... 71
3.2.4 Adding Motion: Applying Animated Effects ... 72
3.2.5 Final Map Layout and Interface Design ... 76
3.3 Presenting the Maps ... 79
3.3.1 Map Variation "A"—Vibration for Representing Earthquake Magnitude ... 80
3.3.2 Map Variation "B"—Pictorial Motion for Representing Earthquake Magnitude ... 80
3.3.3 Map Variation "C"—Pulsation for Earthquake Magnitude ... 81
3.3.4 Map Variation "D"—Static Pictorial Symbols for Earthquake Magnitude ... 81
3.4 Testing and Evaluating the Maps ... 81
3.4.1 How the Online Survey Was Conducted ... 82
3.4.2 How the Eye-Tracking Study Was Conducted ... 86
3.4.3 Participants in the Online Study ... 87
3.4.4 Participants in the Eye-Tracking Study ... 91

**Chapter 4** Understanding How Well the Maps Worked: Results from the Online Study ... 95

4.1 How Accurate Were the Participants ... 95
4.1.1 Finding the Strongest Earthquakes (Using Motion Cues) ... 95
4.1.2 Finding the Most Fatal Earthquakes (Using Symbol Size) ... 97
4.1.3 Identifying the Most Common Earthquake Depth (Using Color) ... 99
4.1.4 Finding the Year with the Most Earthquakes (Timeline Task) ... 101
4.2 How Fast Did Participants Complete the Tasks ... 102
4.2.1 Speed of Finding the Strongest Earthquakes ... 103
4.2.2 Speed of Identifying the Highest Fatalities ... 105
4.2.3 Speed of Determining the Most Frequent Depth ... 106
4.2.4 Speed of Finding the Year with the Most Earthquakes ... 109
4.3 How Participant Characteristics Shaped the Results ... 111
4.3.1 Comparing Responses by Gender ... 111
4.3.2 Does Education Level Matter? ... 119

4.3.3 How Self-Assessed Map Skills Influenced Performance ... 128
4.3.4 The Role of Everyday Map-Use Habits ... 141
4.3.5 How Confident Participants Felt About Their Answers ... 154
4.3.6 What Participants Thought About the Map Designs ... 167

**Chapter 5** Performance and Visual Attention on Animated Maps: Eye-Tracking Results ... 169

5.1 How Accurate Participants Performed ... 169
5.1.1 Finding the Strongest Earthquakes (Motion-Based Search) ... 169
5.1.2 Finding the Deadliest Earthquakes (Size-Based Search) ... 171
5.1.3 Identifying the Most Common Earthquake Depth (Color-Based Search) ... 173
5.1.4 Identifying the Year with the Most Earthquakes (Temporal-Based Search) ... 174
5.2 How Fast Participants Completed Tasks ... 176
5.2.1 Time to Find the Strongest Earthquakes ... 176
5.2.2 Time to Find the Deadliest Earthquakes ... 178
5.2.3 Time to Identify Earthquake Depth ... 179
5.2.4 Time to Identify the Most Active Year ... 180
5.3 How Participant Characteristics Shaped the Results in Eye-Tracking Experiment ... 182
5.3.1 Differences Between Men and Women ... 182
5.3.2 How Confident Participants Felt About Their Answers ... 189
5.3.3 How Participants Rated the Map Styles ... 198
5.4 What Eye Movements Reveal ... 200
5.4.1 How Many Fixations Participants Made ... 201
5.4.2 How Long Fixations Lasted ... 204
5.4.3 How Far the Gaze Moved (Saccade Distance) ... 206
5.4.4 How Fast the Gaze Moved (Saccade Velocity) ... 207
5.4.5 How Much Time Participants Spent Looking at the Legend and Temporal Scale ... 209

**Chapter 6** What We Learned about Animated Symbols: Bringing the Two Studies Together ... 212

6.1 What the Two Methods Taught Us: Online Survey vs. Eye-Tracking ... 212
6.2 How Well Participants Performed: Comparing Accuracy Across Both Studies ... 213

6.2.1 How Well People Found the Strongest Earthquakes ..... 214
6.2.2 How Well People Found the Deadliest Earthquakes ..... 217
6.2.3 How Well People Identified the Most Common Earthquake Depth ..... 220
6.2.4 How Well People Interpreted the Earthquake Timeline ..... 223
6.3 How Fast Participants Worked: Comparing Response Times Across Both Studies ..... 227
6.3.1 How Quickly People Found the Strongest Earthquakes ..... 228
6.3.2 How Quickly People Found the Deadliest Earthquakes ..... 231
6.3.3 How Quickly People Identified Earthquake Depth ..... 233
6.3.4 How Quickly People Interpreted the Earthquake Timeline ..... 236

**Chapter 7** Practical Lessons for Using Animated Symbols on Maps ..... 239

7.1 Accuracy, Speed, and Eye Movements: The Big Picture ..... 239
7.2 Tips for Designing Effective Animated Map Symbols ..... 242
7.2.1 Design Motion That Matches the Meaning of the Data ..... 243
7.2.2 Choose the Right Type of Motion for the Task ..... 243
7.2.3 Motion Performance Depends on Viewing Conditions ..... 244
7.2.4 Motion-Based Symbols Aid Users with Different Backgrounds ..... 244
7.3 What This Study Didn't Cover—and What Should Come Next ..... 245
7.3.1 Sample Size and Participant Diversity ..... 245
7.3.2 Task Type and Domain Specificity ..... 246
7.3.3 Symbol Density and Scene Complexity ..... 247
7.3.4 Motion Type Scope ..... 247
7.3.5 Device and Environmental Variability ..... 248
7.3.6 Integration of Real-Time Eye-Tracking Feedback ..... 248

**Appendix 1** ..... 251

**References** ..... 255

**Index** ..... 275

# Preface

For a long time, cartographers and researchers from related fields have been seeking new methods for presenting spatial data—especially those with a temporal dimension. At the turn of the 20th and 21st centuries, the principles of cartographic design became more clearly defined, and a wide range of mapping methods was established, differentiated by their application to point, linear, or areal phenomena. Yet, among these methods, the potential of animation in cartographic design has remained largely underexplored.

Most existing applications of animation in maps focus on presenting chronological sequences—consecutive days, months, or years—to visualize how phenomena evolve over time and space. However, I began to wonder: what if certain map symbols could be animated independently of the passage of time? Similar techniques have long been used in cinematography, scientific visualization, and computer games, where motion, rotation, vibration, and pulsation are employed to attract attention or convey information.

This idea inspired my first empirical research in 2014, in which I investigated whether the movement of point symbols could effectively communicate information on maps. The results showed that modulating the speed of rotation, for instance, could successfully differentiate symbols representing phosphorus accumulation in benchmark lakes and improve map-reading performance. Surprisingly, although such dynamic methods can represent quantitative data, they have not yet been formally recognized within standard cartographic methodology and remain largely conceptual.

The growing integration of animation into mainstream GIS software has since confirmed the timeliness of this research. Around 2018–2019, ESRI introduced animated symbols in the form of GIFs (Graphics Interchange Format), and similar solutions soon appeared in QGIS (Quantum Geographic Information System). Yet, these tools are still typically used to display *qualitative* rather than *quantitative* differences. For instance, while one can technically prepare sets of animated symbols that vary in speed to represent numerical data, this is not yet a standard design procedure—unlike choosing a color scheme or symbol size.

My subsequent studies focused on defining, classifying, and evaluating several types of dynamic changes, such as flickering, pulsation, and rotation. Together with other researchers, we examined the effectiveness and efficiency of these techniques in cartographic communication, exploring how dynamic point symbols influence visual perception and how the parameters of movement and speed differences can be optimized for clarity and legibility.

This book is the culmination of that research. It presents both theoretical foundations and empirical findings on the use of dynamic point symbols for quantitative visualization, with earthquakes serving as the primary case study. It also seeks to demonstrate that dynamic symbols are not merely a visual novelty but a valuable and, in some contexts, superior alternative to static cartographic representations.

Although the structure of this book is presented in the table of contents, it may also be viewed in a more conceptual way. The first part provides an introduction that defines the main research goal, specific methodological objectives, and the theoretical foundations situating this work at the intersection of geography, cartography, psychology, and visualization. The next part introduces earthquakes—the main phenomenon under study, which serves as a proxy for other point-based phenomena characterized by multiple measurable attributes. This section also analyzes how earthquake maps are currently designed, particularly those published online, and examines the role and place of animation techniques in cartography, highlighting previous studies that have influenced the development of animated visualization methods.

A substantial portion of this book is devoted to the original research procedure. It begins with the conceptual framework and design assumptions, continues through data collection and visualization, and concludes with an evaluation of the cartographic products. The results are then discussed in relation to existing findings in cartography, data visualization, and perceptual psychology. Before concluding, this book synthesizes the most important results, allowing readers to grasp the key insights without referring to individual figures or tables.

The final chapter offers reflections and conclusions on the study as a whole, acknowledging its methodological limitations—common to all empirical and experimental research. It also provides practical guidelines for cartographers who wish to apply dynamic point symbols in their work, as well as suggestions for further scientific exploration of animated mapping.

Ultimately, I hope this book will serve as both a scientific synthesis and a practical guide for researchers, educators, and GIS users interested in the perceptual and cognitive dimensions of map animation.

# About the Author

**Paweł Cybulski, PhD,** is an Assistant Professor in the Department of Cartography and Geomatics, Adam Mickiewicz University, Poznań, Poland. A geographer by education and a cartographer by specialization, he explores how spatial and temporal phenomena can be effectively visualized and understood through maps. His research bridges geography, cognitive science, and visual analytics to explore how users perceive, process, and understand spatial information. He has published widely in leading journals, including *Cartography and Geographic Information Science*, *The Cartographic Journal*, and the *ISPRS International Journal of Geo-Information*. His work explores topics such as dynamic point symbols, animated mapping, preattentive visual processing, and the use of eye-tracking to evaluate map design.

He served as Principal Investigator of the National Science Centre (NCN) project "Preattentive Attributes of Dynamic Point Symbols in Quantitative Mapping" (2021–2024), as Co-Investigator in the German-Polish Foundation for Science (DPWS) project "Development of the Cultural Landscape of German and Polish Industrial Centres" (2016–2018, with Ruhr University Bochum), and as Researcher in the NCN project "Visualisation of the Topographical Space in Europe on Manuscript Multi-Sheet Maps from the 18th Century" (2013–2016).

In his teaching, he leads courses on Animated and Multimedia Cartography, Survey Engineering, Cartographic Generalization, and Experimental Validation of Cartographic Products, integrating perceptual and experimental approaches into geographical education. His teaching reflects a commitment to combining scientific rigor with innovation in map design and spatial communication.

# 1 Animated Symbols in Digital Cartography

## 1.1 WHY MOVEMENT MATTERS IN MODERN MAPS

Earthquakes are among the most dramatic expressions of our planet's dynamic nature. They remind us that the ground beneath our feet is never completely still—its constant, unseen motion can suddenly transform into destructive force. These events can reshape landscapes, destroy infrastructure, and deeply affect human lives. Although we often associate major earthquakes with places like Japan or the Pacific Rim, Europe also experiences frequent seismic activity. Regions such as the Italian Peninsula, the Balkans, the Caucasus, and Iceland regularly record tremors of varying intensity (Stucchi et al., 2013).

To understand and communicate such complex natural processes, geographers turn to maps—not only as static depictions of location but as tools for visualizing change over time. In this chapter, the focus is on how movement and animation can make such dynamic processes more visible and understandable. The discussion begins with the role of animated symbols in representing change on maps, particularly how they help translate temporal or rapidly evolving data—like earthquake sequences—into visual form. The chapter also introduces the broader research context of dynamic cartography, outlining why motion is more than just an aesthetic addition; it is a way of extending cartographic language to capture processes, not just patterns. To illustrate these ideas, consider a sequence of earthquakes unfolding along a fault line. A traditional static map can show where the tremors occurred, but not how they propagated over time. An animated map, by contrast, allows the viewer to follow the temporal rhythm of the events, intuitively grasping their pace, clustering, and intensity. This simple example reflects the core motivation of this book: to explore how motion—when used thoughtfully—can enrich our understanding of spatial phenomena and support more effective geographic communication.

For most people, earthquakes are difficult to imagine until they are experienced firsthand. The sudden shaking of the ground, the rumbling sound, and the sense of instability are all symptoms of deep geological forces at work. In scientific terms, earthquakes occur when stress accumulated along fault lines is released, sending seismic waves through the Earth's crust. These waves can rattle buildings, reshape landscapes, and sometimes trigger secondary disasters such as landslides or tsunamis. To make sense of these invisible processes, scientists and cartographers rely on seismic maps—visual tools that translate complex data into patterns we can see. Such maps show where earthquake epicenters are located, how strong they are, and how many people were affected. In other words, they turn abstract numbers into a spatial story. Through a well-designed map, even someone with no background

DOI: 10.1201/9781003728535-1

in seismology can quickly grasp where earthquakes happen most often, how their strength varies, and what regions are most at risk (Navakanesh et al., 2019).

When the ground begins to tremble, seconds can make the difference between safety and disaster. That's why understanding earthquakes isn't just the task of seismologists—it's something that concerns everyone who lives in a world that never fully stands still. Monitoring seismic activity is only part of the challenge; the real difficulty lies in communicating it clearly to the people who need this knowledge most—citizens, planners, and emergency responders. Here is where cartography becomes more than science—it becomes a language. Traditional printed maps can show where an earthquake occurred, but they freeze the story in a single moment. By contrast, animated and interactive maps have the power to bring that story to life. They can reveal the rhythm of seismic events, showing how one tremor triggers another, or how intensity shifts across regions. These moving symbols do not simply decorate the data; they make invisible processes visible, helping us understand the *when* and *how*, not just the *where* (Karl, 1992). To illustrate this, imagine watching an animation of earthquakes unfolding over a year: at first, a few gentle flickers, then clusters of activity appearing in southern Europe, rippling toward the east. Suddenly, the viewer grasps the scale, the timing, and the pattern—something no static image could convey as clearly. This is the promise of dynamic mapping: it transforms data into experience, bridging the gap between scientific knowledge and public understanding, and in doing so, strengthens both awareness and preparedness.

Seismic activity presents its own storytelling challenge. Unlike many geographic phenomena, earthquakes unfold in both space and time, and their fleeting, sudden nature makes them difficult to convey on a static map. To make sense of these processes, cartographic methods must highlight not only where earthquakes occur but also how they evolve. As the volume of geological data has grown, so too has the need for more expressive visualization techniques—methods capable of capturing motion, rhythm, and intensity in ways that numbers alone cannot. One response to this challenge has been the development of dynamic point symbols, a technique that uses changes in size, color, shape, and movement to represent shifting geographic data (Tinghua, 1998; Lai & Yeh, 2004). These animated symbols enable cartographers to present multiple dimensions of earthquake activity—such as magnitude, depth, and number of people affected—within a single visual framework. But the idea of animation in cartography is far from new. The first experiments appeared long before the digital era. As early as 1918, Martin E. Johnson used film to animate maps in his travel documentaries, like *Among the Cannibal Isles of the South Pacific*, bringing geography to life on screen (Caquard, 2009). Even in cinema, animation served geopolitical storytelling—Walt Disney's 1939/1940 film sequence depicting the invasion of Poland used animated arrows to dramatize military movement (Peterson, 1999; Fidotta, 2014). Norman J. W. Thrower's Animated Cartography (1959) offered one of the first theoretical explorations of how movement could enhance map communication, followed by Tobler's (1970) pioneering "computer movie" simulating urban growth in the Detroit region, and Moellering's (1973, 1976) studies that formally defined animation as an extension of cartographic language.

Today's work on dynamic point symbols continues this creative lineage, transforming early film experiments into systematic, computer-based techniques for visualizing spatio-temporal data. Yet these innovations bring new questions. Because motion naturally attracts human attention, animated symbols can be both powerful and distracting (Wolfe & Horowitz, 2004). The challenge is to use movement not as decoration, but as meaning—to make motion a carrier of information rather than a source of confusion. To explore this balance, the visualization of earthquakes serves in this chapter as a practical and symbolic case: a phenomenon that is both inherently dynamic and rich in measurable attributes, ideal for testing how animated cartographic symbols shape perception, understanding, and decision-making.

Understanding how animated symbols shape our perception of earthquakes requires more than theory—it calls for careful observation and testing. To explore this, one must first look closely at the maps that already exist. A good starting point is the wide range of earthquake maps published on the web. These digital maps reach millions of users, from school students to emergency managers, and play a quiet but powerful role in shaping how society perceives seismic risk. Examining what kind of information they present—and how it is shown—reveals not only current trends in cartographic design but also the underlying assumptions about what people need to know. Once this landscape is understood, new solutions in dynamic and interactive cartography can be proposed, building on both technical innovations and human-centered design. But ideas on paper are not enough; they need to be tested in practice. This is where user studies come in. Evaluating how people interpret and interact with maps provides valuable insight into what truly works. Such assessments should combine objective measures—how accurately and efficiently users complete tasks—with subjective ones, such as their preferences, confidence, or comfort in using different types of maps. After all, as noted by Slocum et al. (2001) and Vondrakova and Vozenilek (2016), the appeal of a map often determines whether it will actually be used. In this sense, studying user perception is not the final step but an integral part of designing maps that inform, engage, and empower.

When cartographers evaluate a map, they have several paths to choose from. Some focus on what people say they like—the subjective approach—while others measure what people do when they use the map, taking an objective or quantitative perspective. There are also qualitative approaches that explore how people describe their experience in their own words. In practice, most modern studies combine these methods, comparing users' preferences with their actual performance. This mixed approach helps reveal whether a map that looks appealing also helps users find information quickly and accurately (Wielebski & Medyńska-Gulij, 2019). Another valuable perspective is the comparative approach, where different maps or visual techniques are tested side by side to see which design communicates best (Bláha, 2009). In recent years, these comparisons have been enhanced by a powerful tool: eye-tracking. By following where users look and for how long, eye-tracking studies reveal which areas of a map capture attention, which are overlooked, and where confusion arises (Krassanakis & Cybulski, 2021). Together, these methods give researchers a fuller picture—not only of how maps perform, but of *how people think and see* when they interact with them.

This kind of comprehensive evaluation is especially useful when studying dynamic and interactive maps, where multiple visual variables—motion, color, size—compete for attention. Earthquakes provide an ideal testing ground for such analysis. Each epicenter can encode several pieces of information at once: its magnitude, depth, number of fatalities, and even the economic impact. These data-rich points make earthquakes a perfect proxy for testing symbol design in other complex phenomena—whether volcanic eruptions, floods, migration patterns, or air quality. In all of these cases, different attributes coexist in both space and time, requiring visualization techniques that balance clarity with complexity. By framing earthquakes in this broader way, the discussion moves beyond a single case study. The real question becomes not only *whether* dynamic map symbols work, but *why* they work. Insights from cognitive psychology—such as visual attention, preattentive processing, and cognitive load—help explain why certain animations guide the eye more effectively than others. In this sense, earthquakes are more than a dataset; they are a lens through which we can understand how people perceive movement, change, and spatial relationships on maps. The lessons learned here extend far beyond seismology. They show how motion, color, and size can be systematically combined to enhance spatio-temporal understanding across many branches of geographic visualization. Ultimately, this connection between perception and cartography makes animated maps not just more dynamic, but more meaningful.

The growing presence of animation in cartography is not limited to research—it is increasingly becoming part of everyday mapping tools. One of the most significant shifts has occurred within Geographic Information Systems (GIS), where dynamic symbols have moved from experimental visualization techniques into mainstream practice. In 2017, the GIS company ESRI (Environmental Systems Research Institute) introduced a new feature in ArcGIS Pro that allowed users to incorporate animated GIFs directly into point symbols. In their announcement, titled "Bring Your Points to Life with GIF Symbols," ESRI showcased how simple animations—like flickering lights or pulsing circles—could draw attention to specific map features and communicate change over time. Open-source software has followed a similar path. The developers of QGIS (Quantum Geographic Information System), for instance, introduced animated symbol capabilities through QField 2.2, making it possible to design dynamic map points that react or pulse on mobile devices (https://www.opengis.ch/2022/08/16/a-new-trick-up-qfields-sleeve-animated-maps/; accessed on 21.11.2023). These advancements represent an important moment in cartography: dynamic symbols are no longer confined to specialized animation or design software. They are becoming part of the standard GIS toolkit, accessible to professionals, educators, and even enthusiasts.

What this means in practice is that creating animated maps no longer requires advanced programming or graphics expertise. Instead, cartographers can use automated workflows within familiar GIS environments to integrate motion directly into their map symbols. This shift not only simplifies the process but also broadens the potential audience for dynamic mapping—making it easier for municipalities, research institutions, or NGOs to communicate changing spatial phenomena quickly and effectively. As animation becomes an everyday feature of GIS, the need for empirical assessment—to understand how users perceive and interpret these moving symbols—becomes all the more essential.

Not everyone agrees that animation always improves maps. Some early researchers, such as Dorling (1992) and Monmonier and Gluck (1994), warned that dynamic symbols can sometimes complicate rather than clarify information. Movement, they argued, adds an extra layer of cognitive work: users must track changes over time while still remembering where things were before. Later, Koussoulakou and Kraak (1992) offered a more nuanced perspective, suggesting that animation may not necessarily make map reading more accurate, but it can make it faster—helping users grasp patterns or messages in less time. Building on this idea, Harrower (2004, 2007) emphasized that animated maps are inherently more demanding on our working memory because they require us to recall spatial configurations and notice subtle shifts in position, color, or size as they unfold on screen. Yet, despite these challenges, animation has found a natural home in the digital age. Online and interactive platforms have made movement an expected part of visual communication, from weather forecasts to data dashboards. When designed carefully, motion becomes not a distraction but a guide—drawing attention to what matters most. In the case of earthquakes, dynamic symbols do precisely that: they capture the user's gaze, signal the unfolding of events, and communicate the rhythm of seismic activity in a way that static maps simply cannot. In this sense, using motion is not just a stylistic choice; it's a meaningful way to help viewers see—and feel—the temporal pulse of the Earth itself.

The motivation behind this research comes from a simple yet pressing question: how can we use maps not just to show data, but to make sense of it? In an age where vast amounts of geographic information are available online, the challenge is no longer collecting data—it's communicating it clearly and meaningfully. This chapter focuses on evaluating specific cartographic methods to determine where and how they can be most effective in expressing change over time and space. The goal is not purely academic; it is also social and practical. By improving how seismic data are visualized, maps can help raise public awareness of the risks posed by natural disasters and support more informed decision-making by authorities and citizens alike. Human cognition is naturally tuned to recognize patterns in what we see. Our brains are remarkably good at detecting movement, noticing differences, and connecting visual cues to meaning. This is why visualizations—when done well—can transform raw data into understanding. As Edward Tufte, one of the pioneers of information design, once observed: "Graphics, at their best, serve as instruments for reasoning about quantitative information. Often, the most effective way to describe, explore, and summarize a dataset—no matter how extensive—is to visualize those numbers" (Tufte, 1983). His words still resonate today, especially in cartography, where visual methods can turn abstract numbers about earthquakes—magnitude, depth, and geographical coordinates—into intuitive and powerful insights. In this sense, effective map design is not just a technical exercise; it is a way of transforming data into shared understanding, helping people grasp the dynamic nature of the world they inhabit.

Studying how earthquakes are presented on maps requires both a systematic and creative approach. One of the most useful techniques is comparative mapping, where different cartographic representations of the same phenomenon are analyzed side by side. In this case, a wide range of earthquake maps—many of them found on the web—serve as the basis for comparison. By examining these existing examples, it becomes possible to identify which kinds of data are most often shown, how they are visually encoded, and which visual variables (such as size, color, or shape) are used most effectively. Once these visualizations are analyzed, researchers can select the most relevant earthquake attributes—such as magnitude, depth, or number of fatalities—and determine the most appropriate methods for displaying them. Comparative studies also allow for the evaluation of different map designs and user interface elements (Dukaczewski & Bielecka, 2009; Horbiński & Cybulski, 2018). Through this process, practical solutions emerge: which types of symbols communicate best, which are most common, which layouts reduce confusion, and which interactions genuinely enhance understanding (Cybulski, 2016; Dawidowicz & Sońta, 2019). But comparing earthquake maps is not just about aesthetics or technology; it is about understanding how different cartographic choices affect comprehension. As Baranowski et al. (2009) and Beaumont et al. (2005) showed, the way data are presented directly influences what users take away from a visualization. For example, some maps rely on proportional symbols to convey magnitude, while others use choropleth shading or colored point symbols to emphasize depth or frequency (Korycka-Skorupa, 2000; Hennig, 2014; Pődör & Kiszely, 2014). Each technique highlights a different aspect of the phenomenon and can lead viewers to focus on distinct spatial or temporal patterns. In the digital era, another crucial dimension must be considered: interactivity. Most contemporary earthquake maps are published online, inviting users to explore data by zooming, panning, or filtering. These interactive tools, as noted by Roth (2013a, 2013b), significantly shape how people retrieve information and engage with spatial data. When designed well, they encourage exploration rather than passive observation—turning map reading into an act of discovery. For this reason, evaluating not only the visual design but also the interactive experience of earthquake maps is an essential step in understanding how modern cartography communicates complex geographic phenomena.

## 1.2 EVALUATING WHAT WORKS—AND WHY

As digital cartography has matured, its focus has evolved from simply making maps look appealing to understanding why certain designs work better than others. Today's cartographers are less concerned with aesthetic appeal alone (which is still important) and more interested in the mechanics of map effectiveness—that is, how well a visual design actually communicates information. This shift has given rise to a rigorous methodology for testing and evaluating the quality of cartographic products. In empirical cartography, one of the key measures is the effectiveness of cartographic communication—a concept that captures how successfully a map conveys its intended message (Elzakker & Wealands, 2007). In everyday language, as defined by the *Cambridge Academic Content Dictionary* (2016), effectiveness is "the ability to be successful and produce the intended results." Within cartographic research, this has long been expressed in terms such as correctness or success rate (Dong et al., 2012; Flannery, 1971). In other words, if users can correctly interpret what a map is trying to show, the design can be said to be effective. Closely related to effectiveness is the concept of efficiency—how quickly users can retrieve information from a map. As Garlandini and Fabrikant (2009) noted, the less time it takes a viewer to find a location, recognize a pattern, or answer a spatial question, the more efficient the map is. This dual focus on accuracy and speed captures the essence of how people interact with visual information: some maps make us pause and think, while others help us see instantly. In this chapter, both dimensions are assessed. Effectiveness is measured by the number of correct answers users provide to map-based questions, while efficiency is measured by their response time—how long it takes to reach those answers. Together, these metrics offer a balanced view of map performance, reflecting not only how well a design communicates, but also how effortlessly it does so.

To measure how effectively and efficiently users interpret maps, researchers rely on structured tasks that mirror real-world interactions. These tasks come from established taxonomies of visualization operations—frameworks that describe the kinds of questions people typically ask when working with visual data (Ogao & Kraak, 2002). For example, participants might be asked to locate a specific symbol on a map, compare two regions, or identify where a phenomenon reaches its highest value. Each of these exercises reflects a different way of reading spatial information and helps reveal how well the map supports perception and reasoning. In the language of cartographic research, such exercises are often categorized as either elementary or general map-based tasks. Elementary tasks involve straightforward identification—spotting one item or value—while general tasks require broader comparison or synthesis across the map (Andrienko et al., 2003). Both types are crucial for evaluating dynamic visualizations because they connect the abstract dimensions of space, time, and attributes—the three pillars of spatio-temporal representation—to the concrete actions users perform while interpreting a map. In this way, task-based evaluation becomes not just a measure of performance, but a window into how people think and see in spatial contexts.

Effectiveness and efficiency do not stand alone—they are part of a larger framework known as usability. In simple terms, usability describes how well a product helps people achieve their goals comfortably, accurately, and with satisfaction. According to the ISO 9241-11: *Guidance on Usability*, usability is defined as "the extent to which a product can meet specific user needs in an effective, efficient, and satisfactory manner." In the context of cartography, this means more than just whether a user can read a map correctly—it also concerns how intuitive, enjoyable, and engaging the experience is. Researchers have since broadened this definition to include qualities that go beyond measurable performance. As Cawthon and Moere (2007) and Štěrba et al. (2014, 2015) suggest, usability also involves how easily a system can be learned, how well it is remembered over time, and even its aesthetic appeal—the sense of clarity, harmony, or visual pleasure that makes a map inviting to explore. After all, a map that is technically accurate but visually confusing is not truly usable. Because of this broader understanding, this chapter complements objective performance measures with subjective user assessments. These evaluations capture the human side of map interaction—how users feel about the experience, which designs they prefer, and what makes a visualization intuitive or frustrating. In doing so, the research embraces usability as both a scientific and human-centered measure of success.

To understand how people interpret earthquake maps—how effectively and efficiently they use them—it is essential to hear directly from the users themselves. For this purpose, the survey method remains one of the most valuable tools in cartographic research. Surveys allow researchers to gather insights from real users, capturing not only what they do, but how they think and feel about a map's design (Cinnamon et al., 2009; Cybulski & Medyńska-Gulij, 2018; Faliszewska, 2012). In this type of study, the number of participants plays a crucial role. The more people take part, the more confidently one can generalize the findings. However, achieving large sample sizes can be challenging in controlled laboratory conditions, where testing is time-consuming and limited to small groups. To overcome this, many modern cartographic studies have turned to online surveys, which make it possible to reach a broader and more diverse audience (Brychtová & Çöltekin, 2017; Cybulski & Medyńska, 2018; Horbiński & Zagata, 2022; Schiewe, 2019). Online environments also reflect the real-world context in which most users now encounter maps—on screens, through interactive interfaces, and often across different devices. In this sense, online surveys do more than just provide a convenient data collection method; they situate the research within the same digital ecosystem where today's maps are experienced. By aligning the research setting with actual user environments, the study enhances both the empirical validity of its findings and their applicability to modern mapping practice.

While online surveys are excellent for reaching large and diverse groups of users, they also come with certain limitations. Researchers have little control over the participants' environment—lighting, screen quality, distractions—and it is difficult to verify who is actually taking part or whether their responses are entirely genuine. For this reason, online experiments are best complemented by laboratory studies, where conditions can be carefully controlled and participants' behavior observed in more detail. Laboratory testing offers a key advantage: it allows for the use of eye-tracking technology, which provides direct insight into how users visually engage with a map (Dong et al., 2014a; Gołębiowska et al., 2017). Eye-tracking works by recording two types of gaze behavior: fixations, moments when the eyes rest on a specific area to process information, and saccades, the rapid jumps between those points of focus (Cassin & Solomon, 1990). Together, these movements reveal how users explore visual information—what attracts their attention first, where they hesitate, and which parts of the map are ignored. In this study, eye-tracking sessions were conducted in controlled laboratory conditions with a smaller number of participants, complementing the broader results of the online survey. This approach makes it possible not only to verify survey findings but also to interpret why users performed in certain ways. By analyzing fixation- and saccade-based metrics—such as time to first fixation, fixation count, or scan path speed (Kiefer et al., 2017)—the research connects *what* users saw with *how* they saw it. In doing so, eye-tracking bridges the gap between statistical performance and human perception, offering a deeper understanding of how animated cartographic symbols guide attention and shape comprehension.

## 1.3 THE STUDY AT A GLANCE: AIMS, DESIGN, AND KEY QUESTIONS

This chapter aims to explore how movement influences the way people read and understand maps. More specifically, it investigates the empirical performance of dynamic point symbols—animated visual markers used to represent changing geographic phenomena—within the context of earthquake visualization in Europe.

The goal is to determine how effectively these symbols communicate spatio-temporal information and how efficiently users can interpret them under real conditions. To achieve this, the study combines objective measures of effectiveness (accuracy of user responses) and efficiency (response time) with detailed eye-tracking analysis, which provides insight into users' visual behavior during map interaction.

A second, equally important objective is to compare dynamic point symbols with more traditional, nondynamic alternatives such as pictorial symbols. By placing these two approaches side by side, the study seeks to understand not only which design performs better, but why. Do animated symbols truly enhance comprehension, or do they merely capture attention? How do they affect object identification, object comparison, and the overall usability of the map? Addressing these questions helps clarify when motion adds genuine value to cartographic communication and when simpler designs may suffice.

Ultimately, the objectives of the study reach beyond earthquakes alone. By using seismic activity as a representative case, the research aims to uncover broader principles for visualizing dynamic geographic data—principles that can inform the design of maps depicting other point-based phenomena, such as volcanic eruptions, weather patterns, or urban events.

The empirical assessment designed for this study is built around several complementary components, each addressing a different aspect of how users interact with dynamic maps. The first and most fundamental element is the evaluation of cartographic effectiveness—how accurately users can extract information from the map. This measure reflects whether the intended message of the visualization is being understood as the cartographer planned. The second component looks at efficiency, which focuses on the speed of information retrieval. Here, both accuracy and response time matter: quick answers are valuable only when they are also correct. By analyzing these two dimensions together—effectiveness and efficiency—it becomes possible to form a more complete picture of how well a map supports perception, decision-making, and comprehension. The third and perhaps most revealing element is the analysis of eye movements. Using eye-tracking data, the study explores how users arrive at their answers, not just whether they are right or wrong. Eye movement metrics such as fixation duration, scan paths, and time to first fixation offer a window into visual strategy—showing where users' attention is drawn, how they navigate the map, and where confusion or hesitation occurs. These insights help explain why certain users or groups respond more accurately or more quickly and identify potential visual or cognitive obstacles presented by animated cartographic designs. Together, these three components—accuracy, efficiency, and visual behavior—form a comprehensive framework for understanding the effectiveness of dynamic symbols, blending measurable outcomes with deeper insights into the cognitive processes that drive them.

Taken together, these measures—accuracy, speed, and visual behavior—offer an objective picture of how well dynamic point symbols perform in conveying spatio-temporal earthquake data. Yet numbers alone cannot capture the full story. To complement these performance metrics, the study also includes subjective user feedback, focusing on participants' impressions, preferences, and perceptions of visual appeal. This combination of quantitative and qualitative perspectives ensures that the evaluation reflects not only how users perform, but also how they experience animated cartographic design.

Beyond its main objective, this study also pursues a series of more specific goals that together form a complete research pathway—from conceptual exploration to empirical testing:

- *Conducting a comparative analysis of earthquake maps available on the Internet* (this initial review provides a foundation for understanding how seismic data are currently represented and what visual strategies are most commonly used in public communication);
- Designing and developing a set of dynamic point symbols specifically tailored for earthquake visualization;
- Creating interactive earthquake maps to visualize specific time periods (illustrating seismic activity);
- Designing and conducting an online survey involving map-based tasks (invites a broad group of participants to perform map-based tasks, providing data on how effectively and efficiently they interpret dynamic representations);
- Conducting a laboratory experiment to record eye movements during map-based tasks (complement to large-scale assessment, offering deeper insight into participants' visual strategies and attention patterns);
- Performing a statistical analysis of both the online survey and eye movement results.

The issues discussed so far lead to a series of guiding questions that shape the direction of this study. Each one addresses a different dimension of how people perceive, interpret, and interact with dynamic cartographic symbols in the visualization of earthquakes.

The first and most general question asks:

- Can dynamic point symbols improve the effectiveness and efficiency of spatio-temporal earthquake maps across different visual variables—such as motion, size, color, and time?

This question gets to the heart of the study, testing whether animation truly enhances understanding or simply adds visual complexity.

The next question focuses more closely on the type of motion itself:

- How do different motion styles—such as pulsation, vibration, or pictorial movement—affect users' visual search performance depending on the nature of the task and the meaning of the data?

Because maps are used in many contexts, the research also considers the role of the user's environment:

- Does performance change when users view maps in a controlled laboratory setting compared to an online, real-world environment?

This question helps determine how stable the findings are across different usage conditions.

Because users engage with maps in diverse ways, individual variability represents a key dimension of cartographic inquiry:

- How do individual user characteristics (gender, confidence, map experience) influence interaction with dynamic symbols?

Finally, to uncover the mechanisms behind these patterns, the study turns to the psychology of vision:

- What are the cognitive mechanisms underlying dynamic symbol processing as revealed through eye-tracking metrics?

Together, these questions define the scope of the research. They link the practical design of animated cartography with the cognitive and perceptual processes that shape how we see, think about, and understand dynamic representations of the world.

# 2 How We See and Study Maps
## *From Earthquakes to Visual Perception*

### 2.1 THE NATURE OF EARTHQUAKES: MAPPING A MOVING EARTH

Earthquakes begin deep beneath our feet, where immense forces slowly build until the Earth can no longer contain them. At their core, they are caused by the sudden movement of two massive blocks of rock—tectonic plates—sliding past one another along a fracture in the Earth's crust known as a fault or fault plane. When these blocks shift, the energy that has been quietly stored for years, sometimes centuries, is released in a burst of seismic waves, shaking the ground above (Braile, 2009; Graniczny & Mizerski, 2017). The exact point underground where the rupture starts is called the hypocenter. Directly above it, on the Earth's surface, lies the epicenter—the location we usually see marked on maps. This is the place where the shaking is often felt most strongly. The process begins with stress accumulating deep in the crust as the plates push, pull, or grind against one another. Eventually, the built-up strain overcomes friction, releasing energy that travels outward in all directions as vibration. Not all earthquakes happen in isolation. Some begin with smaller, subtle tremors known as foreshocks, which occur near the same fault before a larger event. The tricky part is that scientists can only identify a foreshock in hindsight—after the more powerful mainshock has already struck (Gates & Ritchie, 2006). Once the main event passes, the Earth rarely returns to complete stillness. Instead, a series of smaller quakes—called aftershocks—ripple through the same region as the crust continues to readjust. Depending on the strength of the mainshock, these aftershocks can persist for weeks, months, or even years, each one a reminder of the planet's ongoing effort to restore balance. In this constant push and pull between buildup and release, we witness the living dynamics of our planet. Earthquakes are not random acts of nature but part of a continuous cycle of stress and adjustment, shaping the very landscapes we inhabit (https://www.usgs.gov/programs/earthquake-hazards/science-earthquakes, accessed 5.10.2023).

Earthquakes belong to a larger family of natural hazards—one of the five main groups identified by scientists—that arise from the forces working within the Earth rather than upon its surface (Bryant, 2005). These are called geological hazards, or endogenous hazards, because they are born from the restless movements of our planet's interior. Yet not all earthquakes share the same origin. Most of them are tectonic

DOI: 10.1201/9781003728535-2

earthquakes, the direct result of rock masses shifting along vast fractures known as fault lines. These movements occur where tectonic plates meet—at convergent boundaries, where plates collide, and at divergent ones, where they pull apart. Each slip releases pent-up energy, reminding us that the crust is never completely stable. Interestingly, some earthquakes happen far from these boundaries, deep within the continental interiors. These "intraplate earthquakes" show that stress and strain can accumulate and release even in regions we consider geologically quiet. Such events reveal just how complex and far-reaching the interactions within Earth's crust can be, and why seismology remains one of the most intricate sciences of all. The network of tectonic plate boundaries, often shown on maps like the one in Figure 2.1, outlines this invisible engine of global motion. Not every tremor, however, comes from the grinding of plates. Some are born from volcanic activity—the explosive release of magma or the collapse of emptied magma chambers beneath volcanoes (White & McCausland, 2016). These volcanic earthquakes tend to cluster near active or dormant volcanoes, signaling pressure changes below the surface. Others emerge from the collapse of underground cavities, as seen in karst regions, where dissolving limestone leaves fragile ceilings that can suddenly give way. These are sometimes called sinkhole earthquakes (Caramanna et al., 2008; Gutiérrez et al., 2008). And then there are the earthquakes that we cause ourselves. Known as anthropogenic or human-induced earthquakes, they result from our growing influence on the Earth's

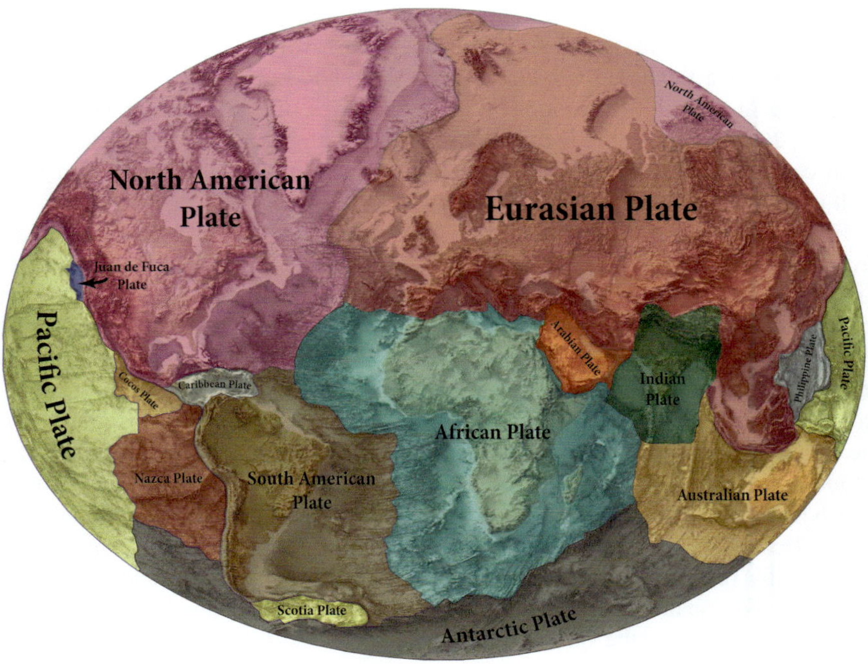


**FIGURE 2.1** Boundaries of tectonic plates based on the USGS tectonic plates of the Earth (based on the United States Geological Survey). (Modified from a map design by Oğuzhan Köse (ORCID 0000-0002-9821-5901), used with permission.)

crust. Activities such as deep mining, the impoundment of large water reservoirs, or the extraction of groundwater can disturb the delicate balance of forces within rocks, triggering small but measurable seismic events (Foulger et al., 2018). Though usually minor compared to tectonic quakes, these human-induced tremors remind us that our actions can reverberate far deeper than we realize.

To study earthquakes, scientists must first find a way to listen to the Earth. That is the role of the seismograph, a sensitive instrument that records the planet's vibrations and translates them into data called seismograms. The basic design of a seismograph is surprisingly elegant. Imagine a heavyweight suspended from a spring and attached to a pen, while the entire device rests on a solid base connected to the ground. When the Earth shakes, the base moves with it—but the suspended weight, thanks to inertia, stays almost still. The difference in movement between the ground and the hanging mass is what the instrument records. In essence, a seismograph measures stillness against motion, transforming invisible ground tremors into visible lines of ink or digital signals. A seismogram—the output of this process—is more than just a squiggly line. It's a timeline of the earthquake itself. Within those waves, scientists can identify different kinds of motion. The first to arrive are P waves, or primary waves, which travel quickly through both solid rock and liquid layers. They're compressional waves, moving particles back and forth in the same direction as the wave travels—like sound waves rippling through air. Following them are S waves, or secondary waves, which move more slowly but with greater destructive potential. These waves cause the ground to move side to side, perpendicular to their direction of travel. Finally come the surface waves, which ripple along the Earth's exterior. Among them, Love waves shake the ground horizontally, while Rayleigh waves roll in an elliptical motion, much like waves on the ocean (Fowler, 2005; Milsom, 2003). Although they travel more slowly than P and S waves, surface waves often cause the greatest damage because they move the ground—and everything built upon it—in complex, overlapping patterns.

Every earthquake tells a story not only through where it happens, but also through how powerful it is. This strength is expressed as the magnitude of the earthquake—a measure of how much energy the Earth released during the event. In practical terms, magnitude is calculated from the readings captured by seismographs, which record the shaking of the ground as seismic waves pass by. By analyzing the height and duration of these waves on a seismogram, scientists can estimate the total energy involved in the rupture. But not all earthquakes are measured in the same way. Over time, researchers have developed several magnitude scales, each designed to capture different aspects of the seismic process. The best-known is the Richter scale ($M_L$), introduced by Charles F. Richter in 1935 (Richter, 1935), which originally measured the amplitude of seismic waves recorded by instruments in southern California. As technology and global data expanded, new scales emerged to handle the variety of earthquake types and distances. For example, the body-wave magnitude scale (mB), developed by Beno Gutenberg in 1945 (Gutenberg, 1945), focuses on waves that travel through the Earth's interior, while the moment magnitude scale ($M_w$)—now the international standard—estimates energy release based on the physical properties of the fault rupture itself (Kanamori, 1977). Other regional systems, such as the K-class magnitude ($M_K$) used in parts of Eurasia (Rautian et al., 2007), are tailored to local geological conditions and data availability. These differences aren't merely technical; they reflect the challenge of describing a complex natural process with a single number. The International Association of Seismology and Physics of the Earth's Interior (IASPEI) has since worked to harmonize these approaches, establishing standardized procedures and equations to ensure that magnitudes are comparable across regions and methods (Bormann & Dewej, 2014). Thanks to this international coordination, an earthquake's magnitude—whether reported from Tokyo, Rome, or San Francisco—now speaks a common scientific language.

## 2.2 HOW MAPS TELL EARTHQUAKE STORIES: VISUAL VARIABLES AND COMMUNICATION

When an earthquake shakes the ground, the world feels it—but it's through maps that we see it. A seismograph records the vibrations, a computer processes the data, but the map is what turns those numbers into meaning. Cartographic visualization bridges the gap between raw seismic data and human understanding. It transforms technical information—coordinates, magnitudes, depths—into visual stories that anyone can read at a glance. In this sense, an earthquake map is more than a scientific tool; it's a shared language between data and perception, between what the Earth does and what we can comprehend. Cartographic visualization has become a cornerstone of modern seismology and geospatial science. It represents a collaboration between human cognition and computer processing, where algorithms handle vast quantities of data, and design principles shape how that data is perceived. Researchers in GIScience, geovisualization, and information visualization have long studied how to make this partnership effective (Andrienko et al., 2010; Dykes et al., 2005; MacEachren & Kraak, 1997). At its heart lies a simple goal: to turn complexity into clarity. Through carefully designed graphics, colors, and motion, maps can reveal hidden patterns—how earthquakes cluster along fault lines, how their magnitudes vary, or how seismic activity evolves over time.

But the power of maps extends far beyond laboratories or scientific journals. Today, digital cartography is part of everyday life. Earthquake maps appear on news sites, emergency dashboards, and mobile apps, guiding not only researchers but ordinary people making real-world decisions. Someone planning a trip might check seismic zones; a city planner might use historical earthquake data when designing safer infrastructure; a family might want to understand the risks of living near an active fault. In each of these cases, maps translate data into awareness—a vital bridge between science and society. As geographic information becomes more accessible and interactive, this communication role of maps has only grown stronger. With modern Geographic Information Systems (GIS) tools, anyone can explore when and where earthquakes occur, compare time periods, or even simulate future scenarios. Communities use these maps to plan evacuation routes, governments rely on them for disaster preparedness, and individuals use them to understand the landscapes they inhabit. In short, cartographic visualization turns information into action. When designed thoughtfully, an earthquake map does not merely display data—it invites exploration, sparks questions, and helps people see their world in motion.

In a graduated symbol map, data values—such as earthquake magnitude—are grouped into a few categories. Each range of values is represented by a symbol of a specific size. For example, if we divide earthquakes into five classes, we might use five circle sizes, from small to large, to show weak to strong events. The result is a map that lets you see patterns at a glance: clusters of tiny circles suggest frequent but minor tremors, while larger circles signal areas of greater seismic energy. This process of classification and scaling helps simplify vast datasets into something visually digestible. By contrast, proportional symbol maps skip the classification step. Instead of fixed symbol sizes, each circle's diameter changes continuously in proportion to the data value it represents. In such maps, a small earthquake might appear as a barely noticeable dot, while a massive one dominates the area around it. This method often feels more organic and expressive—like watching the Earth's pulse drawn directly on the map. Color can add another layer of meaning. For instance, blue tones might represent shallow earthquakes, while deeper shades of red indicate those that occurred far beneath the surface. When size and color work together, the viewer can simultaneously perceive how strong and how deep an earthquake was. A compelling example of this approach can be seen in the interactive Earthquake Live Map https://earth3dmap.com/earthquake-live-map, which displays seismic events on a rotating 3D globe. Each dot varies in both size and color to represent magnitude and depth, turning global seismic activity into an ever-changing constellation of motion and meaning. Through such visual techniques, maps become more than static reference tools—they become instruments of insight, allowing us to see patterns that words and tables could never convey.

The Earth 3D Map (Earthquake Live Map) is a fascinating example of what cartographers call cartographic redundancy or visual suggestiveness (Cybulski & Medyńska-Gulij, 2018; Shortridge & Welch, 1982). In simple terms, redundancy occurs when more than one visual feature—such as color and size—is used to represent the same piece of information. At first glance, this might sound inefficient, but in practice, it often makes maps easier to read. Our visual system thrives on repetition and reinforcement; when different cues point to the same meaning, the message becomes clearer and more memorable. In the Earth 3D Map, for instance, the size of each circular symbol represents earthquake magnitude—the larger the circle, the stronger the quake. At the same time, color reinforces this same information: intense reds and oranges hint at greater magnitude, while softer tones indicate smaller events. Technically, color doesn't add new data; it simply strengthens the signal already carried by size. Yet this visual echo helps users notice patterns faster and with less cognitive effort. It's a subtle design strategy that makes complex information feel intuitive. Interestingly, this particular interactive map does not include a traditional legend—a reminder that in digital environments, interaction often replaces static explanation. Instead of consulting a key, users discover meaning by clicking on symbols. Each click reveals a small information panel showing the earthquake's magnitude, depth, date, and coordinates. This kind of design trades immediate overview for exploratory engagement: the user learns through curiosity, not instruction. However, it's important to note that the map displays earthquakes collectively, rather than in chronological order. It shows what has happened recently, but not the sequence of how events unfolded. This design choice highlights a key challenge in seismic visualization—while spatial patterns are clear, temporal ones can easily disappear. Understanding both requires careful cartographic design, where motion, sequence, and interaction work together to tell the full story.

Another powerful example of how earthquake information is communicated through maps comes from the European Union Emergency Response Coordination Centre (ERCC). On its official website, the Centre regularly publishes maps showing seismic events that have occurred on a given day. At first glance, these maps look dense with information—and they are. Each one combines multiple layers of data: earthquake magnitude, fatalities, tsunami alerts, peak horizontal ground acceleration (in m/s²) as an indicator of seismic hazard, global disaster alerts, and the date of occurrence. When a user selects a specific event, the map reveals even more: the depth of the quake, its geographical coordinates, and even the estimated number of people affected. It's a striking example of how much information can be packed into a single visual environment when design is done carefully. Here again, the principle of cartographic redundancy plays a key role. A single attribute—such as magnitude—is often represented by more than one graphic variable, such as color and size, allowing the message to reach the user through multiple perceptual channels. For example, larger, darker circles might represent stronger earthquakes, while smaller, lighter ones indicate weaker tremors. The repetition across visual cues helps ensure that users can read the map quickly, even under stressful conditions—an important factor when these maps are used by emergency managers or the general public. Figure 2.2, for instance, shows earthquakes with magnitudes of 5.0 or greater on the Richter

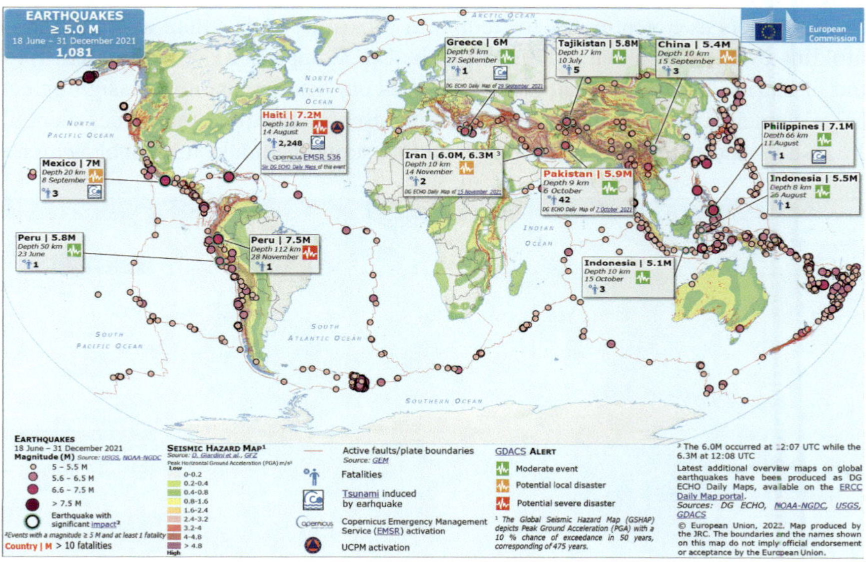


**FIGURE 2.2** Map of earthquakes with magnitude ≥5.0 in a global overview, accompanied by additional information concerning fatalities, seismic hazard, occurrences of tsunamis, etc. (more maps available at https://erccportal.jrc.ec.europa.eu/ECHO-Products/Maps#/maps?pageIndex=1&mt=Daily%20Map). (Source: European Union, 2021. Map produced by the Joint Research Centre (JRC), DG ECHO. Data sources: USGS, NOAA-NGDC, GDACS. Licensed under CC BY 4.0.)

scale. The visualization is designed for global coverage and daily updates, which explains why the temporal dimension—the sequence of how events unfold—is not represented here. The focus is on overview and awareness, not animation or chronology. Each day's map captures a snapshot of global seismic activity, providing an immediate sense of where the Earth has moved most recently. This approach reflects a key aspect of applied cartography: sometimes the goal is not to tell the whole story over time, but to offer a clear, timely picture of the present.

A different design philosophy can be seen in the earthquake maps produced by the United States Geological Survey (USGS)—one of the world's leading authorities on seismic data. Unlike the EU example, these maps avoid redundancy, assigning each visual variable a unique role. The size of each point symbol represents the magnitude of the earthquake, while its color conveys when the event occurred. This separation of meaning allows the viewer to read two dimensions of information—strength and timing—at a glance. In addition to these dynamic point markers, the USGS interface includes contextual layers such as tectonic fault lines, a hazard map showing the likelihood of damaging ground motion, and population density overlays that hint at potential human impact. By clicking on an individual earthquake symbol, users can access precise details in a side panel: the exact time of occurrence, geographical coordinates, and depth of the seismic source. Although the map includes a time component, the temporal dimension here plays a supporting role. Rather than animating the sequence of events, the visualization groups them by recency—distinguishing between quakes that occurred within the past hour, day, week, or month. In this way, color acts as a gentle cue for temporal proximity rather than as a timeline of unfolding events. As shown in Figure 2.3, the USGS approach exemplifies a different kind of clarity. Instead of redundancy, it relies on visual economy—assigning each design element a distinct purpose and trusting users to integrate those cues mentally. The result is a clean, information-rich map that supports quick situational awareness

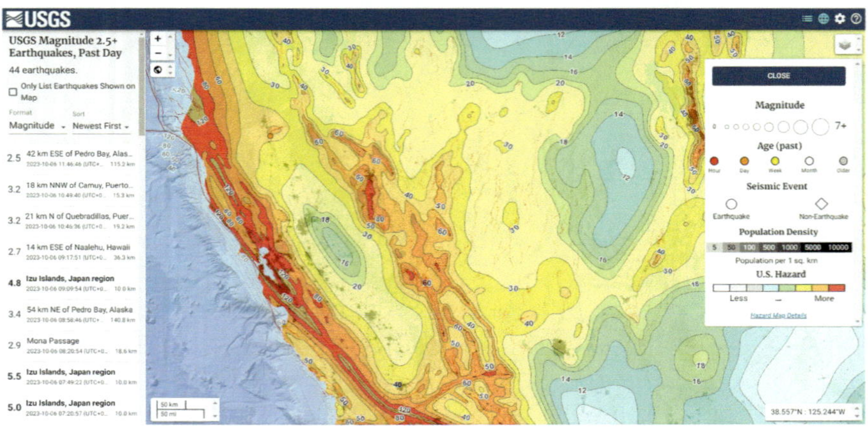


**FIGURE 2.3** The USGS earthquake interactive map presents data regarding earthquake magnitude, age, population density, and hazard levels. Supplementary details such as earthquake depth, spatial coordinates, or precise timing can be accessed through user–map interactions (https://earthquake.usgs.gov/earthquakes/map/?extent=21.0845,-130.16602&extent=52.3756,-59.85352). (Source: U.S. Geological Survey, Department of the Interior/USGS.)

while preserving interpretive depth. It's a reminder that in cartography, there is no single "correct" way to visualize earthquakes; rather, each map reflects the priorities of its maker—whether to inform, to warn, or to invite exploration.

A particularly engaging example of how modern cartography brings earthquakes to life is the Seismic Explorer (https://seismic-explorer.concord.org/). Unlike traditional maps that freeze data in a single moment, this interactive platform lets users experience earthquakes as they unfold across both space and time. The map displays seismic events by magnitude and depth, emphasizing not only how strong an earthquake was but how deep beneath the Earth's surface it originated. This combination offers a richer, more layered understanding of seismic activity—showing that not all powerful quakes are shallow, and not all small ones are harmless. What makes the Seismic Explorer especially compelling is its interactive timeline. Instead of presenting earthquakes all at once, it allows users to select a custom time range—days, months, or years—and then animate the sequence of events. As dots appear and fade across the map, viewers can watch the Earth's crust come alive, revealing the rhythm and clustering of seismic activity. This feature transforms what might have been a static dataset into a temporal narrative, helping users grasp the progression of events that no still image could convey. When a user clicks on an individual earthquake, a detailed information window opens, displaying the magnitude, exact date and time, and depth of the event. Additional controls in the lower panel make the experience even more exploratory: users can filter earthquakes by magnitude, add layers showing volcanic activity or tectonic plates, and even switch map backgrounds to adjust context and contrast. In essence, the Seismic Explorer illustrates how interactivity and animation together can elevate map reading from observation to exploration. Instead of passively viewing data, the user plays with it—testing hypotheses, spotting trends, and watching natural processes unfold before their eyes. As shown in Figure 2.4, this design not only informs but also invites curiosity, demonstrating how digital cartography can transform scientific data into an intuitive, almost cinematic experience of the living Earth.

**FIGURE 2.4** Seismic Explorer is an example of an interactive map utilizing animation techniques to display the chronological sequence of events (https://seismic-explorer.concord.org/). Seismic Explorer screenshot used with permission from the Concord Consortium.

Another fascinating example of how seismic data can come alive through digital mapping is the World EQ Locator (https://nagix.github.io/world-eq-locator/), an interactive globe built on data from the ANSS Comprehensive Earth Catalog. This extensive database collects detailed records of earthquakes from around the world—each entry containing information about an event's location, magnitude, and depth. On the World EQ Locator map, every earthquake is represented as a single colored point, with color and vertical position together indicating how deep beneath the Earth's surface the seismic source lies. As you move your cursor across the map, hovering over a point reveals additional details: the hypocenter depth (that is, the precise underground origin of the quake) and its magnitude. The simplicity of this design makes it easy to scan the globe and quickly spot patterns—clusters of deep tremors in one region, shallower shocks in another. What makes this visualization especially engaging is its three-dimensional perspective. By adjusting the camera angle, users can tilt the globe and "look" beneath the surface, gaining an almost geological sense of how deep or shallow different earthquakes occur. This spatial layering turns the map into a kind of virtual cross-section of the Earth, helping users visualize not just where, but how deep the planet's crust is moving. Another subtle but powerful design feature is the use of dynamic point symbols. In some versions, selected earthquakes are marked by pulsating effects—a rhythmic expansion and contraction of the symbol that naturally draws the eye to the event's precise location (Robinson, 2011). This visual cue transforms what could be a static pinpoint into a living indicator, conveying both significance and immediacy. As shown in Figure 2.5, the World EQ Locator offers a global view of all earthquakes recorded in the database, combining clarity, interactivity, and subtle motion. It's not just a map—it's a window into the planet's restless interior. Blending color, depth, and animation allows viewers to feel the Earth's structure in a way that numbers or text alone could never achieve.

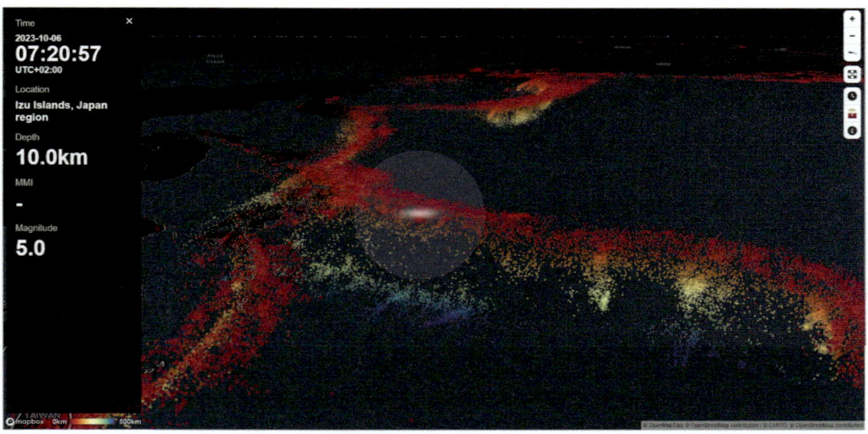


**FIGURE 2.5** An instance of an interactive online map displaying all recorded earthquakes is depicted, with the foremost feature being the presentation of point symbols, all of uniform size, representing the depth of each earthquake (https://nagix.github.io/world-eq-locator/). World EQ Locator visualization by Akihiko Kusanagi (MIT License), using USGS ANSS ComCat data.

One of the most striking examples of dynamic mapping in action comes from the Pacific Tsunami Warning Center (PTWC), part of the U.S. National Oceanic and Atmospheric Administration (NOAA). Here, earthquakes are not simply plotted on a static background—they unfold across the map in chronological sequence, transforming data into an animated story of seismic motion. Each new earthquake appears as a pulsating point symbol, its rhythmic expansion and fading signaling the moment an event is added to the timeline. This simple motion immediately draws the viewer's eye to new activity, giving a sense of when and where the Earth has moved. In this visualization, multiple layers of information come together seamlessly. Each earthquake carries details about its depth, year of occurrence, and magnitude, and, crucially for tsunami monitoring, the amplitude of the resulting wave. These attributes help convey not just the power of the quake but its potential impact on coastal regions. The animation is accompanied by an interactive timeline, allowing viewers to pause, rewind, or focus on specific time periods. Because the visualization is streamed through YouTube, the familiar video interface doubles as a navigation tool—turning what might otherwise be a technical dataset into a publicly accessible educational experience. The PTWC's YouTube channel hosts many similar animations, each tailored to a particular region or time span. Together, these visualizations document the restless pulse of the Pacific, showing how tectonic energy travels across the world's largest ocean basin. The map doesn't just display earthquakes—it tells their story, tracing the invisible connections between seismic events and the tsunamis they generate. As illustrated in Figure 2.6, even a single frame captures the essence of motion-based cartography: clear, rhythmic, and alive with temporal meaning.

Of course, the examples discussed so far represent only a glimpse of the many earthquake maps available today. Across the web, countless variations exist—each

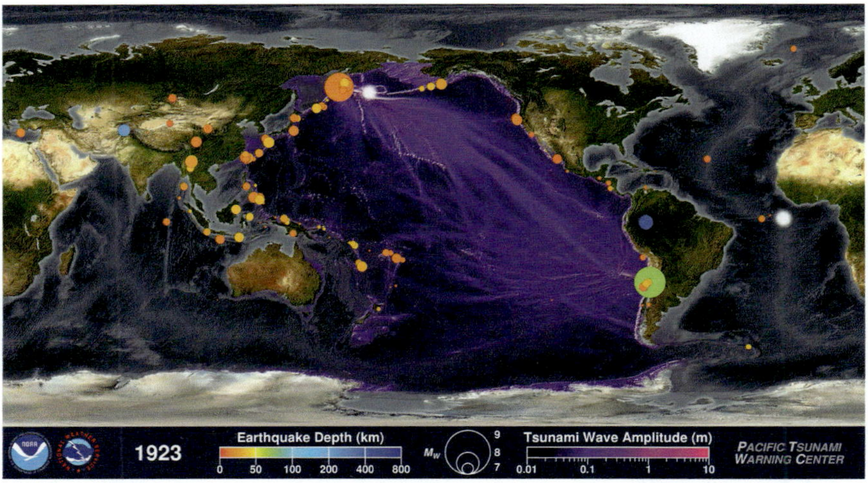


**FIGURE 2.6** Spatio-temporal visualization of earthquakes and their associated tsunamis over a 120-year period (1901–2020). (Source: Nathan Becker, NOAA/NWS Pacific Tsunami Warning Center. Retrieved from https://www.youtube.com/watch?v=ilFEKSZQv5o.)

created for a slightly different purpose, audience, or platform. For the purpose of this chapter, however, the focus has been narrowed to a selection of representative examples—those that best illustrate the dominant trends in how seismic data are communicated to the public. These maps were chosen not only for their visibility and popularity but also because they showcase the most common cartographic techniques used to depict earthquakes in digital media. A comprehensive catalog of all the earthquake maps reviewed for this study, along with details about the types of data they present and the visualization methods they employ, is provided in Appendix 1. This inventory offers a broader overview of the diversity of approaches currently used to map seismic activity, ranging from minimalist point-based visualizations to complex, multilayered interfaces. It's worth noting that earthquake mapping extends beyond online platforms. A review of print media, such as the one conducted by Sarın and Uluğtekin (2019) in their study of Turkish newspapers, reveals that many traditional outlets still rely on pictorial symbols to convey seismic information. These maps, often produced for rapid publication, tend to prioritize clarity and visual appeal over analytical depth. Yet they remain a vital part of how the public encounters geographic information, bridging the gap between scientific data and everyday understanding. By examining both digital and printed examples, we gain a fuller picture of how earthquake cartography operates across different media—each with its own strengths, limitations, and audience expectations. Together, they set the stage for the next discussion: how dynamic symbols can build upon these foundations to make earthquake visualization more expressive, intuitive, and responsive to human perception.

When exploring earthquake maps, one might also encounter a different kind of visualization—the seismic hazard map. Unlike maps that show where earthquakes have already occurred, these maps illustrate where and how strongly the ground is likely to shake in the future. In other words, they are not records of past events but forecasts of potential risk. Seismic hazard maps display the probability that a certain level of ground motion will be exceeded within a given time period—most often a 10% chance within 50 years, a standard benchmark used by hazard assessment programs worldwide (Pagani et al., 2020). At the heart of these maps is a measure known as peak ground acceleration (PGA)—the highest level of ground acceleration recorded at a location during an earthquake. Imagine it as the "maximum jolt" the ground experiences when the shaking is at its strongest. This value is captured using an instrument called an accelerograph, which measures how rapidly the ground moves back and forth. Unlike the Richter or moment magnitude scales, which express the total energy released by an earthquake, PGA focuses on what happens at a specific spot—how intense the shaking feels at ground level (Douglas, 2003; Cua et al., 2010). Because earthquakes generate motion in all directions, PGA is often divided into horizontal and vertical components to reflect how the ground moves side to side and up and down. Local geological conditions also make a significant difference. For instance, the same earthquake will cause stronger shaking on soft, water-saturated soils than on solid bedrock. These microscale variations in the Earth's surface—what geologists call site effects—play a major role in determining the intensity of ground motion (Wang, 2008).

## 2.3 WHEN MAPS BEGIN TO MOVE: DYNAMIC SYMBOLS AND ANIMATED CARTOGRAPHY

The idea that maps could move was once revolutionary. For centuries, cartography was a still image—a frozen view of space. But in the mid-20th century, that began to change. As computers entered the discipline, researchers started asking a daring question: *what if maps could show change as it happens*? This question gave birth to what we now call animated cartography (Thrower, 1959), a field that would later grow into the broader practice of multimedia cartography (Medyńska-Gulij et al., 2021). With the arrival of computational methods in the 1960s and 1970s, maps were no longer limited to static representations. Early pioneers such as Waldo Tobler and Harold Moellering used computers to simulate geographic processes—urban growth, migration, transportation—and to explore animation as a new form of cartographic communication (Tobler, 1970; Moellering, 1976). Their work showed that computers could do more than store and calculate data; they could make geography visible in motion. These early "computer movies" brought time into the language of cartography, extending what maps could say about how the world changes. By the 1990s, this early experimentation had matured into a vibrant field of research. Scholars such as Campbell and Egbert (1990), DiBiase et al. (1992), Dorling and Openshaw (1992), Kraak et al. (1997), and Blok (1997) helped expand animation from a novelty into a tool of scientific visualization. Their work emphasized that animation was not only about catching the eye—it was a way to think. Movement could reveal temporal rhythms, spatial relationships, and evolving patterns that static maps could only hint at. These researchers also helped formalize the practice, developing taxonomies of animated maps and proposing frameworks for empirical evaluation and user-centered design (Lobben, 2003; Ogao & Kraak, 2002).

As computing power grew, so did the possibilities. Animation opened doors to new kinds of interaction—users could now filter, brush, and dynamically compare spatial data, switching between perspectives in real time (Dykes, 1997). This flexibility made animation especially useful for exploratory analysis, where quick shifts of view often lead to new insights. Researchers also began integrating these ideas directly into GIS. Studies explored how dynamic point symbols could behave consistently across keyframes or respond algorithmically to changing data inputs (Xiaofang et al., 2005; Guorui et al., 2006). By the late 2010s, what began as an academic experiment had entered the mainstream. When ESRI's ArcGIS Pro and QGIS (Quantum Geographic Information System) introduced the ability to use animated point symbols—such as pulsating lights or flickering markers—animation officially became part of everyday mapping practice. What started with a few frames of simulated motion in the 1970s has evolved into a core component of modern geovisualization, where movement itself is recognized as a powerful visual language.

These innovations laid the foundation for how we now think about dynamic point symbols—tiny yet expressive markers that bring movement to the map. In essence, a dynamic point symbol represents a geographic event anchored in space but unfolding in time (DiBiase et al., 1992). Imagine each symbol as a living point on the map, its position defined by coordinates, but its behavior—its flicker, pulse, or shift—reflecting change. This idea became especially important once information systems began providing real-time data. Today, maps don't just show what has happened; they show what is happening. Whether tracking a thunderstorm, visualizing air quality, or monitoring earthquakes, the most natural way to represent this ongoing change is through movement itself (Gar-on et al., 2001).

Motion on a map can serve many purposes. It can draw attention to a specific event—like a flashing epicenter during an earthquake—or suggest direction, such as the movement of a ship or migrating animal (Tinghua, 1998). In other cases, motion can reveal transitions: boundaries shifting through history, the advance of ice sheets, or the spread of wildfires (Maggi et al., 2016). Because motion can mean so many things, it becomes crucial to clarify what kind of change a dynamic symbol represents (Cunty & Mathian, 2017). While an animated map might show several temporal movements at once—growth, displacement, fading—dynamic point symbols typically focus on one specific kind of transformation (Kraak, 2025). Their role is to spotlight individual changes without overwhelming the viewer. The power of these symbols lies in how they extend the classic visual variables of cartography—the fundamental "alphabet" of map design first described by Jacques Bertin (1981). Bertin's system defined how attributes like shape, size, value, color, texture, and orientation could communicate different kinds of information. Later researchers built upon this vocabulary. Morrison (1974) expanded the list, while MacEachren (1992, 1995) introduced variables such as focus, transparency, and resolution to handle more nuanced, layered data. Żyszkowska (2000) added concepts like layout, density, and optical weight, emphasizing how spatial arrangement affects visual balance, and Slocum et al. (2023) more recently proposed perspective height as another important variable for modern visualization. When these variables are combined and animated, they open up a rich field of creative possibility. Cartographers can design symbols that grow, blink, fade, rotate, or glide, encoding not just what is happening, but how it happens. Yet with this expressive power comes responsibility. Not every motion improves understanding—some can distract or mislead. As Couclelis (1998), Kraak (1998), and Fabrikant and Goldsberry (2005) remind us, successful dynamic mapping depends on finding a balance between expressive richness and cognitive clarity. The best animated maps do not overwhelm the eye—they guide it. They turn motion into meaning, helping us perceive time and change as naturally as we read space.

The story of dynamic symbols doesn't begin with maps—it begins with the human mind. Long before cartographers made symbols move, psychologists were already asking a simple but profound question: how do we notice movement? The answer, as decades of research revealed, lies in the biology of perception. Certain visual cues—motion, size, color, orientation—are so fundamental that our brains detect them automatically, without conscious effort. These are known as preattentive features. We perceive them in less than a quarter of a second, faster than we can think or decide where to look (Bartram, 1997; Hohnsbein & Mateeff, 1998; Julész, 1981; Treisman, 1985; Treisman & Gelade, 1980; Ware, 2012, pp. 139–177). This ability evolved for survival: a sudden flicker in the corner of your eye could mean danger—or opportunity. In the world of visualization, it means something else entirely: motion instantly draws attention. When used carefully, this makes animation a powerful tool for directing the viewer's gaze without overwhelming them. Researchers such as Lowe and Boucheix (2016) have shown that dynamic displays can help people build more accurate mental models of how systems work—whether it's how a machine functions or how an earthquake spreads through a fault zone. In other words, movement doesn't just attract attention; it helps people understand processes that unfold over time.

However, our perception of motion isn't only bottom-up—driven by automatic visual responses. It's also shaped by what we already know and expect. Studies in cognitive psychology have revealed that prior knowledge and spatial ability strongly influence how effectively we interpret animations (Hegarty & Kriz, 2008; Kriz & Hegarty, 2007; Maggi et al., 2021). An experienced geographer might instantly spot a shifting cluster of epicenters, while a novice might miss it entirely. This delicate interplay between bottom-up perception and top-down reasoning defines how we read animated maps. Comparative studies in information visualization further demonstrate that animation can outperform static alternatives—but only in certain contexts. For example, when all key elements appear together and move in the same direction, viewers detect patterns faster and more accurately (Ware & Bobrow, 2004; Archambault et al., 2011; Archambault & Purchase, 2016). Motion can thus reveal structure, as shown in animated flow visualizations, where streamlets make invisible dynamics—like wind or current—visually tangible (Ware et al., 2016). Yet, as with all design power, there's a trade-off: too much motion risks information overload. Sudden or asynchronous changes can disrupt attention, forcing the viewer to continually reorient (Jiang et al., 2023; Lowe, 2019).

Psychologists have identified two particularly stubborn pitfalls in dynamic displays. The first is change blindness—our surprising inability to notice even large alterations between animation frames if our attention is elsewhere (Fish et al., 2011; Moon et al., 2014; Tse, 2008). The second is inattentional blindness, which occurs when we're so focused on one moving element that we completely miss another (Simons & Chabris, 1999; Tversky et al., 2002). Both remind us that while motion can reveal, it can also conceal. For cartographers, these insights carry an important message: the human visual system is not a passive camera—it's an active interpreter. Effective animation design, therefore, depends not only on technology or aesthetics but on psychology. By understanding how attention works—what draws it, what sustains it, and what distracts it—we can design animated maps that illuminate the user rather than overwhelm it.

The leap from static to dynamic cartography didn't happen overnight—it grew out of decades of experimentation with how visual variables communicate information. Before symbols could move, they had to mean something. This groundwork was laid by Jacques Bertin (1981), who introduced the idea that every map symbol could be broken down into a set of visual variables—such as size, shape, color, and orientation—that together form a visual language. Static point symbols, defined by their geometry and position, were the first building blocks of this language. They allowed cartographers to encode complex data spatially and visually long before computers made animation possible. But something fascinating emerged once researchers began looking at maps through the lens of cognitive psychology. Studies showed that several of Bertin's visual variables—especially size, orientation, and color—are not just design tools but perceptual triggers. The human brain processes them preattentively, meaning almost instantaneously, without deliberate focus

(Cybulski & Krassanakis, 2023; Garlandini & Fabrikant, 2009). This discovery laid the perceptual foundation for dynamic cartography: if certain visual features can be noticed automatically, then adding motion to them could further amplify attention and meaning. Each visual variable, when animated, opens a new expressive dimension. For example, the variable orientation can transform into rotation—a symbol spinning gently around its axis to signal change (Cybulski, 2014a). Similarly, only part of a symbol, such as its outline, can be animated to pulse or shimmer (Ge et al., 2025a). These subtle motions add rhythm and texture to the map, creating a visual experience that feels alive but still informative (Dykes, 2002; Midtbø, 2003). Motion also allows cartographers to encode quantitative differences through speed: faster motion can represent greater intensity, while slower motion suggests a milder phenomenon (Cybulski & Wielebski, 2019).

In practice, these effects are made possible by combining principles of animated cartography with modern coding environments. Today, dynamic symbols can be programmed to react to user input—blinking when hovered over, changing color when clicked, or moving in response to real-time data streams (Medyńska-Gulij et al., 2023). This integration bridges traditional design and computational interactivity, making maps not only visually engaging but also responsive and participatory. The implications go even further. With emerging platforms in virtual and augmented reality, cartographic symbols can now exist in immersive environments (Amorim & Schmidt, 2021). Imagine walking through a 3D earthquake visualization where the symbols pulse around you, their speed and color shifting as seismic intensity changes. Such experiences move beyond viewing a map—they invite you into it. Dynamic point symbols thus mark more than a technical advancement; they represent a shift toward embodied and interactive cartography, where spatial data becomes something users can explore, feel, and respond to directly.

### 2.3.1 Interaction as Exploration: Making Animated Maps Responsive

Animation captures attention—but interactivity gives it purpose. Without the ability to pause, rewind, or explore, an animated map becomes little more than a short film: captivating at first, but quickly frustrating. Viewers have no control over what they see or when they see it. As early researchers like Koussoulakou and Kraak (1992) and Monmonier and Gluck (1994) observed, a lack of interaction can undermine both the effectiveness and efficiency of a map. Users want to do more than watch change happen—they want to navigate it. What makes interaction on animated maps unique is that it unfolds not only in space but in time. The landscape changes before the viewer's eyes, and the user's task becomes managing this temporal flow—moving through data as one would through a story. The most basic form of this control is familiar from everyday media: the ability to start, stop, go forward, or backward the animation (Monmonier, 1990). These simple commands transform a fixed sequence into something flexible, letting users control the rhythm of exploration.

In practice, there are two main ways to build this kind of interaction. The first—and historically the simplest—is to use an external player, such as Windows Media Player or online platforms like YouTube, where the map animation is treated like a video clip. However, this approach has clear limitations. Users can control playback but cannot interact with the map itself. Studies revealed that fewer than one-third (28%) of animated maps available online were designed this way (Cybulski, 2016), reflecting a growing recognition that maps are not meant to be watched—they are meant to be used. The second, more advanced approach is to embed interactive controls directly into the map. Here, animation and user input are integrated through programming languages such as JavaScript, allowing users to manipulate time, adjust parameters, and even combine animation with spatial queries. This approach liberates the map from dependence on an external player—it becomes self-contained, responsive, and adaptable. More importantly, it can function both online and offline, offering the same level of engagement across devices and contexts. When animation and interactivity converge, maps stop being static screens and become exploratory interfaces—spaces where users can play, test, and discover. It is in this transformation that cartography fully enters the era of user experience design.

Once animation and interaction are combined, the next question becomes: *how should users control time*? Embedded interactivity can take many forms, and much of it depends on the design of the graphical user interface (GUI)—the visual tools through which people communicate with the map. Buttons, sliders, and play controls are the cartographer's way of handing over the steering wheel, allowing users to explore not only where things happen, but when (Roth, 2013b). Over the years, cartographers have experimented with countless ways to make this experience intuitive (MacEachren, 1994). Among these, one idea has stood the test of time—quite literally: the time slider. Usually placed along the bottom edge of a map, the slider allows viewers to move through an animation frame by frame, stepping back or forward through the unfolding of events (Kraak et al., 1997; Midtbø & Baroš, 2005; Peterson, 1999). This design feels natural because it borrows from familiar media interfaces, but instead of controlling a movie, it controls the passage of geographic change. In its simplest form, a user drags a button along a scroll bar, watching the map transform as time progresses (Slocum et al., 2000). Yet behind this simple gesture lies a powerful concept: the user is no longer a passive observer but an active participant in temporal exploration. By sliding through time, they can pause to inspect moments of interest, jump to specific events, or replay sequences to detect emerging patterns.

Not all maps treat time as a straight line. In recent years, cartographers have explored more imaginative ways to let users move through time—one of the most intriguing being the circular interface (Harrower & Fabrikant, 2008). Instead of sliding along a horizontal timeline, imagine turning a dial of time. Each segment of the circle represents a different period—hours, days, months, or even years—and users can rotate around it to revisit specific moments or compare multiple points in time (Midtbø et al., 2007). This approach breaks away from the familiar left-to-right logic of temporal navigation. By looping time into a circle, the design echoes natural rhythms—day and night, seasons, cycles of earthquakes or weather patterns. It invites users to see data as something that repeats and flows, rather than something that simply progresses. The result is a more organic and exploratory experience, especially suited to mobile devices, where touch gestures make circular motion intuitive and satisfying (Cybulski et al., 2023).

While innovative interfaces like circular dials invite deeper exploration, most contemporary maps that handle time still rely on something far simpler. Think of the familiar controls found on any video player: forward, backward, play, pause, stop, or replay. These are the tools most users instinctively understand—and they form the backbone of temporal navigation on many web-based maps (Kraak et al., 1997; Popelka et al., 2019). In these cases, interaction is less about uncovering patterns and more about controlling the flow of information. Users can start an animation to watch events unfold or pause it to focus on a specific moment, but they rarely dive into data exploration in a deeper sense (Jiang, 1996). This kind of interface doesn't require specialized knowledge—it's intuitive, universal, and easy to implement, which explains why it remains so widespread. You can think of these controls as the remote control of time. They don't change what the data show, but they let viewers decide when and for how long they want to see it. For simple tasks—like following the spread of earthquakes over a month or visualizing the path of a storm—such straightforward interaction is often all that's needed. Figure 2.7 illustrates these conceptual and practical approaches to time control on dynamic maps, summarizing the most commonly used designs for navigating spatio-temporal visualizations. From basic playback buttons to more sophisticated sliders and loops, each approach reflects a different balance between simplicity, accessibility, and analytical depth.

Beyond simple playback controls, some animated maps take interactivity to a much higher level. These advanced systems—often referred to as geovisualizations—go beyond showing data over time to helping users actively analyze and interpret it (Fabrikant, 2005; Gołębiowska et al., 2020; MacEachren & Kraak, 2001). Instead of a single animated window, they may include multiple panels showing different perspectives of the same phenomenon, or allow users to manipulate data directly through tools like filtering, brushing, toggling, or highlighting (Roth, 2012). Imagine a geoscientist studying seismic events across several regions. Rather than just watching earthquakes unfold on a timeline, they might isolate specific magnitudes, compare temporal clusters, or dynamically link different datasets to uncover hidden

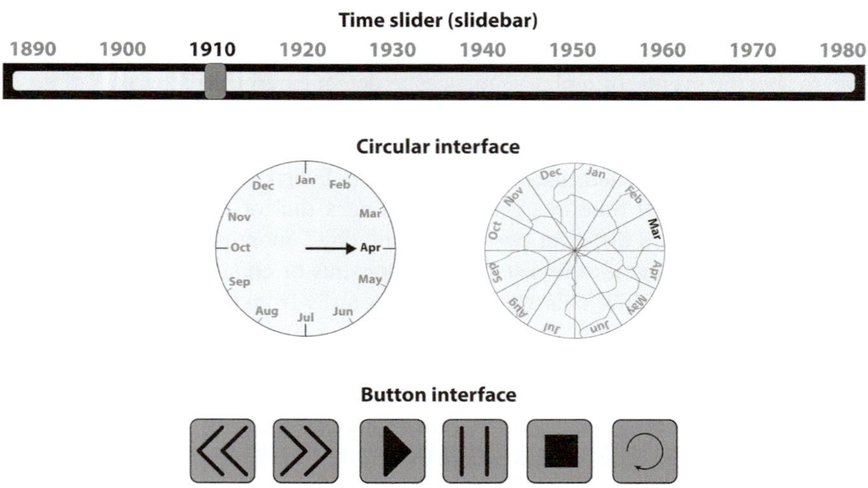


**FIGURE 2.7** Popular temporal navigation interface solutions. The presented examples show conceptual solutions that serve as the basis for specific adaptation to the map design concept.

relationships. These are the kinds of tasks advanced geovisualization systems are designed to support. However, this level of interactivity comes at a cost—it demands both technical skill and analytical intent. As a result, such tools cater to researchers, analysts, and professionals who work with spatial data on a daily basis. In contrast, most public-facing animated maps focus on clarity and accessibility, offering simpler controls that communicate change without overwhelming the viewer. For this reason, the present study centers on user-oriented animated maps, where motion and interaction are designed not for expert analysis but for effective communication—helping users quickly and intuitively grasp the dynamics of a phenomenon like earthquakes.

### 2.3.2 From Vision to Design: Understanding Motion in Cartographic Animation

Human beings are, quite literally, wired to see motion. Our visual system evolved in a world full of movement—of rustling leaves, flowing water, and approaching predators—so it makes sense that we notice motion before almost anything else. In fact, our eyes are astonishingly quick at detecting even the slightest flicker of movement, especially out of the corner of the eye. This sensitivity comes from specialized neural pathways in the brain that are tuned to motion, allowing us to respond rapidly to dynamic changes in our surroundings (Brown et al., 2005; Tynan & Sekuler, 1982). Peripheral vision plays a key role here. While our central vision is responsible for sharp detail, it is the periphery that acts as our early warning system. It's what alerts us when something moves suddenly nearby—even before we consciously look at it (Vater, 2019). But the human brain does more than just detect motion; it also predicts it. When we see a ball flying toward us, for instance, we can estimate its path and catch it at the right moment. This ability to anticipate motion depends on the brain's remarkable knack for combining sensory information with learned expectations (Kwon et al., 2015; Wexler & Klam, 2001). Researchers in cognitive psychology describe this as a dialogue between bottom-up and top-down processing. Bottom-up cues—like speed, direction, or change in brightness—trigger immediate visual responses. Meanwhile, top-down processes use our experience and attention to predict where a moving object will go, even when it briefly disappears from view (Enns & Lleras, 2008; Schmitt et al., 2018). Together, these mechanisms make motion perception one of the most powerful and efficient systems in human vision. It is precisely this sensitivity that makes motion such a potent design tool in cartography.

Cartographic animation gives maps the power to show change—not as a list of numbers or static snapshots, but as a moving story. Its effectiveness, however, depends not just on technology but on how well people can see and interpret motion. The human visual system is remarkably good at recognizing movement, transformation, and rhythm, but it also has limits. Understanding how animation works in mapping, therefore, begins with a basic distinction: some animations show how things change over time, while others simply let us move through space. The first group, known as temporal animations, presents a chronological sequence of events—a kind of visual timeline unfolding before our eyes (Harrower, 2004; Harrower & Fabrikant, 2008). These are the animations we see in weather forecasts or earthquake visualizations, where symbols flicker and shift to mark new events. Temporal animations are especially effective when the pace of change matches the viewer's sense of real time—whether that means minutes for storms, hours for traffic flow, or years for the shifting of political borders (Couclelis, 1998; Cybulski et al., 2023; Peterson, 1993, 1999). When time is well represented, the map doesn't just show movement—it conveys process. The second group, nontemporal animations, operates differently. Instead of showing the passage of time, they create the experience of motion. Think of a fly-by map that lets you soar over a mountain range or zoom smoothly toward a city. Nothing in the map itself is changing—only your point of view is (Cartwright et al., 2007; Fisher et al., 1993; Kraak, 2007). This kind of animation helps users understand spatial relationships and develop a stronger sense of orientation (Lobben, 2003; Multimäki et al., 2015). In this case, animation doesn't show change or movement through space at all—it simply draws attention. A pulsating symbol might emphasize an earthquake's epicenter, or a flashing color could mark an area of risk through dynamic highlighting (Campbell & Egbert, 1990; DiBiase et al., 1992).

Movement on a map does more than make things look interesting—it tells the eye where to look. In both cartography and interactive media, motion serves as one of the most powerful tools for directing attention. For example, animated symbols are often used to highlight important locations—a flashing icon marking a destination, or a pulsating circle showing where an event has just occurred (Robinson, 2011). The same design logic appears in computer games, where players rely on visual cues to navigate vast virtual worlds. Think of the glowing marker that shows your next objective in (1) Age of Empires IV or the flashing waypoint in (2) Need for Speed Underground 2. Similar techniques are used in (3) League of Legends, and (4) World of Tanks, where motion helps players stay oriented and focused on critical points of action (Khan & Rahman, 2018; Zagata & Medyńska-Gulij, 2023). What makes these effects so effective isn't just their visual flair—it's how our brains are wired. Psychological research shows that motion captures attention automatically, without requiring conscious effort (Bartram, 1997; Healey et al., 1995). Even subtle animated effects like a gentle pulse, flicker, or trail of smoke (Cybulski, 2016) tap into what scientists call preattentive processing—a rapid, subconscious mechanism that flags movement as something important. We can notice these signals in less than a quarter of a second, long before we've decided to look at them. (source: 1) https://www.player.one/age-empires-iv-players-can-now-change-size-minimap-update-60878-154474) 2) Need For Speed Underground 2 (source: https://games.tiscali.cz/recenze/need-for-speed-underground-2-recenze-50457), 3) League of Legends (source: https://lol.fandom.com/wiki/New_To_League/Gameplay/Vision), and 4) World of Tanks (source: https://wotbaza.com/mods-wot/120-minimap-tank-directions-for-world-of-tanks).

Among all the visual ingredients that make animated maps effective, speed may be the most underestimated. It's easy to think of animation speed as a mere technical setting—a slider to make things faster or slower—but in fact, it's a powerful cognitive cue. The tempo at which motion unfolds on a map shapes how viewers perceive, interpret, and remember what they see (Harrower, 2007; Łucjan, 2016). Too fast, and the animation races ahead of the mind; too slow, and attention drifts away. Finding the right rhythm is both an art and a science. Research in both cartography and cognitive psychology has shown that animation speed directly influences how efficiently users can process spatio-temporal information. Rapid motion can flood the visual system, making it difficult to register changes in location or value (Hegarty, 2004; Meyer et al., 2010). When symbols flash or move too quickly, the brain struggles to encode their trajectories into memory. On the other hand, when movement is too slow, users often perceive the display as static, even lifeless—a kind of visual monotony that discourages active exploration (Fisher et al., 2008). Studies in visual perception help explain why this balance matters. Colin Ware (2012) demonstrated that motion speed alters the salience of visual stimuli—faster movement draws attention, but only up to the point where clarity begins to suffer. Ric Lowe (2019) showed that pacing in educational animations shapes how mental models are built: the right tempo gives viewers time to connect what they see with what they already know. And Jeff Zacks (2004) found that our brains naturally divide continuous motion into meaningful "event segments"—but only when the speed aligns with our perceptual expectations. In animated cartography, this means that speed isn't just about style; it's about communication. The right tempo helps the viewer feel the rhythm of change—whether that's the pulse of seismic activity, the flow of traffic, or the movement of storms—without cognitive overload. As Roth (2015) and Pratt et al. (2010) emphasize, striking the right balance between pace and engagement ensures that animation enhances understanding rather than distracting from it. In this way, animation speed becomes a kind of visual language—one that must be tuned carefully to the tempo of both the data and the human mind.

In dynamic cartographic design, animation speed serves not only as a perceptual factor but also as a means of quantitative differentiation. Variations in speed can communicate differences in the intensity of a phenomenon: higher animation speeds may indicate greater activity or magnitude, whereas slower changes, in color or size, can suggest lower intensity. This principle is grounded in visual attention research, where both motion and speed are recognized as preattentive features, capable of directing visual attention in less than 250 ms (Chun & Wolfe, 2005; Wolfe, 2020). In data visualization, changes in motion parameters—particularly speed and direction—have been shown to strongly influence perceptual salience and cognitive comprehension. Such variations help viewers detect patterns more readily and retain visual information more effectively (Franconeri et al., 2021; Schill et al., 2020). When combined with other visual variables such as size or color, motion becomes part of a redundant encoding strategy, enhancing perceptual grouping and segmentation in visually complex displays (Gleicher, 2018).

In cartographic practice, these findings correspond closely with the principles of keyframe interpolation, a method that ensures smooth transitions between symbol states and minimizes cognitive strain caused by abrupt visual changes (Oberholzer & Hurni, 2000; Opach, 2006; Liu & Heer, 2014; Chevalier et al., 2014). The process involves defining an initial and a final frame, with the transformation between them automatically interpolated by the software. In the first frame, a point symbol is characterized by a set of visual variables—such as size, color, brightness, or orientation—while in the final frame, one or more of these properties are modified, for instance by enlarging the symbol or changing its hue (Battersby & Goldsberry, 2010; Łucjan & Wojtanowicz, 2018). The interpolation ensures that this transition occurs continuously and perceptibly, maintaining coherence across the animation sequence. Further refinement can be achieved by applying shape hints, which determine how

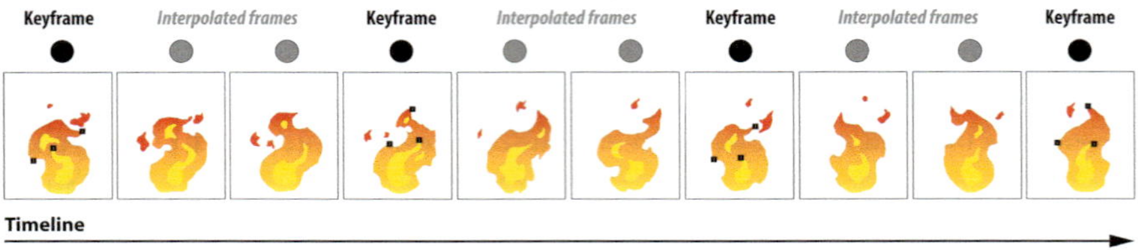


**FIGURE 2.8** Keyframe animation of symbols' shape change during a specific period, illustrating a burning fire icon. Keyframes are designed manually, and frames between them are generated automatically through interpolation. Shape hints (small rectangles within the symbol) help maintain the desired shape interpolation.

and in which direction a symbol morphs during its transformation. These subtle adjustments help maintain visual continuity and preserve the interpretative stability of the animated sequence. Figure 2.8 schematically illustrates this process, showing how dynamic point symbols can be constructed using the shape variable in conjunction with the keyframe interpolation method. Depending on the defined animation speed, the resulting motion can appear fluid and gradual or more pronounced and abrupt, directly shaping the perceptual rhythm of the visualization.

Since different visual variables can be animated, several distinct types of dynamic point symbols can be identified. The manner in which these changes are presented depends largely on the underlying animation structure. Most temporal visualizations employ a frame-based approach, in which a sequence of individual frames—or pages—is displayed in rapid succession to produce the perception of continuous change. Each frame captures a discrete state of the phenomenon, and together they form a temporal sequence depicting its evolution. In contrast, cast-based animation operates on a slightly different principle. Here, each frame contains multiple visual elements, or cells, that can independently change position, shape, or texture to create the illusion of movement. Within this framework, dynamic point symbols function much like foreground cells that periodically flash, shift, or transform against a relatively stable background (Gersmehl, 1990). This approach allows for localized animation—movement confined to specific map elements—while maintaining a coherent overall visual context.

### 2.3.3 Designing Motion: Types of Dynamic Point Symbols

Dynamic point symbols can be categorized in several ways, but one of the most fundamental distinctions concerns their geometry and mode of representation. Building on early typologies (Tyner, 1982), we could divide dynamic symbols into geometric and pictorial forms. Geometric symbols are abstract and schematic, defined primarily by their shape rather than by what they depict. Their movements—such as pulsing, expanding, or rotating—are governed by measurable changes in these visual variables. In contrast, pictorial symbols carry intrinsic meaning through their imagery. They represent recognizable objects, such as a factory, a ship, or a volcano, and therefore their motion must correspond to the real-world behavior of the depicted feature (Board, 1981; Casti, 2005; Ratajski, 1971; Rød, 2004). For instance, a factory symbol might emit puffs of animated smoke or show flickering lights in its windows—movements that reinforce its visual metaphor. Because pictorial symbols are semantically rich, defining standardized motion types for them is far more complex. Their animation depends on both the nature of the object represented and the intended message of the map. In this way, the geometric–pictorial distinction underscores how symbol form and meaning jointly shape motion design, linking visual perception with cartographic communication.

A clear illustration of how animation can reflect the meaning of a pictorial symbol can be found on the Flightradar24 platform, which provides real-time tracking of aircraft and helicopters worldwide (https://www.flightradar24.com/52.40,13.44/11, accessed on 23.11.2023). Each moving icon on the map represents an actual aircraft, continuously updating its position as the flight progresses. The helicopter symbol, for instance, demonstrates two distinct forms of motion: it travels smoothly along its flight path—conveying geographic movement—while its rotor blades rotate in place, visually reinforcing what the symbol represents. This dual animation elegantly combines spatial change with semantic realism: one motion communicates trajectory, the other expresses identity. A similar principle is used on The Map as History, an educational website featuring animated maps of historical events such as the Battle of Verdun (1914–1918) (https://www.the-map-as-history.com/First-World-War-1914-1918/the-battle-of-verdun, accessed on 23.11.2023). Here, pictorial fire symbols are enhanced with subtle smoke animations that billow upward from their centers. The effect is simple but powerful—the added motion resonates with our real-world expectations of fire, making the visualization both more expressive and more immediately legible. In both cases, movement is not decorative; it is meaningful. The animation deepens the connection between the symbol's form and its message, demonstrating how semiotic coherence—when the movement reflects what the symbol stands for—enhances both comprehension and engagement.

In contrast to pictorial designs, geometric dynamic symbols—those built from simple shapes like circles, triangles, or squares—derive their expressive power not from metaphor but from motion itself. Their animations are abstract, driven by changes in basic geometric or visual variables rather than by representational meaning (Cybulski, 2014b). Each geometric symbol possesses an inherent spatial orientation (Dukaczewski, 2014), and by animating this property—such as rotating a triangle continuously around its own axis—designers can create a sense of movement that signals activity or change. The speed of this rotation, like other motion parameters, can be precisely programmed and adjusted in real time. Building on this principle, several distinct types of motion can be defined for geometric symbols (Cybulski, 2022). Rotation emerges from altering a symbol's orientation; pulsation results from rhythmic changes in size, producing the effect of expansion and contraction—a motion style also identified by Wu et al. (2020) as "expanding/shrinking." Flickering occurs when transparency or brightness fluctuates continuously, while vibration is achieved through rapid, cyclical shifts in position, creating a subtle trembling effect. Each of these motion types reflects a different manipulation of fundamental visual variables, offering the cartographer a versatile design vocabulary for encoding dynamic information. The basic movements described here—rotation, pulsation, flickering, and vibration—are illustrated in Figure 2.9.

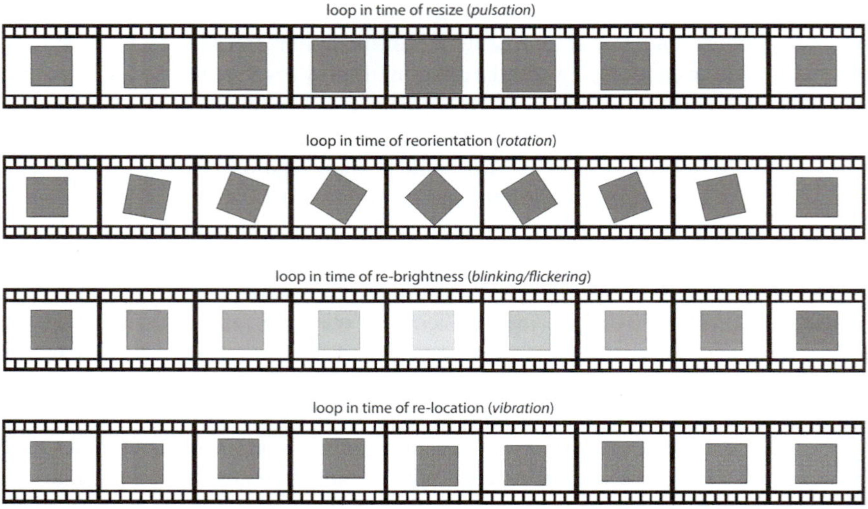


**FIGURE 2.9** Four identified types of alterations based on fundamental geometry (pulsation, rotation, blinking/flickering, and vibration) include shapes like a square, a triangle, a pentagon, or a circle.

Many of the examples discussed so far—whether drawn from computer games or web-based earthquake maps such as World EQ Locator (https://nagix.github.io/world-eq-locator/, accessed 23 November 2023), illustrate how dynamic point symbols can make specific locations stand out or emphasize notable events through movement alone. In these cases, motion primarily serves a qualitative purpose: it highlights, attracts attention, or differentiates one point from another. A pulsing dot, a vibrating icon, or a rotating marker does not necessarily quantify anything; it simply signals significance. Yet, the potential of dynamic symbols extends far beyond these qualitative functions. Motion can also be used to represent quantitative information, linking animation directly to measurable attributes such as magnitude, frequency, or intensity. This introduces a second way of classifying dynamic point symbols—based not on geometry or meaning, but on the type of data they communicate. Within this framework, dynamic motion becomes more than a visual cue; it becomes a variable of representation, capable of encoding numerical differences just as color or size might do on a static map. This expanded perspective underscores the versatility of animated cartography: dynamic symbols can both direct attention and convey data, serving as tools for storytelling and for measurement alike.

When motion is used to convey quantitative rather than purely qualitative information, its design must rely on measurable, controllable parameters. Among the many aspects that can define animation—such as direction, duration, or amplitude—the rate of change proves to be the most effective for representing numerical variation. This parameter allows cartographers to translate data intensity directly into the speed or frequency of a symbol's motion. In practice, faster movement can indicate a stronger or more frequent phenomenon, while slower motion can suggest lower intensity or stability. By manipulating the rate of change, dynamic symbols can visually express fluctuations in magnitude across both time and space. For these differences to be perceptually meaningful, however, the contrast in speed must be clearly distinguishable. Studies in visual perception (Duncan & Humphreys, 1989; Neider & Zelinsky, 2006) have shown that the human eye readily detects differences in rhythmic motion—such as changes in pulsation frequency—when they are separated by a sufficient perceptual threshold. This means that, on a map, users can intuitively associate faster pulsation or rotation with higher values of a given variable, and slower movement with lower ones. The principle is simple yet powerful: perceptible contrasts in dynamic behavior make numerical differences visible. Through calibrated changes in speed, animated symbols turn invisible quantities into observable rhythms, allowing users to grasp variation at a glance. Figure 2.10 illustrates this concept, showing how dynamic pictorial symbols with varying rates of change can be systematically designed to represent quantitative data in a visually intuitive way.

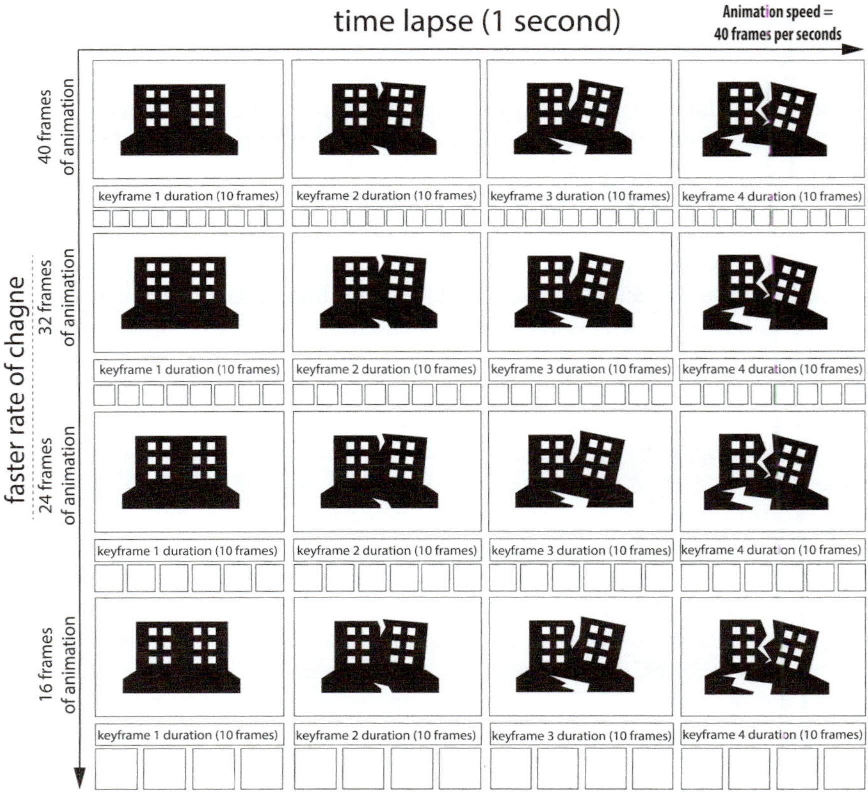


**FIGURE 2.10** Designing dynamic pictorial symbols with various rates of change, which enable the visualization of quantitative spatial data.

## 2.4 SEEING THROUGH THE EYES OF THE USER: EYE TRACKING IN MAP EVALUATION

If the earlier sections of this chapter have shown how maps can move, this one asks a different question: how do people move their eyes when they look at maps? Understanding what users actually see—where their attention goes, how long it lingers, and what is overlooked—is essential for evaluating any visual design. For decades, cartographers relied primarily on indirect measures, such as user surveys or task accuracy, to infer visual attention. Eye-tracking technology, however, makes it possible to observe attention directly. It reveals not only what users look at but also how they look—tracing the invisible choreography of the eyes as they scan, pause, and shift across a map. Eye tracking, in its simplest terms, measures and records where a person's gaze is directed. When we look at a scene, our eyes are in constant motion, performing small, rapid movements called saccades that reposition the gaze from one point of interest to another. Between these movements, our eyes briefly stop to take in detail—these pauses are known as fixations. Together, they form a continuous rhythm of perception, linking attention, vision, and cognition.

In considering the motivation for recording human eye movements, it becomes clear why this method has become so valuable in cartographic research. The primary goal of eye tracking is to capture the intricate patterns of these movements, offering a window into how people visually engage with information. Humans move their eyes to bring specific parts of the visual field into high resolution, allowing them to perceive fine details at the center of gaze (Schiller & Tehovnik, 2001). As this occurs, attentional resources are naturally drawn to the same area, enabling focused processing of the object or region of interest. The motor system coordinates these movements, while the visual system provides a continuous stream of sensory input to guide both vision and action (Land, 2009). By recording these processes, researchers gain insight into what attracts attention, what holds it, and what escapes it altogether. Eye-tracking data therefore, do more than trace lines on a screen—they reveal patterns of perception, moments of uncertainty, and the strategies users employ when interacting with complex visual materials. In essence, eye tracking allows cartographers to see through the eyes of the user, transforming map evaluation from a matter of guesswork into an empirical study of human perception (Duchowski, 2017).

Where we direct our gaze on a map is never random. It results from a subtle negotiation between the world we see and the knowledge we bring to it. Vision, as Wolfe (1994) explains, is guided by two complementary forces: bottom-up and top-down processing. In bottom-up processing, attention is driven by the stimulus itself—the brightness of a color, the sudden motion of a symbol, or the sharp contrast between features. These elements capture the eye automatically, even before we consciously decide to look. In contrast, top-down processing begins in the mind. Here, prior knowledge, goals, and expectations steer perception: we look where we think important information will be, or where we have learned to find it (Kinchla & Wolfe, 1979). Both mechanisms operate together whenever we read a map. The eye may first be drawn to a flashing epicenter through bottom-up salience, but it lingers or returns through top-down reasoning—because we know it marks the main earthquake. This interplay between sensory input and cognition, long recognized in psychology (Cavanagh, 1991), lies at the heart of cartographic perception: maps do not just show; they invite the mind to interpret.

When we talk about how the eyes move, we are really describing a collection of distinct yet coordinated actions. The main types are fixations, saccades, smooth pursuit, and nystagmus (Tatler & Wade, 2003). Each plays a specific role in how we explore visual scenes such as maps. During a fixation, the eyes hold relatively still, allowing detailed inspection of one location—perhaps a legend entry, a cluster of epicenters, or a symbol that caught our attention. Yet even in these moments of apparent stillness, the eyes are never completely motionless. They perform tiny, involuntary movements known as tremor, drift, and microsaccades, which help maintain visual stability. These microsaccades are extremely small—typically only 1 or 2 min of arc in amplitude—and occur in seemingly random directions. You can think of them as the visual system's quiet hum, a background noise generated as it tries to keep the image of an object stable on the retina (Carpenter, 1977). This "noise" is not a flaw but a feature: it prevents the image from fading due to neural adaptation and ensures that even a stationary object remains perceptually vivid. In the context of fixation analysis, these subtle fluctuations usually stay within a narrow window of about five degrees of visual angle, forming the fine-grained texture of what we call steady gaze.

Another key type of eye movement is the saccade—a rapid, jerking shift of the eyes that repositions the fovea, the small central area of the retina responsible for sharp vision, to a new location in the visual field. The term saccade comes from the French word meaning "flick of a sail," an apt description of how quickly the eyes jump from one point to another (Gregory, 1990). These movements can be both voluntary, such as when we deliberately look from one map symbol to another, and reflexive, triggered automatically by sudden changes in the visual scene. Each saccade lasts only about 10–100 ms—so brief that, during the movement itself, we are effectively blind (Shebilske & Fisher, 1983). The brain simply suppresses visual input until the eyes land on their new target. What feels like seamless vision is, in fact, a sequence of rapid, coordinated jumps stitched together by perception. On a map, these tiny leaps are what allow us to scan from legend to content, from one epicenter to another—building understanding one glance at a time.

While saccades move the eyes in quick jumps, smooth pursuit does something very different—it allows us to follow motion continuously. When a target begins to move, the oculomotor system calculates its speed and direction, then adjusts the eyes so that the gaze moves in perfect synchrony with it (Robinson, 1968). In other words, the eyes "lock on" to the moving object and track it smoothly, matching its velocity. This coordination is remarkably precise: the visual and motor systems work together to keep the image stable on the fovea even as it moves across space. On a dynamic map, for example, smooth pursuit occurs when a viewer follows an animated symbol—perhaps a pulsating point representing an earthquake—across the screen. Unlike saccades, which leap between points, smooth pursuit creates the feeling of continuous visual flow, letting us perceive motion rather than snapshots of change.

A more complex type of eye movement is nystagmus, which appears as a rhythmic, repetitive motion of the eyes. One well-studied form, optokinetic nystagmus, occurs when the eyes track a moving visual field—such as passing scenery from a train window. In this case, the eyes alternate between two distinct phases: a slow phase, during which they smoothly follow the motion, and a quick phase, where they snap back in the opposite direction to begin tracking again. When recorded, this pattern produces a characteristic sawtooth-like waveform, combining smooth pursuit and saccadic components (Robinson, 1968). A related phenomenon, known as vestibular nystagmus, operates on a similar principle but originates from head movement rather than from the motion of visual stimuli. Here, the eyes move to compensate for the head's rotation or tilt, stabilizing what we see even as our body moves. In both forms, nystagmus represents a remarkable partnership between the visual and vestibular systems—an automatic coordination that keeps our gaze steady and our vision clear during motion (Duchowski, 2017).

To capture these intricate eye movements, researchers use a device known as an eye tracker—the central instrument in modern studies of visual attention. In essence, an eye tracker measures where a person is looking and how their gaze moves over time. To achieve this precision, the system must separate movements of the eyes from movements of the head. The most straightforward way to do this is to stabilize the head, ensuring that any change in gaze corresponds directly to eye rotation. However, in many research or applied settings—especially those involving natural viewing—participants are allowed to move freely. In such cases, the device must track several features of the eye simultaneously to distinguish between head motion and true eye movement. Two features are particularly important: the pupil center and the corneal reflection, the latter created by a small, harmless infrared light source directed at the eye (Duchowski, 2017). By continuously measuring the relative position of these two points, the system can calculate the point of regard—that is, exactly where the viewer is looking. Calibration plays a crucial role in this process: before data collection begins, the viewer follows a sequence of dots on the screen, allowing the software to link eye position with specific screen coordinates. Once calibrated, the eye tracker can determine gaze location with impressive accuracy, transforming subtle movements of the eye into precise data about visual attention and behavior (Duchowski, 2017).

Calibration lies at the heart of every accurate eye-tracking study. It is the process that allows the system to translate raw eye movements into precise coordinates on a display. In recent years, this seemingly simple step has become a central topic in research on human–computer interaction and gaze-based systems, as scientists strive to make calibration faster, more robust, and more adaptive to different users and contexts (Garde et al., 2021). In practice, calibration involves asking participants to look at a series of points—typically arranged in patterns of 5, 9, or 13 targets—distributed across the screen. As each point appears, the participant is instructed to keep their gaze fixed on it while the system records the corresponding eye features. By comparing the measured gaze positions with the known coordinates of these calibration points, the eye tracker can compute an accurate mapping between the user's eye movements and the display surface (Kasprowski et al., 2014). In effect, calibration teaches the system how to "read" each person's unique eye geometry, ensuring that every subsequent fixation and saccade is measured with precision.

Modern eye trackers—such as the Tobii Pro Fusion, Gazepoint GP3, or EyeLink 1000+—represent a remarkable leap in precision and speed. These devices can record the position of the gaze in real time, delivering instantaneous (x, y) coordinates at sampling rates reaching up to 2,000 Hz. To put that in perspective, this means the system captures as many as 2,000 measurements every second, allowing researchers to follow even the smallest and fastest eye movements with exceptional accuracy. Such temporal resolution is essential for studying the subtle dynamics of visual attention—how quickly the gaze shifts, how long it lingers, and how it reacts to changing stimuli on a map. Thanks to this capability, eye trackers have become indispensable tools in cartographic research, enabling scientists to analyze not only where users look, but how their eyes—and, by extension, their minds—move through visual information (Ooms et al., 2015; Ooms & Krassanakis, 2018).

### 2.4.1 What Eye Movements Reveal: Fixations, Saccades, and Scan Paths

Eye-movement metrics offer a unique window into how people see, think, and decide. When we look at a map, our eyes are not merely gathering information—they are performing a complex selection process known as visual attention. This process allows us to identify a target object, such as a specific earthquake symbol, among many potential distractors scattered across the display (Cybulski, 2020; Wolfe, 2000). Each fixation, saccade, or pause reflects a moment of cognitive decision-making: where to look next, what to ignore, and what information to process in detail. In this sense, eye-tracking data serve as both a record of movement and a trace of thought. Beyond attention, eye-movement metrics have also been linked to emotional arousal and stress—factors that influence how people perceive and process information (Skaramagkas et al., 2021). Although these dimensions play a vital role in education, psychology, and cognitive science, they fall outside the scope of this cartographic study. Here, our focus remains on the perceptual and cognitive mechanisms that guide how users explore maps. Nevertheless, it is worth noting that the same physiological data used to analyze map reading can also provide insight into the viewer's cognitive workload—that is, the mental effort required to interpret complex visual information (Skaramagkas et al., 2021). Understanding this workload is key to designing maps that not only communicate clearly but also respect the limits of human attention and memory.

A fixation occurs when the eyes pause briefly to focus on a single point, allowing detailed visual processing to take place. During this moment of stillness, the brain gathers the information needed to recognize symbols, read labels, or interpret spatial relationships on the map. Researchers use several fixation-based metrics to describe how this process unfolds (Bylinskii et al., 2017). Among the most common are the number of fixations, which indicates how often the gaze stopped during a task, and the number of fixations within each area of interest (AOI), showing where attention was concentrated. The total fixation duration—that is, the sum of all fixation times on a given AOI—reveals how much effort or interest that region demanded. Other important measures include the average fixation duration, which can suggest how deeply the user processed information, and the time to first fixation, reflecting how quickly attention was drawn to a specific target. Researchers may also examine fixation density (how closely fixations cluster together) and repeat fixations, which occur when the eyes return to the same place for a second look. Together, these metrics form a detailed map of visual attention, showing not only where users look, but also how they look and for how long.

For example, the average duration of fixations provides valuable insight into how attention operates over time. The number of fixations per second can be interpreted as a measure of attentional speed—how rapidly the eyes move from one point to another. These two measures are closely related: when fixations become longer, they usually occur less frequently, indicating that the viewer is spending more time processing complex or demanding information. In other words, an inverse relationship typically exists between fixation duration and fixation count—longer fixations often signal a heavier cognitive load or greater mental effort (Keskin et al., 2018).

Another important class of eye movements is the saccade—a rapid, ballistic shift of the eyes from one fixation point to another. These movements act as the visual system's way of repositioning attention, allowing us to scan a scene efficiently. The amplitude of a saccade can vary considerably: small saccades occur when reading text or inspecting details within a single area, while larger ones help the eyes jump between distant objects or regions on a map. To describe these movements quantitatively, researchers rely on several saccadic metrics, including the number of saccades, their velocity and amplitude, the saccade rate (how often they occur per second), and their duration (Rucci & Poletti, 2015). Patterns in these measures can reveal a great deal about a user's cognitive and emotional state. For instance, an increase in average saccade velocity has been associated with higher stress levels or greater task complexity, often reflecting a drop in sustained concentration. Conversely, higher cognitive load—when the brain is working harder to interpret information—is typically linked to shorter saccades combined with faster eye movements (Behroozi et al., 2018). In this way, saccadic behavior not only shows where the eyes move, but also how the mind is coping with the demands of the task.

In addition to fixations and saccades, other physiological indicators—such as pupil size and blinking—offer valuable insight into visual and cognitive processes. The pupil diameter is particularly sensitive to both internal and external factors. It varies naturally with a person's emotional state, age, and level of mental effort (Bradley et al., 2008). A larger pupil often signals heightened cognitive or emotional engagement (Skaramagkas et al., 2021). This response becomes especially apparent during demanding map-based tasks, such as performing a focused search for a specific feature or planning an optimal route across a complex landscape (Kiefer et al., 2016). However, pupil size is also affected by environmental conditions—lighting, temperature, or humidity—which must be carefully controlled or accounted for during experiments (Winn et al., 1994). Similarly, blinking—the rapid closing and reopening of the eyelids—is another measurable aspect of eye behavior. Although semi-involuntary, it carries meaningful information. Researchers often track blink duration and blink rate (typically expressed as blinks per minute) to infer levels of attention, fatigue, or discomfort. Like pupil dilation, blinking is also influenced by external factors such as air dryness or ambient temperature (Ousler et al., 2008). When analyzed alongside fixations and saccades, these subtle physiological signals contribute to a fuller understanding of how users visually and cognitively engage with maps.

Recent research has shown that physiological eye metrics can serve as subtle indicators of mental effort during map use. For instance, Cheng et al. (2023) found that an increased blink rate often accompanies a rise in cognitive load during mobile map navigation. However, blinking frequency alone cannot fully capture the complexity of cognitive processing—it becomes truly informative only when interpreted alongside other metrics such as fixation duration or pupil dilation. In related work, Kiefer et al. (2016) observed systematic differences in mean pupil diameter across various types of map-based tasks. Their findings revealed that free exploration tends to produce the lowest cognitive load, while tasks requiring search, polygon comparison, or line following generate moderate load. The most demanding activities—route planning and focused search—evoke the greatest pupil dilation, reflecting intense mental engagement. Taken together, these results highlight the importance of integrating multiple eye-movement indicators when evaluating map usability. They also suggest promising directions for designing adaptive cartographic interfaces that respond intelligently to a user's cognitive state, tailoring map complexity and visual feedback to maintain optimal comprehension (Fairbairn & Hepburn, 2023).

### 2.4.2 From Gaze to Insight: Eye Tracking in Cartographic Research

Although eye tracking has a long history in psychology and cognitive science, its systematic use in cartography is a much more recent development. In psychology, eye-movement studies have helped reveal how people read, recognize faces, or process visual scenes. In cartography, the same method opens a new way of understanding how users engage with maps—tools that represent the complexity of geographic space. By recording where, when, and for how long the eyes rest on specific elements, eye tracking allows researchers to quantify visual attention in precise spatial and temporal terms. Each fixation becomes a data point in a larger behavioral map, showing how the viewer's gaze travels across legends, symbols, and labels. This ability to measure attention directly has made eye tracking a valuable bridge between perception research and map design. Today, it is widely used to investigate questions in spatial cognition, Geographic Information Science (GIScience), and cartography, helping researchers move from observing what maps show to understanding how they are truly seen (Kiefer et al., 2017).

One of the most dynamic areas where eye tracking has deepened our understanding of spatial behavior is map-based navigation. In this domain, researchers have examined how users visually interact with maps across a variety of contexts—from traditional mobile navigation (Dong et al., 2021) to animated route displays (Cybulski, 2021) and immersive virtual reality (VR) environments (Zagata et al., 2021). During such tasks, participants engage in activities that mirror real-world challenges, including orientation, wayfinding, and self-localization within an unfamiliar space. Eye-tracking analyses have revealed which symbol types are recognized most quickly and which help users construct more detailed mental representations—or cognitive maps—of their surroundings (Franke & Schweikart, 2017). The studies also confirm the central role of landmarks in navigation. When participants attempt to determine their own position, their gaze consistently gravitates toward distinctive geographic features—such as rivers, railway tracks, or road junctions—indicating that these elements anchor spatial understanding (Wenclik & Touya, 2023). Interestingly, modern mobile mapping applications increasingly rely on pictorial symbols that can be interpreted intuitively, without the need for consulting a legend. Across different platforms, a degree of symbolic consistency has emerged—icons for transportation, food, or accommodation tend to share visual forms. This standardization helps users transfer symbol knowledge between applications, streamlining interpretation and reducing cognitive load (Piątek & Gołębiowska, 2022).

Eye-movement tracking has also become a standard method for investigating how users search for and recognize symbols on both two- and three-dimensional maps (Yang & Li, 2021). In simple terms, such visual search tasks help researchers understand which map elements most effectively draw and hold attention. According to the general theory of vision (Wolfe, 2021), certain visual properties—such as color, orientation, or shape—can be detected almost instantly, even before conscious attention is directed toward them. These are known as preattentive features, and their role in cartographic design has been confirmed through eye-tracking experiments (Cybulski & Krassanakis, 2023). Research by Pomerantz (2003) and Krassanakis (2013), for example, demonstrated that symbols containing distinctive visual cues—such as a small hole or internal gap—are identified more rapidly than those lacking such characteristics. Similarly, Stachoň et al. (2020) found that the addition of color further accelerates the search process, improving detection speed for both simple and pictorial symbols compared to their black-and-white counterparts. Moreover, the shape of a cartographic symbol has long been recognized as a key factor in facilitating its identification (Michaelidou et al., 2005). Taken together, these findings bridge modern empirical results with classic cartographic theory—specifically, Jacques Bertin's (1981) concept of visual variables, which emphasized that differences in shape, size, and color enable effective visual distinction. Through eye-tracking research, these foundational principles of symbol design have gained concrete experimental validation, linking perception science directly with the practice of mapmaking.

When map-reading tasks become more complex, the visual demands placed on the user also increase. In such cases—like comparing symbol sizes in a graduated point symbol map (Cybulski, 2020) or evaluating color values in a cartogram or chorochromatic map (Brychtová & Çöltekin, 2016, 2017)—the spatial distance between the elements being compared becomes a critical factor influencing how effectively users can make visual judgments. Simply put, the farther apart the two features appear on the map, the greater the cognitive effort required to compare them accurately. Another key aspect that shapes visual search performance is the location of the target symbol itself. Early research by Lloyd (1997) demonstrated that objects placed near the center of the map are found more quickly and with higher accuracy than those located at the edges or corners. This finding aligns with how human vision naturally favors central areas, where visual acuity and attentional focus are strongest. Of course, not all users approach these tasks in the same way. Experience level plays a significant role in how efficiently information is extracted from maps. Studies have consistently shown clear differences between experts and novices (Çöltekin et al., 2017; Keskin et al., 2019). Experts tend to apply more systematic and targeted visual strategies, often anticipating where important information is likely to appear. Novices, by contrast, exhibit less structured gaze patterns and rely more heavily on trial-and-error exploration (Çöltekin et al., 2010; Keskin et al., 2020).

These differences highlight that eye-tracking data can reveal not only how well users perform but also how they visually think their way through a map.

Without revisiting in full the comprehensive overviews provided by Kiefer et al. (2017) and Krassanakis and Cybulski (2021), it is worth emphasizing that eye tracking has become central to understanding how users interact with animated cartographic displays (Dong et al., 2014b). Dynamic maps, composed of multiple changing elements, can be powerful tools for visual communication—but only when designed with careful attention to human perception. Well-constructed multicomponent animations have the potential to guide users' attention efficiently, supporting faster and more accurate cognitive processing (Opach et al., 2011, 2014). However, the capacity of users to divide attention across several simultaneously moving components remains a point of debate (Opach & Nossum, 2011). When too many elements change at once, the visual system may struggle to register subtle differences—a phenomenon known as change blindness. In cartographic animation, this occurs when users fail to notice a modification in a map's visual state, such as a shift in color intensity on a choropleth map (Cybulski & Krassanakis, 2021; Goldsberry & Battersby, 2009; Moon et al., 2014). This limitation partly explains why user performance can vary between different types of map-based tasks, such as filtering, tracking movement, or detecting anomalies (Gołębiowska et al., 2017). Interestingly, research by Griffin et al. (2006) showed that animated maps can outperform static small multiples in communicating spatial patterns, especially when the animation pace is properly adjusted. Slower transitions may help users perceive gradual spatial changes, while faster sequences can highlight strong or abrupt patterns. These findings underscore a key principle: the success of animated cartography depends not merely on motion itself, but on how motion is tuned to the limits and strengths of human visual attention.

Eye-tracking research has also proven invaluable for uncovering difficulties users face when interpreting maps and for highlighting the differences between experts and novices (Ooms et al., 2014). Studies comparing these groups reveal that experience fundamentally shapes both the speed and strategy of visual processing. In dynamic and interactive map tasks, experts tend to locate place names or symbols significantly faster than novices—a pattern reflected in their eye-movement metrics. Specifically, experts show shorter fixation durations and a higher fixation rate per second, suggesting that they can extract and integrate more information from the map within the same amount of time (Ooms et al., 2012). This results in greater efficiency and smoother visual search patterns, as experts rely on well-developed mental schemas that guide their attention effectively across the map. However, later findings by Keskin et al. (2020) offer a more nuanced perspective. In tasks rated as moderately difficult or complex, expert users actually required more time to complete the activity—but their accuracy was higher. This suggests that experts may engage in more deliberate and strategic information processing, verifying details rather than relying solely on intuition. In other words, expertise does not always translate to speed; it often reflects a different quality of attention—one that balances efficiency with precision. Together, these studies illustrate how eye tracking not only measures performance but also reveals the underlying cognitive strategies that distinguish how people, from novices to experts, truly see a map.

## 2.5 WHAT WE ALREADY KNOW AND WHAT WE STILL DON'T: RESEARCH ON DYNAMIC SYMBOLS

The development of dynamic point symbols marks a significant milestone in the evolution of cartographic visualization. Early foundations can be traced to the pioneering work of Tinghua (1998), who extended Jacques Bertin's (1981) visual variable framework into the temporal domain. Building upon the conceptual groundwork established by DiBiase et al. (1992), Tinghua introduced a set of new dynamic visual variables—including duration, rate of change, and rhythm—to describe how motion could function as an integral component of cartographic representation. His study outlined three principal uses for dynamic point symbols. The first was highlighting specific spatial locations, often achieved through blinking or flashing symbols that immediately capture a viewer's attention. The second use involved expressing object attributes, extending the logic of quantitative cartography into animation—where dynamic properties like speed or pulsation encode information such as intensity, magnitude, or category. The third and perhaps most transformative function was representing real-world temporal change: the unfolding of events over time, from weather patterns to historical processes. In this sense, dynamic point symbols served as an animated and interactive extension of traditional mapping methods, enabling the visualization of temporal phenomena in a way that static maps could only suggest. Through motion, the language of cartography expanded—making time itself a visible and interpretable dimension.

An important early step in refining the concept of dynamic symbols was taken by Xiaofang et al. (2005), whose work explored both the development and functional role of animation in map symbol design. Their study sought to consolidate existing knowledge about dynamic visual variables and establish them as a conceptual foundation for constructing animated symbols. In doing so, they proposed a design procedure based on interpolation between keyframes—a method borrowed from computer animation that allows smooth transitions between defined visual states. Xiaofang and colleagues identified two fundamental types of symbol animation. The first type is closely tied to real-time dynamics, where the animation reflects actual temporal progression in the data—such as the movement of a storm system, the spread of an earthquake sequence, or the evolution of traffic flow. This approach conveys the procedural change of phenomena as time advances, maintaining a direct relationship with the real-world temporal dimension. The second type, by contrast, focuses on continuous visual motion designed primarily to attract attention, regardless of real-time correspondence. Here, animation functions as a perceptual cue rather than a literal depiction of temporal change. Its purpose is to draw the viewer's eye, emphasizing certain map elements or categories without implying the passage of time. Together, these two forms of animation—temporal and attention-driven—highlight the dual role of dynamic symbols: they can both represent change and guide perception.

This dual classification of animated map symbols underscores the multifaceted nature of dynamic representation. On one hand, animations that correspond to real-time dynamics emphasize temporal accuracy—their purpose is to mirror genuine changes in the represented phenomenon, preserving the integrity of the time sequence. Such animations strengthen the analytical dimension of a map, allowing users to observe how spatial patterns evolve naturally. On the other hand, attention-driven animations serve a different but equally valuable role. Their strength lies not in temporal fidelity but in visual communication—they are designed to capture the viewer's gaze, emphasize key features, and sustain engagement. Recognizing this distinction is essential for the effective design of dynamic symbols. It reminds us that animated maps can serve both informational and perceptual goals: they can communicate what changes over time, but also guide how users see and interpret those changes.

In concluding their study, Xiaofang et al. (2005) outlined a wide spectrum of applications for maps employing dynamic symbols, illustrating how animation could expand the communicative potential of cartography. They proposed that such maps could be especially effective in visualizing historical events, earthquakes, floods, and street traffic—phenomena defined by motion, transformation, and temporal progression. Through animation, historical maps could move beyond static depictions, allowing viewers to experience temporal sequences dynamically, witnessing how events unfold rather than merely reading about them. Similarly, in the context of natural hazards such as earthquakes or floods, animated maps provide a means to visualize evolving impacts over time, transforming complex geophysical processes into intuitive, time-based narratives. Beyond these core examples, Xiaofang and colleagues emphasized the broader promise of dynamic visualizations across domains such as environmental monitoring, urban planning, and disaster response. In these fields, the integration of animated maps not only enhances analytical understanding but also heightens public awareness—inviting users to engage actively with changing spatial phenomena. Their conclusions underscored a vision of cartography where motion becomes more than aesthetic enhancement—it becomes an essential medium for communicating the living, changing nature of the world.

Another important contribution to the study of dynamic point symbols came from Lai and Yeh (2004), who focused on how animation could be used to represent nontemporal and qualitative changes on maps. Their work sought to evaluate how effectively flickering symbols—both points and lines—could convey information and capture attention in comparison to static representations. To assess this, they measured two key performance indicators: accuracy and reaction time. Accuracy was determined by the number of trials participants required to correctly identify the target symbol, while reaction time reflected the speed of recognition. The experiment was carefully structured into three phases, beginning with a series of visual tests designed to determine whether there were statistically significant differences in reading times between static and blinking point symbols. Several variables were systematically tested, including symbol size, the number of symbols displayed, and foreground–background contrast—all factors known to influence visual detection. Within this stage of the study, the researchers addressed several specific questions:

a. Are blinking symbols more visible than static ones?
b. How does symbol size affect visibility?
c. Is there a threshold beyond which too many blinking symbols become visually distracting?

d. How do different symbol classifications or combinations perform when numerous points are shown simultaneously?
e. Finally, does blinking improve visibility when the contrast between foreground and background is poor?

Through these questions, Lai and Yeh's study established an early experimental framework for understanding how motion—specifically flickering—affects perception in cartographic displays, bridging the gap between attention research and applied map design.

The second phase of the Lai and Yeh (2004) study focused on how users perceive different blinking speeds in animated point symbols. The researchers sought to determine whether participants could accurately distinguish between varying flicker rates, effectively testing the limits of perceptual discrimination in temporal variation. This experiment was not merely about visibility—it explored how finely the human eye can detect differences in rhythm, an essential factor when designing maps where motion encodes meaning. The third phase expanded the investigation to animated line symbols, introducing additional visual variables such as line length, thickness, and animation style. The central question here was whether these factors influenced how users perceived the speed of animation. This part of the study was particularly relevant for applied contexts, as the researchers aimed to extract practical insights for the design of online traffic congestion maps—visualizations in which varying line animations represent different levels of traffic flow. Two distinct animation methods were tested. The first, known as the marquee or interlaced flashing technique, involved lines that appeared to shimmer or flash sequentially along their length. The second method, called grow-out or incremental lengthening, simulated lines that extended progressively within defined bounds, mimicking movement or accumulation. Both techniques were tested under combinations of three animation speeds and variations in line length and thickness, allowing the researchers to evaluate how each factor affected visual clarity and perceptual differentiation. Through this structured approach, Lai and Yeh's study provided one of the earliest systematic examinations of motion perception in linear cartographic symbols, paving the way for later research on dynamic map design and user-centered animation principles.

The results of the visual experiments largely confirmed what might be intuitively expected. Dynamic symbols—those that blinked or flashed—proved to be more visible than their static counterparts, even under challenging conditions with low foreground–background contrast. Motion, in this case, acted as a powerful visual cue, allowing users to detect symbols more readily and with greater accuracy. However, the study also revealed that this advantage came with limits. While flickering enhanced visibility, the researchers were unable to identify a precise threshold for how many blinking symbols could appear on the screen before the display became visually overwhelming. As the number of animated elements increased, participants' reading times grew significantly, suggesting a gradual rise in cognitive load and visual interference. Importantly, Lai and Yeh observed that user performance varied considerably depending on individual ability. Some participants were able to manage dense, animated displays with relative ease, while others struggled as the visual complexity increased. This finding highlighted a key principle for cartographic design:

the effectiveness of dynamic symbols is not absolute but depends on both map design parameters and the perceptual capacities of users.

The findings of Lai and Yeh's (2004) study led to several meaningful conclusions that continue to inform the design of dynamic maps today. The researchers noted that while symbol classification remains a logical and effective approach in traditional cartography—particularly when managing large numbers of elements on a single display—animated or blinking symbols can provide an equally, if not more, discernible alternative. Motion, it seemed, could substitute for classification when used thoughtfully. Participants across different levels of cartographic experience were generally able to distinguish between varying animation speeds, though less experienced users showed greater difficulty in doing so, reflecting the influence of expertise on perceptual sensitivity. Interestingly, the results also showed that larger differences in animation speed made perception easier and faster. When the contrast between blinking rates was pronounced, users required less time to correctly identify the variations. Even when the display contained up to 15 blinking point symbols, participants still reported the visualization as acceptable and manageable, suggesting a relatively high tolerance for dynamic clutter before it becomes distracting. Another key insight concerned the role of contrast. While high contrast is critical for static maps, animation proved to be an effective compensatory mechanism in low-contrast environments, significantly enhancing symbol visibility. Moreover, symbol size and thickness were shown to influence efficiency: larger symbols (4 mm) and thicker lines (9 pixels) resulted in faster and more accurate reading times. Taken together, these outcomes demonstrated that when designed carefully, animated symbols can improve perceptual clarity and user performance, especially under visually demanding conditions.

Building on the foundations laid by earlier studies, Guorui et al. (2006) advanced the discussion of dynamic symbol design by proposing a technical framework for constructing animated map symbols. Their approach was grounded in established theories of dynamic visual variables, yet extended them into a more computational and systematic form. The researchers introduced a multi-transform assembly method, which provided a structured way to control symbol motion and transformation through defined parameters. At the core of their framework was a keyframe-based model, in which each keyframe represented a specific state of the symbol—defined by attributes such as size, shape, color, position, and orientation. Between these keyframes, intermediate states were generated through interpolation, creating smooth transitions over time. The model incorporated both geometric and nongeometric properties, allowing the animation to evolve not only in form and position but also in more abstract visual dimensions. Geometric attributes, in particular, were managed using two-dimensional morph transformation matrices combined with affine transformations, enabling precise control over how symbols scaled, rotated, and deformed as they animated. This work represented an important step toward formalizing the design of dynamic symbols—moving from conceptual discussions of motion to a practical computational system capable of producing consistent, repeatable animations for cartographic applications.

Guorui et al. (2006) emphasized that dynamic visual variables should not replace traditional cartographic principles but rather evolve from them. In their view, motion in maps could be expressed through transformations of static visual variables—extending the classic framework of Bertin (1981) into the temporal domain. To achieve this, the authors introduced a multilayered symbol structure, enabling multiple visual components to animate simultaneously. Each animation sequence was anchored by keyframes, which defined the start and end points of a transformation, while interpolation functions governed how symbols transitioned between these states. These functions could simulate various dynamics—such as acceleration, deceleration, or periodic oscillation—closely mirroring the logic of film animation and aligning cartographic design with broader principles of motion graphics. The most innovative contribution of the study, however, was its symbol storage system, which employed a database-driven structure to manage dynamic symbol components efficiently. This allowed symbols to be not only designed but also stored, retrieved, and manipulated in ways compatible with real-time rendering. The authors further proposed a symbol transformation function matrix capable of handling geometric transitions, such as morphing a circle into a square. By combining geometric morphing with attribute interpolation, their framework offered a powerful foundation for implementing dynamic map symbols within GIS environments, bridging conceptual design and computational execution.

Cybulski (2014a) conducted one of the first studies to focus exclusively on a single type of dynamic visual variable—rotation. In this experiment, two distinct symbol designs were tested. The first relied on a simple hexagonal geometry, while the second incorporated an additional graphic line element extending from the hexagon. This line served as a visual cue, helping participants more easily perceive the angle and speed of rotation. The dynamic symbols were used to represent unemployment rates in Poland over a 10-year period (2002–2012), allowing the researcher to test how effectively motion-based encoding could convey changes in a statistical dataset. The experiment involved 34 participants, who viewed a time series of animated maps in which each year was displayed for 5 s, followed by a 5-s interval showing the corresponding date. During the experiment, participants were asked to answer questions related to the content of the map, enabling the researcher to measure both effectiveness (accuracy of responses) and efficiency (response time). The study's findings offered valuable insights into the perceptual behavior of map users. Cybulski concluded that increasing the difference in rotation speeds among symbol classes significantly improved the clarity of interpretation. In other words, users were better able to distinguish between categories when rotational speeds were more noticeably differentiated, regardless of the specific number of class intervals used. This suggested that perceptual separation—not just the visual design of the symbol—plays a key role in ensuring the legibility of dynamic quantitative maps.

During the experiment, participants were observed spending additional time studying the rotational speeds of symbols displayed in the map legend before turning their attention to the main map. This behavior reflected a natural strategy: users first sought to understand the meaning of motion before interpreting the broader spatial patterns. As a result, Cybulski suggested that map legends should themselves include animated examples of the symbols used, allowing viewers to familiarize themselves with motion cues prior to the full animation of the map. Such a design practice can significantly enhance comprehension and reduce cognitive effort during dynamic map reading.

Cybulski and Wielebski (2019) extended the investigation of dynamic point symbols by examining how different types of symbol motion influence the effectiveness and efficiency of map-based communication. Their research explored how animation within the symbol itself—rather than merely over time—affects users' ability to interpret quantitative data. Three forms of symbol variability were tested, each based on a simple geometric shape: blinking, pulsation, and rotation. These were compared against a classic animation approach, where only the temporal sequence changes while the symbols remain static. This design allowed the researchers to isolate the perceptual contribution of internal symbol dynamics to map comprehension. The study engaged 160 participants, who viewed a series of maps presented in several animation variants. In one experimental condition, each map depicted data for a single year during the time series, allowing users to focus on the behavior of the symbols themselves rather than temporal transitions between frames. Participants viewed either static or animated symbols for a fixed duration, after which the map disappeared, and they were asked to identify the value represented by a specific symbol. Beyond accuracy and response time, the researchers also examined how long participants spent reading the animated legend before beginning the task, as well as their self-reported confidence in the answers they provided. This multifaceted design enabled the authors to assess not only what participants perceived, but how certain they felt about their interpretations—offering a more comprehensive picture of how dynamic symbols influence both cognitive processing and user assurance in map reading.

The results of Cybulski and Wielebski's (2019) study demonstrated that dynamic point symbols can be just as effective as their static counterparts in communicating quantitative information—provided users are given slightly more time to process them. Specifically, the researchers found that participants required an additional 3 s to analyze animated symbols before reaching a level of performance comparable to static maps. After this short adjustment period, accuracy levels approached 90%, indicating that motion-based encoding does not inherently hinder comprehension but simply demands a brief period of perceptual adaptation. A particularly noteworthy finding was the strong correlation between the amount of time participants devoted to studying the animated legend and the correctness of their responses. In other words, the more time users spent familiarizing themselves with how motion represented data, the more accurately they interpreted the map. This observation aligns with earlier conclusions by Buziek (2000), who found that users of noninteractive animated maps rely much more heavily on legends than users of static visualizations. The legend, in such cases, serves as a key interpretive tool for decoding dynamic variables. An exception to this pattern was observed for rotation-based symbols. Participants required minimal time to understand their meaning, suggesting that rotational motion is more intuitive and immediately interpretable than other forms of animation, such as blinking or pulsation. This insight reinforces the idea that not all types of motion are equal in their perceptual clarity—some, like rotation, may inherently align better with users' visual expectations and cognitive processing.

Further insights into the perception of dynamic point symbols were provided by Cybulski and Krassanakis (2023), who examined whether motion in map symbols could be processed preattentively—that is, detected automatically, without the need for conscious visual search (Hohnsbein & Mateeff, 1998). The study involved 103 participants, divided into two groups, whose eye movements were recorded using an EyeLink 1000 Plus eye-tracking system. This setup allowed the researchers to capture precise data on how users' gaze interacted with animated map elements. The core research hypothesis proposed that the velocity of motion in dynamic point symbols constitutes a preattentive visual feature, meaning that users could recognize differences in motion speed almost instantly, even without actively searching for them. To test this, the experiment compared several types of symbol motion—rotation, blinking (or flickering), and pulsation—each representing a distinct form of temporal variation. The goal was not only to determine whether these motion types attract attention automatically but also to identify the optimal motion parameters, specifically how symbol speed should vary across class intervals to achieve maximum perceptual clarity. This research thus advanced earlier studies by linking the theoretical concept of preattentive processing directly with empirical eye-tracking evidence, providing a deeper understanding of how motion-based visual variables function in the perception of animated maps.

The study also found that the distribution of motion velocity among symbols played a critical role in perception. When symbol speeds followed a logarithmic distribution, participants had greater difficulty identifying the fastest-moving elements, indicating that such scaling does not favor preattentive detection. In contrast, the linear distribution of velocity appeared to facilitate quicker and more accurate recognition. Among all tested motion types, the rotating square emerged as the most effective stimulus for preattentive detection. Regardless of how the motion speeds were distributed, participants consistently identified rotating symbols faster and with greater accuracy. This finding underscores that certain geometric forms and motion types, particularly rotation, possess inherent perceptual advantages, making them especially suitable for attention-guiding tasks in animated maps.

One of the most recent investigations into dynamic point symbols was conducted by Ge et al. (2025a), who explored how different encoding types, animation speeds, and modulation areas influence user performance in visual search tasks. Their study compared three forms of motion-based variation—flashing, pulsation, and lightness modulation—to determine which combination of parameters most effectively guided attention. The results revealed that, under specific conditions, flashing at certain animation rates produced higher effectiveness than pulsation, suggesting that temporal rhythm and frequency play a crucial role in perceptual efficiency. Perhaps the most striking finding was that contour-based pulsation, in which only the outline of the symbol is animated, performed better than animating the entire symbol area. This design reduced visual fatigue while maintaining strong attentional capture.

Interestingly, these conclusions were not entirely consistent with those from a companion study by Ge et al. (2025b), which found that color-changing point symbols can also serve as highly effective dynamic encodings. Moreover, while earlier studies—such as those by Cybulski and Wielebski (2019) and Cybulski (2024)—indicated that higher animation speeds tend to improve detection, Ge and colleagues observed the opposite effect: a lower animation rate led to both greater accuracy and shorter response times. Despite these differences, both studies converged on one important insight—that pulsation, when applied selectively to the symbol's contour, can successfully convey information without overwhelming the user. This subtle form of motion appears to balance perceptual clarity with aesthetic restraint, pointing toward new directions in the design of user-friendly dynamic symbols.

### 2.5.1 Where the Gaps Remain: The Need for Research

Dynamic point symbols, whether pictorial or geometric, represent one of the most recent advancements in GIS—introducing motion as a new dimension in cartographic communication. Like traditional static symbols, these dynamic forms are attribute-driven, meaning their behavior is directly tied to data values rather than arbitrary animation. This allows them to represent quantitative information not only through conventional visual variables such as size, color, and shape, but also through dynamic variables—for example, motion speed, direction, or rhythm. By integrating these temporal elements, modern cartography moves closer to depicting how geographic phenomena actually change over time, transforming maps from static images into living visual systems that mirror the dynamics of the world they represent.

While visual variables have long been supported by a well-established methodological framework for representing quantitative differences, the situation is far less mature when it comes to dynamic variables. In static cartography, the relationships between data variation and visual encoding—such as changes in color value, size, or shape—are grounded in decades of empirical testing and theoretical refinement. By contrast, the use of motion as a quantitative indicator remains comparatively underexplored. Only a handful of studies, including those discussed earlier, have undertaken systematic empirical investigations into how variations in motion parameters—such as speed, frequency, or rhythm—relate to data class differences. Even so, the existing body of research offers several important insights, suggesting that:

- Recognizable motion differences follow perceptual similarity principles. For users to reliably distinguish between varying speeds of symbol movement, the differences must adhere to the rules of Similarity Theory (Duncan & Humphreys, 1989). In practice, this means that the distribution of motion velocities across data classes should follow an arithmetic or exponential progression, which allows viewers to detect the fastest-moving symbols preattentively (Cybulski & Krassanakis, 2023).
- Dynamic symbols have demonstrated empirical validity and practical potential. A growing number of experimental studies, even those conducted with fictitious datasets, have shown that dynamic symbols can be both effective and efficient in map interpretation tasks. Their successful implementation in GIS environments—including platforms such as ArcGIS Pro and QGIS (with QField 2.2)—further confirms their potential for spatio-temporal visualization across both qualitative and quantitative mapping contexts.
- Different motion types and geometries vary in perceptual effectiveness. Research consistently indicates that certain forms of motion—such as rotation or pulsation—are more easily perceived than others, and that symbol geometry can significantly influence recognition accuracy (Lai & Yeh, 2004; Cybulski & Wielebski, 2019).

Taken together, these findings point to a clear research gap in the application and empirical evaluation of dynamic symbols using real-world data. While experimental studies with controlled or fictitious datasets have provided valuable theoretical insights, there remains a need to validate these principles in authentic cartographic contexts. Future research must focus on developing dynamic maps of real phenomena that incorporate optimized motion parameters—particularly speed differentiation—and empirically tested symbol geometries and motion types. Only through such applied studies can we fully assess how dynamic variables perform under real data conditions and translate experimental promise into practical cartographic solutions.

# 3 How the Study Was Done

## *Combining Online Surveys and Eye-Tracking*

Before examining the results, it helps to understand how the study was planned, structured, and carried out. This chapter walks the reader through the entire research process—from the first ideas about how earthquakes should be shown on a map, to the creation of animated symbols, and finally to the evaluation of how people interact with these maps.

The study uses a mixed-methods approach, meaning that it combines statistical analysis with user-based experiments. Statistical methods are used to check the research hypotheses and to measure how well participants performed during the tasks. Such an approach is common in contemporary cartographic research, especially in studies that examine how people read and understand maps (Cybulski, 2020; Dong et al., 2018a; Liao et al., 2017).

The research process unfolded in several key stages:

The first stage (Section 3.1) establishes the foundation of the study. Here, the goals of the research are clarified, the spatial data sources are selected, and early decisions are made about how earthquakes should be represented on the map. This includes choosing which visual and dynamic variables to use, and ensuring that the planned visualization will be clear, readable, and appropriate for user testing.

In the second stage (Section 3.2), the conceptual ideas are turned into actual maps. The dynamic earthquake symbols are designed, motion is added in a way that makes sense cartographically, and differences in speed between data classes are defined. The symbols are then placed onto the map according to the spatial data. At the same time, a simple graphical user interface (GUI) is created so that users can interact with the visualization, especially by controlling the timeline.

The third stage (Section 3.3) focused on producing the final map visualizations—fully functional versions that combined the spatial data, the designed animations, and the interactive elements. At this point, the maps were also published online through direct links, allowing participants and readers to explore how they worked in real time.

The fourth stage (Section 3.4) involved evaluating these maps. This stage explains how the user studies were carried out, both through large-scale online surveys and controlled laboratory sessions with eye-tracking. It also outlines the equipment used, the way data were collected, and the statistical methods applied. In addition, it introduces the performance and eye-movement metrics used to assess how effectively and efficiently participants interacted with the maps.

DOI: 10.1201/9781003728535-3

Together, these stages outline a structured yet flexible methodological framework that connects cartographic design, data visualization, and empirical evaluation—showing not only what was done, but why each step was important.

## 3.1 PLANNING THE STUDY

The first step (Section 3.1.1) in the research involved establishing how earthquakes are typically visualized on maps and identifying which data attributes are most frequently represented. To build this foundation, an analysis was conducted on 30 earthquake-mapping websites, each offering a distinct approach to presenting seismic events. The results of this review revealed a consistent pattern: in almost every case, location and magnitude were the two dominant features visualized, while depth, time, and number of fatalities appeared as secondary but still common attributes. Together, these variables form the essential components of how earthquakes are most often communicated cartographically.

Based on these findings, a cartographic framework was developed specifically for this study. In this framework, magnitude is expressed using dynamic point symbols, where the speed of motion functions as the key visual variable. Depth is represented through color, and fatalities through symbol size. This multidimensional approach allows users to interpret different aspects of the earthquake data simultaneously through motion, color, and size, while maintaining visual clarity and reducing cognitive strain. The design process also incorporated several cognitive and perceptual considerations. The number of classes was limited to avoid information overload, and color schemes were chosen to be colorblind safe, ensuring accessibility for a wider audience. To enable direct comparison, static map versions were included alongside the animated ones, providing a clear reference point for evaluating how motion affects map reading, comprehension, and overall user experience.

The second stage of conceptualization (Section 3.1.2) concerned the data sources and spatial coverage of the study. Two main datasets were selected to ensure both global reliability and temporal precision. The first source, EM-DAT (The Emergency Events Database), provides comprehensive spatiotemporal data on global disasters, including historical records of earthquake events. The second source, the United States Geological Survey (USGS), offers real-time monitoring of seismic activity, delivering detailed information about the location, magnitude, and depth of earthquakes. As the research focuses specifically on the European region, the dataset was spatially filtered to match this extent. However, the definition of Europe's eastern boundary presents a well-known cartographic and geopolitical ambiguity. To address this, the study adopted a pragmatic approach, extending the spatial range to include the entire territory of Turkey. This decision was supported by both tectonic considerations, given Turkey's active seismic zones, and cultural–geographical reasoning, which often associates the region with broader European frameworks in scientific studies. The earthquake epicenters were plotted using the geographical coordinates provided in the databases, while administrative boundaries served as the contextual background. To maintain visual clarity and minimize interference between thematic color coding—used, for example, to represent depth—and background map elements, the base map was designed in grayscale. This neutral visual foundation ensured that the dynamic and color-coded earthquake symbols remained perceptually dominant and easy to interpret.

The third part of the conceptualization stage (Section 3.1.3) focused on the practical design decisions related to both the experimental maps and the user interface. These decisions were guided by the need to maintain experimental control while ensuring that participants could interact with the visualization in an intuitive and consistent way. A time slider was implemented as the sole interactive element, allowing users to manually explore temporal changes in the earthquake data. Importantly, automatic playback was intentionally excluded. This design choice aimed to eliminate confounding factors such as individual differences in preferred animation speed or pacing, ensuring that all participants viewed the temporal sequence at a comparable rhythm. The layout of interface elements followed established cartographic conventions and empirical findings. The timeline was positioned along the bottom edge of the map, aligning with the natural left-to-right reading direction and supporting smooth temporal navigation. The legend, placed on the right side of the map, served a particularly critical role in this study. Because dynamic point symbols are still a relatively new cartographic technique, the legend acted as the key interpretive tool for users to understand motion types and their corresponding data classes. Its position was determined based on previous research highlighting the efficiency of right-side legends in supporting faster and more accurate map reading. The research design also incorporated a dual methodological approach to capture both the breadth and depth of user responses. A large-scale online survey was conducted to gather quantitative data from a wide participant group, while a controlled laboratory study employing eye-tracking technology provided detailed insights into users' visual attention and cognitive processing. Together, these two approaches created a balanced and comprehensive framework for evaluating how animated map symbols influence perception and performance.

The final stage of the study's conceptualization (Section 3.1.4) defined the tasks assigned to participants and the metrics used to evaluate their performance. Drawing on established frameworks from cartography and spatiotemporal data analysis, the tasks were categorized into four fundamental types: identification, comparison, grouping, and ordering. Each of these could involve either spatial relationships—focusing on location and distribution—or temporal relationships—focusing on sequence and change over time. In this chapter, earthquake epicenters were treated as temporal map objects, meaning that each existed only at a particular moment and geographic location. This approach allowed participants to engage directly with how spatial and temporal dimensions intertwine in animated visualizations. To evaluate user performance, two core indicators were used. Effectiveness was measured by the accuracy of participants' responses, while efficiency was assessed by the time required to complete each task. Additional data on user preferences and confidence levels were collected using Likert-scale questionnaires and, in some cases, spatial ability tests, providing insight into individual differences in map-reading skill. In the laboratory setting, eye-tracking provided a deeper layer of analysis. Key metrics included fixation counts and durations, which reflect patterns of visual attention; pupil size, used as an indicator of cognitive load; and saccades or scan paths, which reveal search strategies and how users visually explore the map. Together, these measures created a detailed picture of how participants perceived, interpreted, and interacted with both static and animated map representations, bridging the gap between performance outcomes and underlying visual behavior.

### 3.1.1 How Earthquake Features Were Visualized on Maps

There are many different cartographic methods used to visualize earthquakes (Pődör & Kiszely, 2014; Willmes et al., 2008). At the conceptual stage of this study, it was essential to understand how earthquakes are commonly represented and what kinds of related data appear most often in such visualizations. This step provided the groundwork for later design decisions and ensured that the experimental maps reflected realistic and recognizable mapping practices. To explore this, a systematic review was conducted of websites that publish earthquake maps, all of which are listed in Appendix 1. The purpose of this review was to identify consistent trends in both data representation and visualization methods. Specifically, three key aspects were analyzed:

- the types of data most frequently displayed on earthquake maps (such as magnitude, depth, time, and number of fatalities);
- the cartographic methods and symbolization techniques used to represent these phenomena; and
- the interactive tools and controls provided to users for navigating or adjusting the map's temporal content.

This overview helped clarify how earthquake data are most often presented in practice and revealed opportunities for improving existing approaches—especially through the inclusion of dynamic elements like motion and interactivity, which can enrich the way users perceive and understand seismic activity.

The analysis of 30 earthquake-mapping websites revealed several consistent trends in how seismic data are presented. The most common attribute visualized was the magnitude of the earthquake, which appeared on 93% of all maps reviewed. In nearly all of these cases, the moment magnitude scale ($M_w$) was used as the standard measure of earthquake strength. This consistency reflects both its scientific reliability and its widespread acceptance as the modern replacement for the older Richter scale. The second most frequently represented element was the geographical location of each event. However, it is important to distinguish between general location information and precise coordinates. While most maps depicted the approximate area of the epicenter, only a small fraction—around 10% of all analyzed maps—allowed users to read exact coordinates interactively. In most cases, this spatial information was expressed through a point symbol placed on the map, designed to reflect the relative geographic relationships between earthquake events rather than precise positional accuracy.

Other attributes commonly shown on earthquake maps included information about the depth of the earthquake, typically expressed in kilometers. This value provides important context, as deeper earthquakes tend to cause less surface damage than shallow ones. The date of occurrence was also a frequent element, most often presented in a simplified format showing only the year of the event. More detailed temporal information—such as the exact day and time—was generally available only through interactive tools, for example, by clicking or hovering over a specific event. A smaller number of maps, about 16% of those analyzed, also displayed fatality counts, indicating the number of people who lost their lives as a result of the earthquake. Although this information adds a valuable human dimension to the visualization, it appeared less consistently, likely because such data are not always immediately available or comparable across sources.

When examining the methods used to visualize earthquake data, three main cartographic approaches were identified. The first method employs circular point symbols of uniform size, differentiated by color to indicate the depth of the earthquake. This approach resembles a simple point map, where color functions as a differential visual variable (Żyszkowska et al., 2012), or what can also be described as a point choropleth map (Pieniążek & Zych, 2017). For the purposes of this study, this technique is referred to as the color point map. The second common method is the use of proportional or graduated point symbols (Slocum et al., 2023). In this technique, the size of each symbol corresponds directly to a quantitative attribute—most often, the earthquake's magnitude or the number of fatalities—allowing users to quickly compare relative intensities across different locations. Finally, the most widespread visualization strategy combines both color and size variation. In this method, symbol size represents magnitude, while color indicates depth. This dual encoding is particularly effective because it communicates two critical aspects of the phenomenon simultaneously: the strength of the earthquake and its subsurface depth, both of which are essential for interpreting seismic impact.

In this study, the depth of earthquakes was visualized using color, following conventions observed in most existing earthquake maps. Typically, these maps employ a bipolar color scheme, ranging from red—used to represent shallow earthquake hypocenters—to blue, which denotes deep-focus earthquakes. Adopting a similar approach in this work ensured visual consistency with common mapping practices and helped users interpret the visualization intuitively. However, two important design considerations must be addressed when applying this color scheme. The first concerns color vision deficiency, which affects approximately 4% of the population (National Eye Institute: https://nei.nih.gov/health/color_blindness/facts_about; accessed on 15.02.2024). For these users, the red–blue spectrum can be difficult or even impossible to distinguish accurately. To minimize this accessibility issue, the color scheme—regardless of its color model (RGB, HSV, or others)—must be colorblind safe. Fortunately, tools such as ColorBrewer.org provide tested and accessible color palettes that help cartographers design effective and inclusive thematic maps (Harrower & Brewer, 2003). Using such resources allows designers to maintain both visual clarity and universal accessibility, ensuring that the map communicates equally well to all viewers.

While color accessibility is a key consideration, another important aspect concerns the use of animation in cartographic communication. Animated maps that display temporal data require users to retain information from one animation frame and compare it with subsequent frames. This process imposes an additional cognitive demand compared to reading a static map, since the viewer must rely on working memory rather than simultaneous visual inspection (Harrower, 2007; Lloyd, 2005). For this reason, it is advisable to limit the number of color classes used in animated visualizations. Following Miller's principle (1956)—which suggests that the human brain can effectively process about 7 ± 2 units of information at a time—this study adopted a simplified five-class system to reduce cognitive load and maintain legibility during animation playback. An exception was made for the motion-based classification used in dynamic point symbols. Because animated symbols are a relatively new addition to cartographic design—introduced in mapping software such as ESRI ArcGIS Pro and QGIS only in 2018—the decision was made to employ four motion classes instead of five. This more restrained classification ensured that users could perceive differences in motion speed clearly, without becoming overwhelmed by excessive variability.

The second key characteristic commonly represented in online earthquake maps is the magnitude of the event. In this study, magnitude was visualized through the use of dynamic point symbols, introducing motion as a new visual variable. Specifically, the velocity of motion—that is, the speed at which a symbol moves or changes—was employed to represent differences in earthquake magnitude. In this approach, faster moving symbols correspond to stronger earthquakes, while slower movements indicate weaker ones. This design decision was made to establish a direct perceptual association between the intensity of motion and the intensity of the earthquake itself. From a cognitive perspective, faster motion naturally draws the viewer's attention and conveys a sense of greater energy or impact, making it a more intuitive and associative representation of seismic strength than static visual variables alone.

Given the study's objectives, it was also necessary to include a static mapping method to serve as a reference for comparison. This approach allowed for a direct evaluation of how dynamic symbols perform relative to traditional static representations. In this design, static point symbols were used to depict earthquake magnitude on a separate, nonanimated map. This static version functioned as a control condition, providing a stable visual baseline against which the effectiveness and efficiency of the dynamic point symbol method could be assessed. By comparing both approaches—one static, one dynamic—the study aimed to determine whether the introduction of motion as a visual variable truly enhances users' ability to interpret quantitative earthquake information, or whether static designs remain equally effective under certain conditions.

The third key attribute identified in the analysis of online earthquake maps is the number of fatalities—that is, the count of deaths resulting from the earthquake's effects. To represent this feature, the study employed the visual variable of size. This choice follows long-standing cartographic convention, as symbol size is one of the most effective and perceptually intuitive means of conveying quantitative differences (Garlandini & Fabrikant, 2009). Larger symbols naturally draw attention and imply greater magnitude or severity, making size particularly suitable for representing human impact. Moreover, using size to indicate fatalities aligns this map with common practices in seismological visualization, where circle scaling has traditionally been used to communicate relative differences in earthquake strength or damage. In this study, the integration of symbol size, color, and motion together provided a coherent, multivariable system—allowing users to perceive at a glance not only how strong an earthquake was, but also how deep it occurred and how severe its consequences were.

Earthquake maps typically display geographical locations as approximate positions on the map surface. However, in several of the analyzed examples, interactive functions allowed users to access the exact coordinates of earthquake epicenters.

While this feature can enhance analytical precision, the primary objective of this study was not to develop a highly interactive mapping tool, but rather to evaluate how different visualization methods influence user perception and performance. For this reason, the maps created for the study display only the geographical positions of earthquakes—without providing detailed coordinate readouts. This design choice simplifies the user experience and keeps the focus on symbol interpretation rather than data retrieval. The resulting conceptual framework for data representation, based on the most commonly used mapping approaches identified in the review, is illustrated in Figure 3.1. This figure summarizes how magnitude, depth, and

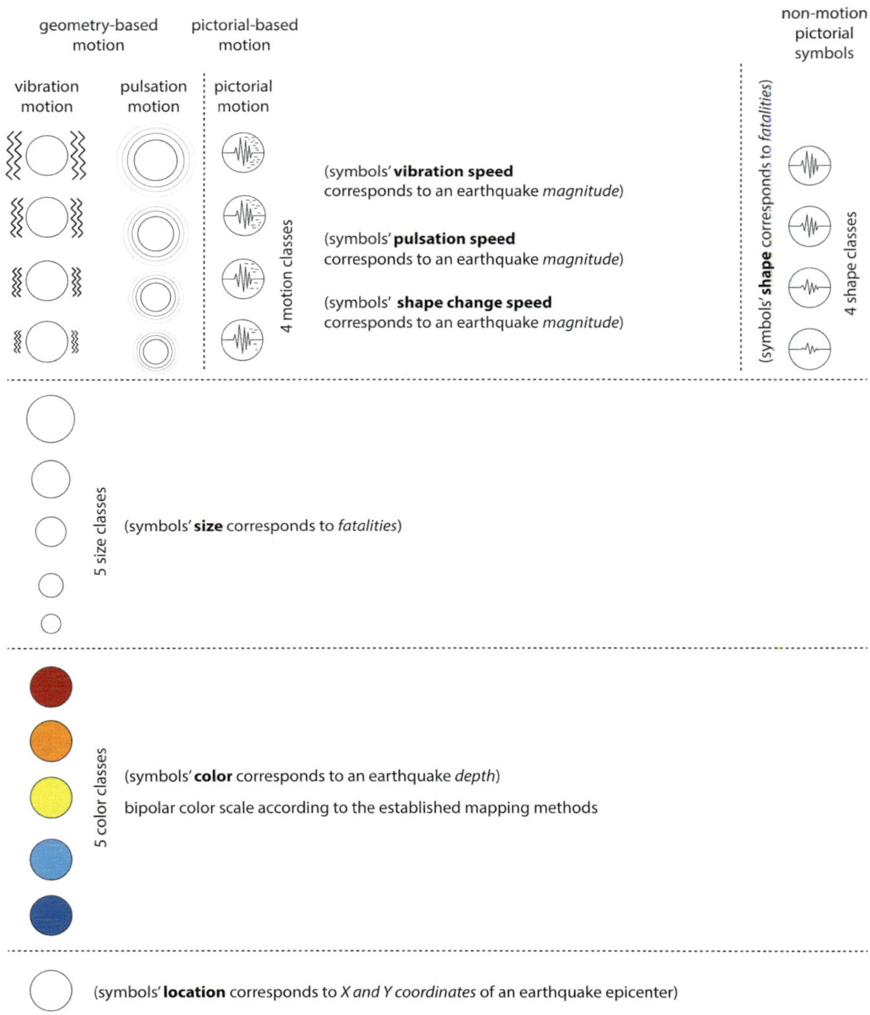


**FIGURE 3.1** Visual variables (location, color, size), motion features (speed), and corresponding earthquake features.

fatalities are visualized across both static and dynamic symbol designs, establishing the foundation for subsequent stages of map development and experimental testing.

### 3.1.2 Where the Earthquake Data Came From

To create the earthquake maps for this study, the dataset needed to include a set of core attributes that accurately describe each seismic event. These attributes comprised the magnitude of the earthquake, its depth, the number of fatalities, the spatial location of the epicenter, and the year of occurrence. Together, these variables form the essential descriptive framework for understanding both the physical characteristics and the human consequences of earthquakes. Each of these data features was selected not only for its scientific relevance, but also for its cartographic utility—allowing for effective visualization through the combined use of size, color, and motion as visual variables. In this way, the dataset provided a solid empirical basis for assessing how different types of map symbols—not only those used in seismic visualizations but also in many other types of geographical data—communicate quantitative and temporal information to users.

The first dataset used in this study was obtained from the EM-DAT International Disaster Database, a global repository that provides spatiotemporal information on major natural and technological disasters from 1900 to the present. The database is maintained by the Centre for Research on the Epidemiology of Disasters (CRED) and aggregates information from multiple reputable sources, including United Nations agencies, nongovernmental organizations, reinsurance companies, research institutions, and press agencies. The EM-DAT database is freely available for non-commercial academic use (https://www.emdat.be/; accessed on 16.02.2024). For the purpose of this study, the dataset was downloaded in CSV (comma-separated values) format, which facilitated subsequent data cleaning, processing, and import into GIS software for spatial visualization.

The second spatiotemporal data source used in this study was the USGS. The USGS plays a central role in the monitoring and communication of earthquake activity worldwide, providing reliable and regularly updated data on seismic events. Its mission extends beyond observation: the agency also evaluates earthquake impacts, estimates associated hazards, and contributes to the National Earthquake Hazard Reduction Program (NEHRP)—a U.S. initiative focused on improving public safety through research, preparedness, and risk assessment. For public access, the USGS maintains an extensive Earthquake Catalog, which allows users to search and download records of past seismic events according to custom parameters such as time, location, and magnitude. In addition, the organization offers an interactive online map displaying earthquakes that have occurred within the previous 24 h, making real-time seismic information easily accessible to both scientists and the general public.

Because this study focuses specifically on earthquake activity in Europe, the collected data were spatially filtered to correspond to the European continent. However, defining the exact boundaries of Europe presents a well-known cartographic challenge, as the continent's limits have been interpreted differently across historical, cultural, and geographical contexts. The northern, southern, and western borders of Europe are relatively unambiguous, following the natural coastlines of the Arctic Ocean, the Mediterranean Sea, and the Atlantic Ocean, respectively. The eastern boundary, however, remains the most debated. Over the centuries, scholars and geographers have proposed various delineations, reflecting both tectonic divisions and cultural–historical interpretations of where Europe ends and Asia begins.

In geographical terms, the Ural Mountains are often regarded as the natural boundary separating Europe from Asia. However, the contemporary definition of Europe extends beyond geography alone—it also reflects cultural, economic, and political dimensions that shape how the continent is understood. The greatest ambiguity arises along the southeastern frontier, particularly in the case of Turkey, where a smaller portion of the territory lies within the European continent, while the larger part is geographically situated in Asia. From a tectonic perspective, this area represents a complex region where the Eurasian and Arabian plates converge, and where the Anatolian Sub-Plate forms a distinct zone of seismic activity (Özalp et al., 2016). For this reason, and because earthquakes are inherently spatial phenomena whose impacts often extend across political boundaries, the entire territory of Turkey was included in this study. Moreover, both the EM-DAT and USGS databases provide data referenced by geographic coordinates, without assigning earthquakes to specific continents. This makes the inclusion of Turkey not only justified but necessary to maintain spatial and analytical consistency. The resulting study area, encompassing Europe as defined for this research, is illustrated in Figure 3.2, which presents the approximate spatial extent of the data used for the subsequent stages of analysis.

**FIGURE 3.2** The administrative boundaries of the study area.

The earthquake epicenters were plotted according to their geographical coordinates obtained from the previously described databases and positioned on a physical–geographic map that includes the administrative boundaries of individual countries. This type of background map is widely used in online cartographic visualizations of earthquakes, as it provides essential spatial context while maintaining familiarity for general audiences. However, displaying earthquake depth using a bipolar color scale—for instance, transitioning from red for shallow events to blue for deeper ones—introduces a potential challenge. When such colors are applied over a full-color physical–geographic background, the resulting map can become visually complex, making it difficult for users to distinguish thematic colors from the base map itself. To address this, this study adopts a grayscale base map, ensuring that thematic colors representing earthquake depth remain clear, legible, and perceptually distinct. This design decision aligns with established principles of visual hierarchy, which state that the most important information should stand out clearly from the background, and with the principle of cartographic clarity, which helps participants focus on the mapped data rather than the background details.

### 3.1.3 Key Technical and Design Decisions

Animated maps are designed to show how geographic phenomena change over time, and in the case of earthquakes, this means depicting variations in magnitude, depth, location, and fatalities across multiple years. To help users interpret these temporal patterns, a timeline element must be visible on the map interface and, when possible, provide basic interaction—for example, allowing the user to move through time or pause the animation. With this in mind, several key technical and design decisions were made regarding the GUI and the visual form of the animated maps. These decisions concern both the layout and functionality of interface components.

A fundamental requirement for the animated earthquake maps is that users must be able to control time. This means allowing them to select specific moments they wish to view or navigate freely across the temporal range of the data. The most common and effective way to achieve this is through a time slider, a familiar interface element that both visualizes change over time and enables user control. The slider displays the progression of time and includes a draggable handle that lets users move backward and forward across the timeline at their own pace. In this study, interaction is deliberately limited to this basic level of temporal control. The animation operates through embedded interactions rather than an external video player, meaning that users can manually explore changes by adjusting the time slider but cannot automatically play the animation. This design ensures a consistent frame-by-frame comparison, helping users focus on temporal differences without being influenced by animation speed or playback dynamics.

At this point, it is important to clarify the decision to exclude automatic playback of the animated map. This omission was deliberate and motivated by the desire to avoid perceptual variables that lie outside the scope of this study. Allowing the map to play as a continuous animation would introduce additional factors—most notably, the speed of playback—that can significantly influence how users process and interpret visual information. Previous research has shown that animation speed can greatly affect both effectiveness and efficiency in information acquisition. Several studies have identified the existence of an optimal animation speed, or a range of speeds, that best supports user comprehension and performance (Acevedo & Masuoka, 1997; Jobard & Lefer, 1997; MacEachren & DiBiase, 1991). More recent approaches even allow users to manually adjust playback speed according to their preferences (Roth et al., 2015). However, to eliminate the potential confounding influence of playback speed on user performance, this study employs a nonplayable animation design. In this configuration, users can explore the temporal dimension only by manually moving the time slider, ensuring that any differences in task performance arise from the map's symbol design and content, not from variability in animation speed.

The next design consideration concerned the placement of the interactive timeline within the map interface. Both observations from existing animated earthquake maps and insights drawn from the cartographic literature suggest that the bottom of the map frame is the most effective and intuitive location for this element. This placement allows users to maintain a clear view of the main map content while easily accessing temporal controls within their natural reading and viewing path. Positioning the timeline at the bottom also aligns with common design conventions in dynamic visualizations, where temporal navigation tools are typically located beneath the main display area.

Another important element of the map interface that required careful consideration was the placement of the legend. On some animated maps, the legend is omitted entirely (Cybulski, 2016), which may be acceptable when users can easily interpret the symbols. However, in the case of dynamic point symbols, this is not advisable. These symbols are relatively new to cartographic practice—with early research dating back to the early 2000s and their first appearance in GIS software around 2018—and are therefore less intuitive for many users (Cybulski & Wielebski, 2019). For this reason, including a clear and accessible legend becomes essential. The placement of the legend should not be based on aesthetic preference alone but instead guided by empirical evidence. A recent study by Edler et al. (2020), which combined traditional map-based tasks with eye-tracking analysis, demonstrated that placing the legend on the right side of the map leads to faster learning of the textual definitions of map symbols. Participants were able to understand and memorize the same amount of information in significantly less time compared to when the legend was positioned on the left. Based on these findings, the legend in this study was placed on the right-hand side of the map interface.

In line with the study's objectives, the designed maps were intended to be evaluated by participants through structured user testing. In traditional cartographic research, particularly in studies involving eye-tracking methods, the number of participants is often limited to a few dozen due to the technical and procedural constraints of laboratory-based experimentation (Brychtová & Çöltekin, 2016; Dong et al., 2018b). Such controlled studies provide high-quality behavioral data but are inherently restricted in sample size. In contrast, online survey research offers the possibility of reaching a larger and more diverse audience, enabling broader statistical analyses and insights into user performance across different demographic groups (Cybulski & Medyńska-Gulij, 2018; Horbiński & Zagata, 2022; Woźniak et al., 2024). For this reason, the study adopts a dual methodological approach. The first component consists of a large-scale online survey, designed to capture a broad spectrum of user responses and general trends. The second component involves laboratory-based eye-tracking experiments, which serve to validate the online findings under controlled conditions and provide in-depth insights into users' visual behavior during map interaction.

### 3.1.4 Designing and Analyzing the Map Tasks

Maps that represent spatiotemporal phenomena—even those without animation—can display a wide range of changes over time and space. These may include thematic changes (such as variations in attribute values), existential changes (the appearance or disappearance of geographic features), or spatial changes (including movement, displacement, or height variation). Because of this diversity, users can perform many different types of map-based tasks, each targeting a specific aspect of change (Andrienko et al., 2003, 2010). At the most basic level, these tasks can be classified as elementary searches, involving questions related to space ("where"), time ("when"), and object ("what"), or combinations of these components (Pequet, 1994). For example, a user might look for where an earthquake occurred, when it happened, or what its magnitude was. When combined, these dimensions form more complex analytical tasks that require users to integrate spatial, temporal, and thematic information simultaneously.

Each type of information on a spatiotemporal map can be explored at different levels of detail—that is, by focusing on a single element, a group of elements, or the entire phenomenon represented by all elements together. However, some researchers, such as Andrienko et al. (2003), recommend simplifying this classification into two main levels: elementary search, which targets individual features or events, and general search, which concerns broader patterns or relationships among many features. In this context, map objects can differ in their temporal characteristics. Some features are spatial objects, meaning they have a fixed location and exist continuously throughout the entire time period of analysis—for example, cities, rivers, or mountain ranges. In contrast, other features—such as earthquake epicenters—are considered temporal map objects or events, because they occur only once at a specific time and location (Andrienko et al., 2008). Given these characteristics, earthquake data are best understood as a collection of spatial events. Each epicenter represents an object defined by precise coordinates and a single moment in time, making it an ideal phenomenon for analyzing how users interpret temporal point data on dynamic maps.

Taking into account all the approaches discussed above, it becomes clear that map-based tasks can be designed to address both the spatial and temporal dimensions of earthquake events. In this context, users may be asked to locate a specific earthquake epicenter in terms of where it occurred on the map and when it took place in time (Beitlova et al., 2023; Board, 1978).

Within this general framework, several distinct categories of map-based tasks can be identified, each emphasizing different aspects of spatial and temporal reasoning. The most relevant for this study include:

- Identification task

  These tasks involve recognizing and selecting specific objects or time moments that meet given criteria.
  a. Spatial identification—identifying and selecting spatial–temporal objects with particular features or combinations of features.
  b. Temporal identification—identifying and selecting specific moments in time when a feature occurs or is absent.
- Comparison task

  Comparison tasks focus on examining differences or similarities between objects or between points in time.
  a. Spatial comparison—comparing two or more spatial–temporal objects based on their shared or differing characteristics.
  b. Temporal comparison—comparing two or more time moments in terms of the occurrence, absence, or change of temporal map objects.
- Grouping task

  Grouping tasks involve finding and assembling objects or times that share common features.
  a. Spatial grouping—selecting a set of temporal objects that exhibit a common property or characteristic.
  b. Temporal grouping—selecting a set of moments in time that share a specific attribute or pattern related to temporal objects.
- Ordering task

  Ordering tasks emphasize arranging objects or times according to specific rules or attributes.
  a. Spatial ordering—arranging temporal objects in a defined spatial sequence.
  b. Temporal ordering—arranging moments in time according to the characteristics of the corresponding temporal objects.

It is important to note that the map-based tasks described above can also be combined or expanded to create more complex analytical challenges. For instance, a single activity may involve both identification and comparison, or it may require participants to identify target objects, count them, and then compare the results to another group of objects. Such combinations represent a higher level of task complexity, going beyond basic identification to engage more advanced cognitive processes, such as reasoning and pattern recognition. When additional measurement techniques are introduced—most notably eye-tracking—these tasks gain another dimension related to visual perception. In such settings, object recognition can be understood in two complementary ways: through visual detection, as recorded by the eye tracker, and through manual confirmation, such as clicking on the object (Kiefer et al., 2013; Qin et al., 2024).

In this study, each map-based task will be evaluated in terms of both effectiveness and efficiency. Effectiveness will be determined by the accuracy of participants' responses, expressed as the percentage of correct answers, while efficiency will be measured by the time required to complete each task. These two indicators will serve as objective measures of user performance, providing a clear understanding of how accurately and quickly participants interpret cartographic information. In addition to these objective measures, subjective evaluations will also be collected to better understand users' opinions, expectations, and preferences regarding the maps. Such evaluations will help identify whether participants find the visualizations intuitive, engaging, and visually clear. To capture these insights, Likert-scale questionnaires will be applied, following established practices in user experience and cartographic research (Joshi et al., 2015). Participants will also be asked to self-assess their map-reading skills, a standard approach used in empirical cartography studies to contextualize performance results (Biland & Çöltekin, 2017; Vincent et al., 2019; Wilkening & Fabrikant, 2011). While this study will not include formal psychometric testing, such as the Santa Barbara Sense of Direction (SBSOD) scale (Hegarty et al., 2002), that method is referenced here as an example of how spatial ability can be objectively assessed in future research.

In this chapter, eye-tracking analysis will be used alongside performance and preference measures to gain a deeper understanding of how users visually engage with maps. Several eye-movement metrics will be applied to examine participants' visual behavior, decision-making strategies, and cognitive load while they perform specific map-based tasks. First, users' attention will be assessed through the analysis of fixation coordinates and their frequency, indicating where the gaze remains relatively stable on the map. These data will show which areas attract users' attention and how their focus shifts during task execution (Dobson, 1979; Steinke, 1987). Second, to evaluate cognitive load, metrics such as fixation duration, dwell time, and pupil size will be examined, as these measures tend to increase when mental effort rises (Holmqvist et al., 2011). Finally, patterns of saccadic movement—the rapid eye shifts between fixations—will be analyzed to reveal users' search strategies and visual efficiency. Measures such as scan path length, saccade amplitude, and search speed will help identify how participants explore the map and how effectively they transition between relevant areas of interest (Kiik et al., 2017).

## 3.2 CREATING THE DIGITAL MAPS

Transforming raw seismic data into a meaningful cartographic experience requires more than technical precision—it demands a thoughtful balance between science and design. In this stage of the research, the focus shifts from conceptual assumptions to practical implementation: the development of a digital map environment capable of representing earthquakes dynamically and interactively. The process begins with data. The map to be designed will combine a grayscale physical–geographic base layer, ensuring clarity and visual hierarchy, with vector-based symbols that convey the magnitude, depth, and human impact of each earthquake. These symbols, crafted in a graphic design environment, will later gain motion—pictorial motion, blinking, or pulsating to express seismic intensity. To make this possible, the cartographic workflow integrates both GIS processing and animation software, uniting analytical rigor with creative flexibility. For geographic data processing, ESRI ArcGIS Pro (version 2.4.0) and ArcMap 10.6.1 will be used to manage spatial data and locate epicenters according to their precise coordinates. Once processed, the data will be exported and enhanced in Adobe Animate CC (version 20.5.1), where the visual form, interactivity, and animation of dynamic point symbols will be developed. By merging these tools, the resulting maps will do more than display information: they will invite interpretation. Users will not only see where earthquakes occurred but will also sense their temporal rhythm, perceiving shifts in frequency, depth, and magnitude through controlled animation.

To conduct the empirical experiment, the maps were designed for display on a desktop computer monitor. A BENQ 21-inch monitor with a resolution of 1920 × 1080 pixels was selected to ensure consistent visualization conditions and maintain adequate detail and readability of cartographic content during the study.

### 3.2.1 Preparing the Map Background and Earthquake Data

The base map, titled "Physical Location Map Europe," was obtained from Wikimedia Commons under the Creative Commons Attribution–Share Alike 3.0 license. The map was derived from the ETOPO Global Relief Model, made available by the National Centers for Environmental Information (NCEI). This dataset provides gridded topographic and bathymetric data organized into 15 × 15-degree tiles (latitude/longitude) with a geographic resolution of 15 arc seconds (Amante & Eakins, 2009; National Centers for Environmental Information, 2022). To enhance the representation of terrain features, SRTM3 (shuttle radar topography mission) data were also incorporated, offering global coverage in 1-degree blocks at a spatial resolution of 90 m, which corresponds to a map scale of approximately 1:200,000 (Prasannakumar et al., 2011). In addition, the Natural Earth raster dataset (https://www.naturalearthdata.com/, accessed on 15.10.2024) was used to refine continental outlines and administrative boundaries. All spatial data were projected using the Lambert Azimuthal Equal Area projection (EPSG:3035), a coordinate system widely applied in European-scale cartographic projects due to its ability to preserve area relationships across large regions.

Since one of the main design goals was to avoid visual confusion between the background and the earthquake symbols, the colors of the original map were converted to grayscale. This change helped ensure that the dynamic and colored symbols would remain clearly visible and easy to interpret. The adjustment was made using Adobe Photoshop CS6 (version 13.0). To ensure the map fit the experimental display setup, the original publicly available physical map of Europe was also slightly adjusted to match the aspect ratio of the 21-inch BENQ monitor used in the study. This ensured that participants could view the entire map in full-screen mode without any distortion or cropping. The base map used as the cartographic background for earthquake visualization is presented here https://commons.wikimedia.org/wiki/File:Europe_blank_laea_location_map.svg#/media/File:Europe_relief_laea_location_map.jpg, accessed on 14.11.2024.

To display earthquake occurrences and their characteristics on the map, each point symbol needed to be accurately positioned according to its geographical coordinates. To make this possible, the raster base map was georeferenced, meaning that real-world coordinate values were assigned to individual pixels. This step ensured that the earthquake data could be precisely overlaid onto the cartographic background. The georeferencing process was performed in ESRI ArcGIS Pro, using four control points to align the base map within the GIS coordinate system. A 1st Order Polynomial (Affine) transformation was applied, resulting in a root mean square (RMS) error of approximately 18,380.91 m. While this value may appear large, it must be interpreted relative to the map's continental scale—where one pixel represents roughly 5 km in height and 3 km in width. In practice, this level of inaccuracy corresponds to less than a millimeter of deviation on the final map, and therefore does not affect the accuracy or readability of the visualization.

Once the cartographic background was spatially aligned, attention turned to the earthquake data itself. The datasets obtained from the EM-DAT International Disaster Database and the USGS contained highly detailed records—sometimes listing more than 40 events per day. To maintain visual clarity and avoid information overload, only earthquakes with a magnitude of 4 Mw or higher were selected for display. This threshold reduced the dataset to a manageable level, typically ranging from a few to several dozen events per year. This approach allows participants to analyze the data for each year without spending excessive time on visual analysis, making the empirical experiment feasible within a reasonable timeframe.

Because complete and consistent information was not available for every year, the timespan selected for analysis extended from 2000 to 2022. This period ensured that the dataset included comprehensive records of earthquake depth, fatalities, and magnitude. Data from the USGS provided accurate coordinates, magnitudes, and depths for each event, while the EM-DAT International Disaster Database supplied information about fatalities. In cases where fatality data were missing, supplementary information was retrieved from credible news sources, such as The Guardian (https://www.theguardian.com/world/2011/may/12/spain-earthquake-havoc-questions, accessed on 08.04.2023). After verifying and merging both datasets, all earthquakes with magnitudes of 4 Mw or greater were imported into ESRI ArcGIS Pro and displayed on the georeferenced map using the *Display XY Data* tool for CSV files. This ensured precise placement of epicenters on the cartographic background and provided a unified foundation for the subsequent design of static and dynamic visualizations.

With all earthquake points correctly placed on the georeferenced map, the next step was to prepare the data for visualization by assigning each event to an appropriate class interval. Thanks to the unified dataset created in the previous stage, it became possible to classify earthquake attributes automatically using standard data-classification techniques. However, when working with spatiotemporal data, classification cannot be performed independently for each year. Each year contains its own local minimum and maximum values for magnitude, depth, or fatalities. If these values were classified separately for every year, the resulting class intervals would change annually. As a consequence, the same symbol size or color might represent a completely different value from 1 year to the next, making comparisons across time misleading or even impossible. To maintain consistency, a single classification scheme was applied across the entire 2000–2022 dataset. This means that class intervals for fatalities, earthquake magnitude, and depth were calculated using

**TABLE 3.1**
**Results of Earthquake Feature Classification for the Entire Time Period**

| Earthquake Feature→ | Earthquake's Depth[a] | Fatalities[a] | Earthquake's Magnitude[b] |
|---|---|---|---|
| 1st class | 0.0–7.0 km | 0–3 people | 4.0–4.9 $M_w$ |
| 2nd class | 7.1–15.0 km | 4–10 people | 5.0–5.9 $M_w$ |
| 3rd class | 15.1–28.0 km | 11–17 people | 6.0–6.9 $M_w$ |
| 4th class | 28.1–69.0 km | 18–51 people | 7.0–7.9 $M_w$ |
| 5th class | 69.1–151.0 km | 52–296 people | Not applicable |

[a] Corresponds to Jenk's natural breaks classification method.
[b] Refers to equal intervals.

all available data from the full 23-year period. By doing so, the visual meaning of each class remains stable over time, allowing participants to interpret symbols reliably as they move through the timeline.

With the full temporal extent of the data established, the next step was to define the class intervals for each mapped attribute. Based on the 2000–2022 dataset, five-class intervals were created for earthquake depth and five intervals for the number of fatalities. In contrast, four class intervals were used for earthquake magnitude. This reduced number of magnitude classes was intentional: since magnitude will be represented through dynamic point symbols, keeping the classification simpler helps ensure that differences in motion speed remain clear and easy for users to interpret. To generate the class intervals, the Jenks natural breaks method was applied to depth and fatality data, as this technique identifies naturally occurring groupings within unevenly distributed datasets (Jenks, 1967; Jenks & Coulson, 1963). For earthquake magnitude, an equal-interval classification was chosen to maintain consistent spacing between classes, which supports the design of motion-based symbol differences. The resulting classification thresholds are summarized in Table 3.1, providing the basis for all subsequent visual encoding in the experimental maps.

Once the class intervals were defined, the precise placement of each earthquake on the map became essential. Accurate point locations ensured that, in the next stage, the corresponding dynamic symbols could be inserted into their correct geographic positions within the graphic design environment. This step enabled the animated symbols—created later in Adobe Animate CC—to align seamlessly with the georeferenced data (Lepetiuk & Ostroukh, 2021).

### 3.2.2 Setting the Time Frame for the Earthquake Data

Before any animation could be created, the cartographic background had to be transferred into the animation software. The workspace was set to 1920 × 1080 pixels, matching the monitor that participants would later use. This ensured that what you see during the experiment is exactly what the map was designed for—no scaling, stretching, or distortion.

The next step was to decide how time would function inside the map. Because this study looks at earthquakes over more than two decades, each year needed to be treated as a separate "moment" in the animation. To accomplish this, every year from 2000 to 2022 was assigned its own keyframe—a fundamental building block in animation. You can think of a keyframe as a snapshot of everything that appears on the map at a given point in time. By giving each year its own keyframe, the map can display a completely different set of symbols for each time-step, reflecting the exact earthquakes that occurred, along with their depth, magnitude, and fatalities. As we (or a study participant) move the time slider from 1 year to another, the map updates instantly, showing only the earthquakes relevant to that specific year. In other words, time becomes an interactive dimension, and the map responds dynamically to the user's actions (Pearce, 2008).

With the temporal structure in place, the next task was to populate each keyframe with the actual earthquake locations. Using the coordinates prepared in the GIS software, a simple dot symbol was placed at every recorded epicenter for each year. This step transformed the timeline into a meaningful sequence of spatial events—now, when you move from 1 year to the next, you're not just flipping through empty frames but watching earthquakes appear exactly where they occurred. In total, 215 earthquakes with a magnitude greater than 4 Mw were mapped across the 23-year period. This means that the animation consists of 23 separate time steps, each showing a different number of earthquakes depending on that year's seismic activity.

To give a clearer sense of this variability, Table 3.2 summarizes the number of earthquakes recorded per year. On average, there were 9.3 earthquakes per year (±1.5 standard error), with a standard deviation of 7.4, indicating noticeable year-to-year fluctuation. The median value was slightly lower at six earthquakes per year. Some years were exceptionally quiet—only two events occurred in 2005—while others were remarkably active, with 2021 and 2022 each recording 25 earthquakes, the highest

**TABLE 3.2**

**Each Year of the Spatiotemporal Visualization with the Exact Number of Earthquakes with Magnitude Higher Than 4 Mw**

| Year | Earthquake Count | Year | Earthquake Count |
|---|---|---|---|
| 2000 | 5 | 2011 | 4 |
| 2001 | 3 | 2012 | 4 |
| 2002 | 11 | 2013 | 3 |
| 2003 | 6 | 2014 | 6 |
| 2004 | 14 | 2015 | 4 |
| 2005 | 2 | 2016 | 12 |
| 2006 | 3 | 2017 | 14 |
| 2007 | 6 | 2018 | 10 |
| 2008 | 7 | 2019 | 20 |
| 2009 | 4 | 2020 | 23 |
| 2010 | 4 | 2021 | 25 |
| | | 2022 | 26 |

counts in the dataset. This variability is important because participants will experience years with sparse activity and years with dense clusters of events, allowing the study to examine how symbol design performs under different levels of visual complexity.

### 3.2.3 Choosing and Designing the Static Map Symbols

With all earthquake locations placed across the yearly keyframes, we can proceed to the next stage: designing the cartographic symbols that will represent the data. Before introducing any movement, we start by creating a complete set of static point symbols, which later serve both as a reference and as a basis for comparing static and dynamic symbology. Our first design decision is to use size as the visual variable representing earthquake fatalities. Based on the five previously defined class intervals, we assign five distinct symbol sizes—each one corresponding to a different fatality range. This approach ensures that differences in the number of deaths are immediately visible on the map. The exact dimensions, expressed in pixels and in millimeters on the target display, appear in Table 3.3. To keep the symbol geometry consistent, we develop all static point symbols inside the Adobe Animate CC library. Once the symbol set is ready, we place each symbol at its correct location on the map. Working in this order—design first, placement second—allows us to maintain a clear and controlled workflow: the symbology remains standardized, and the spatial arrangement accurately reflects the underlying dataset.

Another key visual variable in our design is color, which we use to represent earthquake depth. This choice follows the patterns we identified in our earlier review of online earthquake maps, where depth is most commonly shown using a diverging color scheme—typically shifting from red for shallow hypocenters to blue for deeper ones. To remain consistent with established practice and to make the map easier to interpret, we adopt the same general approach.

However, we also need to ensure that the map is accessible to users with different types of color vision. For this reason, we rely on ColorBrewer2.org, a widely used resource for selecting cartographically sound and colorblind-safe palettes. Using this tool, we choose a five-class diverging scheme that maintains contrast while preserving interpretability for all users (Brewer et al., 2003; Harrower & Brewer, 2003; Robinson, 2014). The exact color values assigned to each class interval are listed in Table 3.4.

**TABLE 3.3**
**Actual Size of the Symbols for Each Class Interval**

| | Size [Points] | Size [mm] |
|---|---|---|
| 1st class | 13 | 3.0 |
| 2nd class | 21 | 4.5 |
| 3rd class | 29 | 6.5 |
| 4th class | 37 | 8.5 |
| 5th class | 45 | 10.0 |

*The size in points represents Adobe Animate CC parameters, while the size in millimeters refers to the target display monitor.*

**TABLE 3.4**
**Actual Color Selection for Map Symbols According to Specific Class Intervals**

| | HEX | RGB | CMYK |
|---|---|---|---|
| 1st class | #d7191c | 215,25,28 | 15,90,80,0 |
| 2nd class | #fdae61 | 253,174,97 | 0,32,55,0 |
| 3rd class | #ffffbf | 255,255,191 | 0,0,25,0 |
| 4th class | #abd9e9 | 171,217,233 | 33,3,0,0 |
| 5th class | #2c7bb6 | 44,123,182 | 85,30,0,0 |

*The choices were supported by ColorBrewer2.org.*

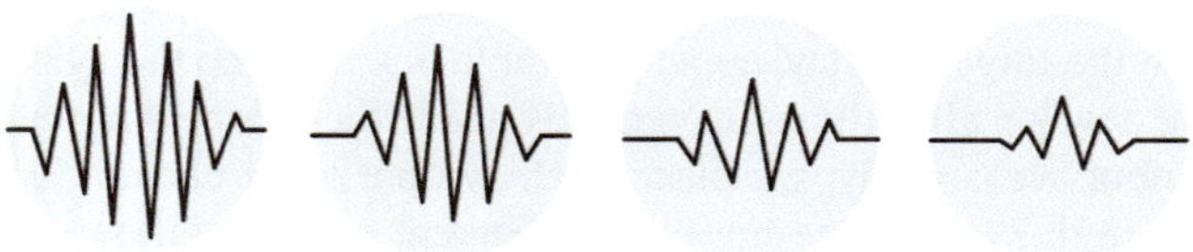

**FIGURE 3.3** Pictorial symbols for presenting the earthquake's magnitude. The larger the pulse, the stronger the earthquake. Circle size varied according to the number of fatalities, and specific color replaced gray background of the symbol according to the earthquake's depth.

According to our conceptual framework, we also need a static pictorial symbol that can serve as a reference when comparing user performance on static versus dynamic maps. To keep the comparison fair and focused, we base this static symbol on the same design used for the dynamic version—a circle containing an inner pulse element. The only distinction is that, in the static version, the pulse remains frozen, without any animation applied. This approach allows us to examine whether it is the motion itself—or simply the symbol's geometry—that influences user performance. Figure 3.3 illustrates the design scheme for each class interval. As with the other symbol sets, we assign size to represent fatalities and color to represent depth, replacing the neutral gray fill in the base pictorial symbol with the appropriate visual variables.

### 3.2.4 Adding Motion: Applying Animated Effects

With the static visual variables (size and color) already defined, the next step is to introduce motion as the third variable—used here to represent earthquake magnitude. In this study, we work with three types of dynamic point symbols:

- Vibration,
- Pulsation,
- Pictorial motion (a symbol containing an animated internal element).

To animate any of these symbols in Adobe Animate, each design must first be converted into a movie clip instance. This transformation allows us to define frame-by-frame motion and control animation properties independently for each symbol. To ensure consistency across all motion types, we establish a reference point: the fastest symbol is designed to complete its full animation cycle in 12 frames. With a global playback rate of 24 frames per second (FPS), this means the fastest-moving symbol repeats its motion twice every second, making it clearly distinguishable during visual search tasks.

At this stage, we must determine how different classes of magnitude should vary in speed. Drawing on prior empirical findings on dynamic point symbols, we adopt an exponential distribution of motion speeds. Earlier studies demonstrated that exponential spacing—rather than linear spacing—produces clearer perceptual differences between classes and improves visual search effectiveness (Cybulski, 2024). Following this approach, each magnitude class receives a distinct animation speed according to an exponential progression, applied consistently across all three symbol types. The exact speed values assigned to each class are presented in Table 3.5, forming the basis for the dynamic encoding used in the experiment.

The first type of dynamic symbol we designed represents vibration, achieved by introducing a subtle, frame-by-frame shift in the symbol's location. When these small positional changes are displayed continuously, the symbol appears to tremble or shake—an effect that users can easily associate with seismic activity. In Adobe Animate, this effect was created by adjusting the *X* and *Y* coordinates of the symbol's center across the 12 animation frames. To make the movement pattern transparent during the design phase, a small cross was placed at the center of the symbol. This cross served as a visual reference, allowing us to track the exact displacement applied in each frame. The resulting design is illustrated in Figure 3.4, which shows both the sequence of shifted positions and the magnitude of displacement. Because depth classes were encoded through color, this vibration effect had to be prepared separately for each color variant used in the experiment. This ensured that all dynamic symbols—regardless of class—shared the same motion logic while preserving their visual distinctions.

To control how quickly each vibrating symbol appears to move, the duration of individual frames is adjusted according to the motion parameters listed in Table 3.5. This adjustment is applied separately for every color class, resulting in 20 unique symbol instances in total—four motion-speed classes multiplied by five color classes. Early experiments with a keyframe-based approach showed that this method

**TABLE 3.5**
**Total Number of Frames of One Animation Instance**

| | Animation Instance Frames |
|---|---|
| 1st class | 96 |
| 2nd class | 48 |
| 3rd class | 24 |
| 4th class | 12 |

*These values are related with the overall display speed which was set to 24 FPS. This means that one second of animation presents 24 frames.*

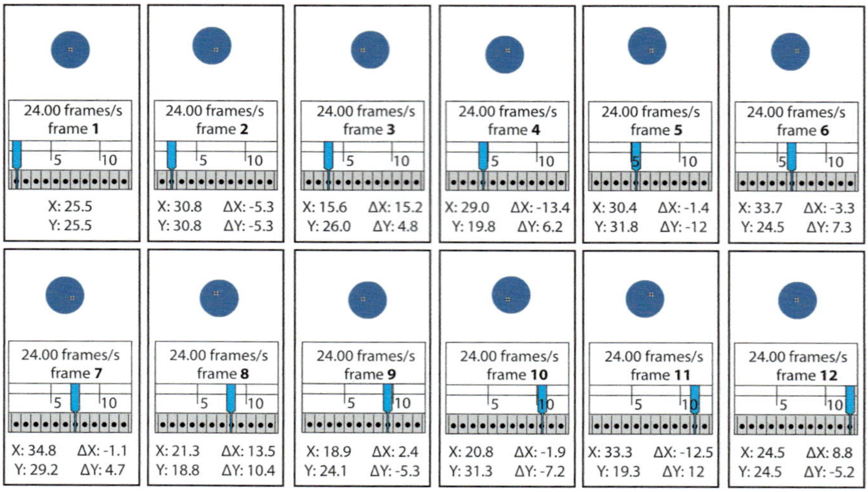


**FIGURE 3.4** Each step of dynamic symbol design, in this case vibration. The crosshair on the circular shape refers to the central position of the initial location. On each frame, there is a slight change in the symbols' location reflected the local display coordinates and the difference related to the center.

produced movement that was too smooth: the symbol gradually interpolated toward each new position rather than shifting abruptly. Because the goal of the vibrating effect is to create a sharp, jittery motion, such smooth transitions are not desirable. For this reason, the animation relies on a frame-by-frame technique, which allows each positional change to occur instantly and preserves the abrupt motion required for a convincing vibration effect.

The next type of dynamic symbol uses pulsation, understood here as a rhythmic change in symbol size. In contrast to vibration, which relies on a frame-by-frame approach, the pulsation effect is created using keyframe animation in Adobe Animate. In the timeline, this difference is visible: keyframe-based animation is shown with an horizontal bar, whereas frame-by-frame animation appears in gray. For each pulsating symbol, there is no need to design twelve separate frames. Instead, only two keyframes are created: one with the symbol at its original size and a second with the symbol briefly enlarged by +10 pixels, scaled in proportion to the base size that represents the number of fatalities. These two states are then repeated across 12 frames, producing a smooth, regular pulsing effect. The detailed structure of this design is illustrated in Figure 3.5. As with the vibrating symbols, color is applied according to earthquake depth, so that pulsation, size, and color work together to communicate multiple aspects of the data.

The final type of dynamic symbol builds on a pictorial design, using a pulse-shaped icon as the base. In its static form, this symbol contains a single pulse line. To animate it, a masking technique is applied in Adobe Animate: the mask gradually reveals the pulse line, creating the impression of movement. The speed of this reveal is directly linked to the earthquake's magnitude—stronger earthquakes are represented by symbols that unfold more rapidly. The full structure and animation process for this pictorial symbol is shown in Figure 3.6. As with the other dynamic symbols, both the size and color of the pictorial symbol correspond to the earthquake's fatality count and depth, ensuring that all visual variables work together to convey the underlying data clearly and consistently.

Each dynamic and static symbol was then assigned to the appropriate class interval based on the classification scheme presented in Table 3.1 in Section 3.2.1. In total, four

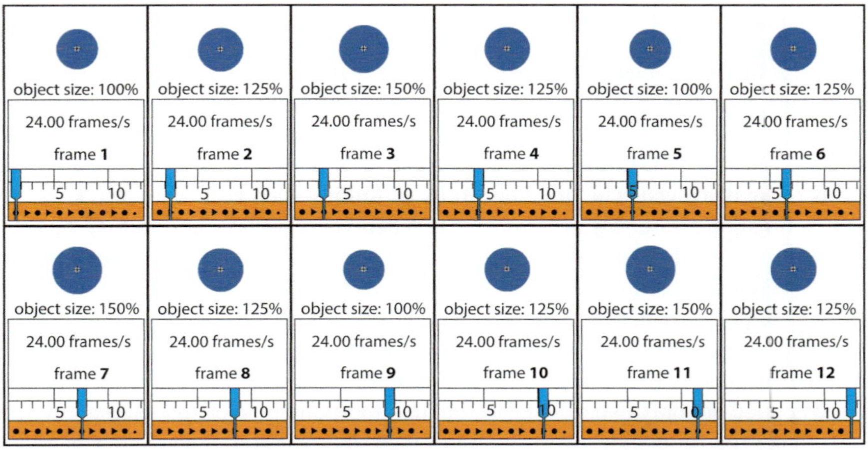


**FIGURE 3.5** The effect of pulsation designed in every frame of a keyframe animation.

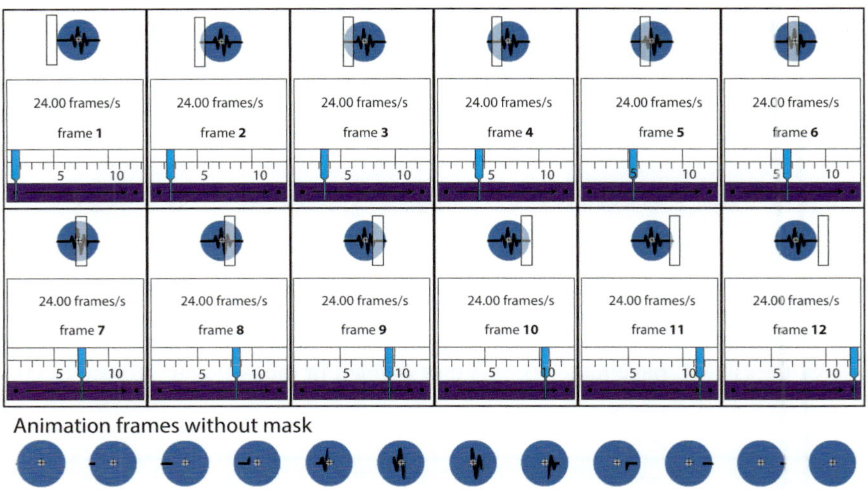


**FIGURE 3.6** Detailed design of a dynamic pictorial symbol that was based on a static symbol. Arrow tween between the first and the last keyframe suggests that this is a movement interpolation of the mask. On the bottom side of the figure, the final animation without a mask is presented.

map variants were prepared: three featuring dynamic symbols (vibration, pulsation, and pictorial motion) and one using the static pictorial symbol as a reference.

For each version of the map, every point representing an earthquake—already positioned according to its coordinates and assigned year—received the correct symbol type and class-specific appearance. This ensured that all four map variants displayed the same underlying data but differed only in how the symbols communicated magnitude, depth, and fatalities.

### 3.2.5 Final Map Layout and Interface Design

To let users move through time easily, the map includes an interactive timeline covering every year from 2000 to 2022. Each year is placed at an equal distance along the slider, creating a clear and predictable layout. All years appear in gray, while the currently selected year turns black so the user can immediately see which moment in time is being displayed. The timeline works by linking each year to its own keyframe in the animation (as shown in Figure 3.7). When you click a year, the map instantly updates to show only the earthquakes that occurred during that period. This approach makes it possible to jump freely between years without any imposed order—ideal for exploring temporal patterns or comparing specific time periods. Although the timeline is powered by a series of keyframes inside the animation software, these technical elements are hidden from view. What the user sees is a simple, clean time slider that can be applied not only to earthquakes, but to almost any type of spatiotemporal map.

To make the time slider work with the map, the position of the slider handle—a small gray dot—had to be linked to the keyframes representing each year. This required a short piece of JavaScript added inside Adobe Animate. Before the code could be applied, the slider graphic was placed at the top of the layer structure so that it remained visible and interactive above all other map elements. Once positioned, the following script was attached to the slider component:

```
document.body.style.backgroundColor = lib.properties.color;
createjs.Touch.enable(stage);

slider1 = new TimelineSlider(this.slider1, 0.0, function()
{
        ;
});
```

This script performs two essential functions. First, it ensures that the background color of the HTML page matches the color defined in the Animate project. Second,

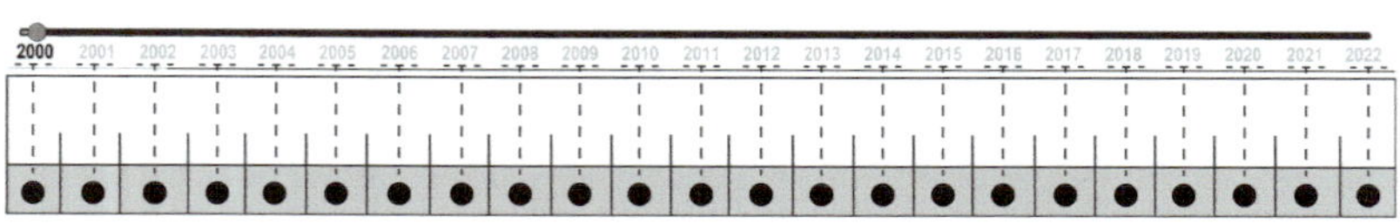


**FIGURE 3.7** The time slider designed for temporal navigation on the earthquake map. It was the same for all map variations.

it enables touch support, which allows the slider to work not only with a mouse but also with touch-based input. Finally, the *TimelineSlider* object links the visual position of the slider to the timeline of the animation. In practice, this means that when users drag the slider handle, the visualization jumps to the corresponding keyframe, displaying the selected year instantly. At this point, the slider exists and can be interacted with, but its detailed behavior—how its movement changes the displayed year—is defined in the next part of the script.

In the next step, additional JavaScript was added to the global script of the project. These lines were responsible for linking the slider's position to specific keyframes in the animation. With this mechanism in place, the user can drag the slider to any year on the timeline, and the map automatically updates to display the earthquakes from that year. The full code is shown below:

```
window.TimelineSlider = function(target, ratio, onUpdate)
{
        this.target = target;
        this.target.loop = false;
        this.ratio = ratio || 0;
        this.onUpdate = onUpdate;
        this.gotoRatio();
        this.target.hit.on("mousedown", this.move.bind(this));
        this.target.hit.on("pressmove", this.move.bind(this));
};

window.TimelineSlider.prototype.move = function(e)
{
        var mouseX = this.target.globalToLocal(e.currentTarget.
stage.mouseX, e.currentTarget.stage.mouseY).x;

        this.ratio = Math.min(Math.max(mouseX /
e.currentTarget.nominalBounds.width + 0.47, 0), 1);
        this.gotoRatio();
};

window.TimelineSlider.prototype.gotoRatio = function(ratio)
{
        this.target.gotoAndStop(Math.floor(this.target.
totalFrames * (ratio || this.ratio)));

        if (this.onUpdate)
        {
        this.onUpdate = this.onUpdate.bind(this);
        this.onUpdate();
        }
};
```

In simple terms, this script translates the slider's horizontal position into a numerical ratio between 0 and 1. That ratio is then mapped to the total number of frames in the animation—each frame representing 1 year between 2000 and 2022. When the user drags the handle, the script calculates the new ratio, jumps to the corresponding frame, and updates the map instantly. As a result, temporal navigation becomes smooth, intuitive, and responsive, allowing users to explore the data year by year with a single motion of the slider.

The final part of the script focused on making the map responsive to different screen sizes. Since the map would be shared through an online survey, participants could open it on a wide range of devices—large desktop monitors, smaller laptops, or even tablets. To ensure that the visualization remained readable and the interface behaved consistently, the project needed a mechanism that automatically adjusted the map to the size and proportions of the viewer's screen. To achieve this, the following JavaScript code was added:

```
var page_body = document.getElementsByTagName("body")[0];
page_body.style.backgroundColor = "#FFFFFF";
page_body.style.overflow = "hidden";
page_body.style.position = "fixed";

var page_canvas = document.getElementsByTagName("canvas")[0];
stageWidth = page_canvas.width;
stageHeight = page_canvas.height;

var viewport = document.querySelector('meta[name=viewport]');
var viewportContent = 'width=device-width, initial-scale=1.0,
minimum-scale=1.0, maximum-scale=1.0, user-scalable=0';

if (viewport === null) {
 var head = document.getElementsByTagName('head')[0];
 viewport = document.createElement('meta');
 viewport.setAttribute('name', 'viewport');
 head.appendChild(viewport);
}

viewport.setAttribute('content', viewportContent);

function onResize() {
 var widthToHeight = stageWidth / stageHeight;
 var newWidth = window.innerWidth;
 var newHeight = window.innerHeight;
 var newWidthToHeight = newWidth / newHeight;
 //
 if (newWidthToHeight > widthToHeight) {
 newWidth = newHeight * widthToHeight;
 page_canvas.style.height = newHeight + "px";
 page_canvas.style.width = newWidth + "px";
 } else {
 newHeight = newWidth / widthToHeight;
 page_canvas.style.height = newHeight + "px";
 page_canvas.style.width = newWidth + "px";
 }
 scale = newWidthToHeight / widthToHeight;
 stage.width = newWidth;
```

```
 stage.height = newHeight;
 page_canvas.style.marginTop = ((window.innerHeight -
newHeight) / 2) + "px";
 page_canvas.style.marginLeft = ((window.innerWidth -
newWidth) / 2) + "px";
}

window.onresize = function () {
 onResize();
}

onResize();

var root = this;
```

In simple terms, this script checks the viewer's available screen space and scales the map so that it always fits without distortion. It preserves the aspect ratio of the original map, centers it within the browser window, and prevents unwanted scrolling or stretching. Every time the window is resized—for example, if a participant rotates a tablet or adjusts a browser window—the script recalculates the layout and redraws the canvas. By using this approach, the map remains usable and visually consistent across all devices, ensuring that every participant sees the same interface and can interact with the timeline, symbols, and animations as intended.

Since the map relies on dynamic point symbols—a relatively new and still unfamiliar technique in cartographic visualization—a clearly visible legend was essential. Following the design assumptions established earlier, the legend was placed in the upper-right corner of the map, where users can access it quickly without obscuring key geographic content. Its structure guides the reader through each visual variable step by step. At the top, the legend explains the dynamic (or, in the static variant, pictorial) symbol used to represent earthquake magnitude. In the middle section, the meaning of symbol size is clarified, helping users interpret the number of fatalities associated with each event. Finally, the lower portion of the legend describes the color scheme used to represent earthquake depth. To complete the layout, the map title is placed in the upper-left corner, creating a balanced and easily readable composition.

## 3.3 PRESENTING THE MAPS

With the maps fully designed and animated, the next step was to make them accessible for user testing. Each map variant was published online, accompanied by a direct hyperlink that allows readers and study participants to open the visualization in a web browser and interact with it freely. To make this possible, every map was exported from Adobe Animate as an HTML5 file supported by a JavaScript script. These files were then uploaded to the server of the Department of Cartography and Geomatics at Adam Mickiewicz University in Poznań. By hosting the maps online, we create a setting where user feedback could be gathered efficiently and under consistent technical conditions. This approach also mirrors real-world scenarios, where web-based maps are the primary way many people explore geographic information. The collected responses play an important role in the field of cognitive cartography, which studies how different map designs influence perception, visual attention, and comprehension. These insights are particularly valuable in areas where people must quickly make sense of spatial information—such as crisis management, hazard communication, or mapping of disaster events.

For clarity in the later stages of analysis, each map variation was assigned a simple label:

"A" for maps using vibration to represent earthquake magnitude,
"B" for maps using pictorial motion,
"C" for maps using pulsation, and
"D" for maps using static pictorial symbols as a reference.

These four versions form the basis for comparing how different types of motion—or the absence of motion—shape users' understanding of the data. Introducing animated elements such as vibration, motion, and pulsation offers a fresh direction in cartographic design of seismic events.

### 3.3.1 Map Variation "A"—Vibration for Representing Earthquake Magnitude

In this map version, earthquake magnitude is expressed through a vibrating point symbol. Vibration serves as a natural visual metaphor for ground shaking, making it an intuitive choice for representing seismic intensity. Because the human visual system is highly responsive to rapid, irregular motion, differences in magnitude become immediately noticeable. At the same time, this technique must be used carefully, as excessive vibration can lead to visual fatigue during longer viewing sessions. In Map Variation A, the vibrating symbol indicates the strength of the earthquake, while the size of the symbol represents the number of fatalities. This combination allows users to quickly assess both the intensity of each event and its human impact within a single, integrated visual design.

https://kartografia.amu.edu.pl/Badanie_PC/trzesienie_ziemi_A/GRUPA_01.html

### 3.3.2 Map Variation "B"—Pictorial Motion for Representing Earthquake Magnitude

In this map version, earthquake magnitude is shown through the speed of a changing pictorial symbol. Because motion speed is a preattentive feature, it can quickly draw the viewer's attention, even before conscious reading or interpretation begins. Although this method has shown promising results in a few empirical studies, it is not as intuitive as traditional visual variables, so some users may find it less immediately clear. In Map Variation B, the faster the symbol changes, the stronger the earthquake. As in the previous map, the size of each symbol shows the number of fatalities, and color represents the earthquake's depth. This combination allows users to compare three key characteristics at once—magnitude, impact, and depth—while exploring an alternative motion-based method of presenting seismic data.

https://kartografia.amu.edu.pl/Badanie_PC/trzesienie_ziemi_A/GRUPA_02.html

### 3.3.3 Map Variation "C"—Pulsation for Earthquake Magnitude

Pulsation—where a symbol rhythmically grows and shrinks—can visually echo the repeating pattern of seismic waves. Because of this, pulsating symbols may be well-suited for showing earthquake magnitude, and some empirical studies suggest that this type of motion can be effective for presenting spatiotemporal and quantitative data. In Map Variation C, the speed of the pulsation reflects the earthquake's magnitude: the stronger the quake, the faster the pulse. As in the previous map versions, the size of the symbol represents the number of fatalities, and color is used to show the depth of the earthquake. This variation allows users to explore how rhythmic motion influences the perception of seismic intensity.

https://kartografia.amu.edu.pl/Badanie_PC/trzesienie_ziemi_A/GRUPA_04.html

### 3.3.4 Map Variation "D"—Static Pictorial Symbols for Earthquake Magnitude

Static symbols represent the most traditional way of showing information on a map. They do not move or change, which makes them stable, familiar, and often easier to read—especially for people who may find dynamic symbols visually demanding. Although static symbols lack the immediacy and emphasis created by motion, they provide a clean reference point for comparison. In Map Variation D, earthquake magnitude is shown using static pictorial symbols, with each icon representing one of the predefined magnitude classes. As in the dynamic versions, the size of each point indicates the number of fatalities, and the color shows the depth of the earthquake. This map serves as the baseline against which all animated versions are compared, helping us understand what motion adds—or fails to add—to the user's interpretation of the data.

https://kartografia.amu.edu.pl/Badanie_PC/trzesienie_ziemi_A/GRUPA_03.html

## 3.4 TESTING AND EVALUATING THE MAPS

To understand how well each map variation works for visualizing earthquakes, we need to test them through an empirical study. This makes it possible to measure effectiveness (how accurately people complete the tasks) and efficiency (how quickly they do so). In this research design, each map variation is tested by a separate group of participants, so that feedback reflects the specific qualities of each visualization method. At the same time, it is important that these groups remain as similar as possible—for example, in terms of age, experience, or background—so that the results can be fairly compared across all map types.

Evaluating maps designed for earthquake visualization involves two complementary stages. The first stage centers on an online survey, a widely used method for collecting user responses to different cartographic designs. This approach makes it possible to gather data from a large number of participants in a relatively short time. The second stage involves a controlled eye-tracking experiment, where participants complete the same map-based tasks while their visual behavior is recorded. By combining these two methods, we can both broaden the reach of the study and strengthen the reliability of the findings. The online survey provides large-scale, diverse user feedback, while the eye-tracking sessions offer detailed insights into how people actually look at and process the maps. Because eye-tracking requires calibration, specialized equipment, and individual testing sessions, the number of participants is naturally smaller. Together, however, these two approaches create a balanced evaluation strategy—one that captures both the breadth of user reactions and the depth of their visual and cognitive processes.

The study was carried out in full compliance with the ethical standards set by the institution's Ethics Commission. All participants were informed about the purpose of the research, the type of data being collected, and their rights during the study. Consent was obtained before participation, and all procedures—both in the online survey and the eye-tracking laboratory—were designed to respect privacy, safety, and participant well-being.

### 3.4.1 How the Online Survey Was Conducted

The online survey was created using the LimeSurvey.org platform. When participants opened the survey, they first saw an introductory screen explaining the purpose of the study and providing information about the researcher. This page also reassured participants that the survey was anonymous, voluntary, and that they could withdraw at any moment without giving a reason. Before moving on, each participant had to confirm that they had read this information, understood the study's goals, and agreed to take part voluntarily. Only after providing this consent did the survey continue to the next section, where participants were asked a few basic demographic questions, such as their age, gender, and level of education. These details helped ensure that the participant groups were comparable across map variations.

The next section of the survey included several subjective questions about participants' opinions and everyday habits. All of these items were presented as radio-button questions, meaning participants could select only one answer from the available options. The first question focused on general map-use experience:

1. *How often do you use maps, in any format (including paper maps, digital maps, mobile maps), in your daily life?*

   And the possible answers here where:
   - I use them every day;
   - I use them couple time a week;
   - I use them couple time a month;
   - I use them less than couple time a month.

   The next item asked participants to reflect on their own map-reading abilities. This was a simple self-assessment question, commonly used in cartographic user studies to capture how confident individuals feel when interpreting maps:

2. *Assess your own map-reading ability*:
   - Very good;
   - Good;
   - Average;
   - Bad;
   - Very bad.

After participants completed the introductory questions, they were shown a set of instructions explaining how to use the map interface. These guidelines clarified that the map displayed earthquakes that occurred in Europe between 2000 and 2022, but only events with a magnitude of 4 or higher were included. The instructions also described how to interpret the visual elements on the map. Circular points indicated the locations of earthquakes, while the year shown on the timeline at the bottom marked when the events took place. Participants were informed that the size of each circle represented the number of fatalities, and that a bipolar color scheme was used to show earthquake depth. Finally, depending on the particular map variation assigned to them, the instructions included an additional explanation of how the dynamic symbols worked and what form of motion (vibration, pulsation, or pictorial change) they would see during the tasks.

For each map variation, participants received a short explanation tailored to the type of symbol they would see. In map variant A, they were told that earthquake magnitude is shown through a vibrating (shaking) point symbol—where faster vibration means a stronger earthquake. In map variant B, the instructions clarified that magnitude is represented using a dynamic pictorial pulse symbol, and the pulse changes more quickly for stronger earthquakes. For map variant C, participants learned that magnitude is shown through a pulsing point symbol that repeatedly grows and shrinks, with the speed of this pulsation reflecting the earthquake's strength. Finally, for map variant D, the static version, participants were informed that magnitude is represented using different pulse shapes inside each circle, where taller pulse lines indicate stronger earthquakes at that location.

The final part of the instructions explained how to interact with the map. Participants were told that a time slider appears at the bottom of the screen and that they can move it back and forth to display each year separately. The map does not include zoom or pan options, so the entire interaction focuses on the timeline. Before answering any questions, participants were asked to spend a moment exploring the map—moving the slider, observing how the symbols change from year to year, and getting familiar with the dynamic effects. When they felt ready to begin the task, they could press the spacebar (or double-tap the screen on a mobile device) to proceed to the first map-based question. At that moment, the first map-based question appeared, selected at random and accompanied by clear instructions on what to do.

Participants were asked to answer four map-based questions, which appeared in a random order to ensure that no particular question benefited from its position in the sequence (the so called learning effect). Three of the questions focused on key descriptive aspects of the earthquakes shown on the map. The fourth question shifted attention to the temporal dimension. These questions were selected to test how well participants could interpret the key visual and dynamic variables on the map. To avoid influencing their responses, the exact wording of the questions was presented exactly as shown below (and in each participant's native language):

1. *How many times did the strongest earthquake (earthquake magnitude) occur?*
2. *How many times did the earthquakes with the highest number of fatalities occur?*
3. *At what depth do earthquakes most often occur in Europe (depth)?*
4. *In which year were there the most earthquakes?*

Before each question was shown, participants received a short instruction explaining what would happen next. They were told that the map would reappear and that their task was to examine it specifically with the upcoming question in mind. The map could be viewed for as long as needed, and when participants felt ready to respond, they could continue by pressing the spacebar (or double-tapping the screen on mobile devices). During this examination phase, the system automatically measured how long it took each participant to search the map and gather enough information to answer the question. Only after they chose to proceed were they shown the answer options, along with the question repeated for clarity.

The questions and answer formats were as follows:

*Ad. 1. How many times did the strongest earthquake (earthquake magnitude) occur?:*

A. 3 times
B. 4 times
C. 5 times
D. 6 times
E. 7 times

*Ad. 2. How many times did the earthquakes with the highest number of fatalities occur?:*

A. 3 times
B. 4 times
C. 5 times
D. 6 times
E. 7 times

*Ad. 3. At what depth do earthquakes most often occur in Europe (depth)?:*

A. 0.0–7.0 [km]
B. 7.1–15.0 [km]
C. 15.1–28.0 [km]
D. 28.1–69.0 [km]
E. 69.1–151.0 [km]

*Ad. 4. In which year were there the most earthquakes?:*

- Dropdown list with each year that was presented on the map

  This structure ensured that every participant followed the same procedure: read the question, examine the map at their own pace, and then choose an answer from clearly defined options.

  After each response, participants were asked to reflect on how confident they felt about the answer they had just given. This self-assessment helped capture not only performance but also how certain users felt while interpreting the map. To measure this, the survey used a Likert scale—a common questionnaire format where participants choose from a set of ordered options that range from very positive to very negative feelings. It is widely used in research because it allows people to express degrees of confidence, agreement, or preference rather than giving a simple yes/no answer. In this study, confidence was measured using a five-point Likert scale with the following options:
  - very confident
  - a little confident
  - neutral
  - a little unsure
  - very unsure

  At the end of the survey, participants were asked one final question about their overall impression of the symbol style they had just worked with. While earlier questions focused on performance and confidence, this last item aimed to capture their subjective attitude toward the visual design itself. Understanding whether users find a symbol style appealing—or off-putting—is important, because even highly effective visualizations can be less successful if users dislike or resist their appearance. The closing question was:

*Assess the style of symbol presentation on the maps you observed:*

- I definitely like it
- I like it
- I have no opinion
- I do not like it
- I definitely don't like it

The online survey was distributed across several channels to reach a broad and diverse group of participants. It was shared through research survey platforms such as SWPanel.pl, Greenpanthera.com.pl, and Opinie.pl, as well as through social media, online forums, and the author's personal network of contacts. Using multiple platforms made it possible to gather responses from people with different backgrounds and levels of map-reading experience, which in turn strengthened the reliability of the results.

### 3.4.2 How the Eye-Tracking Study Was Conducted

The eye-tracking experiment followed the same general structure as the online survey but required several additional steps because of the specialized equipment involved. To avoid repeating information already covered in Section 3.4.1, this part focuses only on the extra procedures needed for the laboratory setting. The first major difference appeared immediately after participants provided their basic demographic information. Before they could begin working with the map, a calibration procedure had to be performed between the eye tracker and the participant's dominant eye. Calibration ensures that the device can accurately record where the participant is looking on the screen.

Because eye-tracking sessions are longer and more demanding than online surveys, the procedure needed to remain as efficient as possible. Extending the session with lengthy dominance tests was not feasible, as it could increase fatigue and reduce data quality. Instead, a simplified method was used: the participant's handedness (left- or right-handed) was taken as an indicator of eye dominance. This approach is supported in earlier research, particularly by McManus (1999), and allowed the study to move smoothly into calibration without unnecessary delay.

During calibration, the study used the EyeLink® 1000 Plus eye tracker (SR Research Ltd., Ottawa, Canada), which was paired with a chin and forehead rest to keep the participant's head stable. While this setup may feel different from everyday map use—and is less "natural" because people normally move freely—it is a necessary step in laboratory research. The stabilization allows the device to record eye movements with extremely high precision. With this configuration, the EyeLink can reach sampling rates of up to 2,000 Hz, meaning it records up to 2,000 measurements of the eye's position every second. Its accuracy can be as precise as 0.15° of visual angle, and on average, it captures a gaze measurement every 1.4 ms (with very low variability, SD $< 0.2$ ms). These parameters make it possible to analyze even the quickest eye movements in a reliable and scientifically meaningful way (EyeLink® 1000 Plus Technical Specifications: https://www.sr-research.com/wp-content/uploads/2017/11/eyelink-1000-plus-specifications.pdf, accessed on 19.07.2024).

The experimental procedure itself was built using WebLink, a tool designed to present and interact with HTML-based materials—making it ideal for showing the same earthquake maps used in the online survey. Before participants began the map tasks, the eye-tracking system needed to be calibrated, which is a crucial step in ensuring accurate gaze measurement. During calibration, the participant is asked to look at a series of dots—nine in this setup—appearing one after another in different parts of the screen. After this, a validation step checks how accurate the calibration is, often reporting the error in degrees of visual angle. In this study, a second validation was also performed mid-experiment to ensure the device was still tracking correctly.

Once calibration and validation were complete, the experiment continued in the same order as the online survey, starting with questions about map-use frequency and self-assessed map-reading skills. All maps in the eye-tracking study were displayed on a 21-inch monitor (46.49 cm × 26.15 cm) with a resolution of 1,920 × 1,080 pixels, matching the display used earlier in the project. The eye tracker, positioned just below the screen rather than mounted onto it, was placed about 55 cm from the participant's eyes. Participants themselves sat roughly 90 cm away from the monitor, a setup that ensured both comfort and the high spatial accuracy needed for reliable eye-movement recording.

## 3.4.3 Participants in the Online Study

### 3.4.3.1 Map Variation "A"—Vibration for Earthquake Magnitude

A total of 1,167 people took part in the online survey. For map variant A—the version that used vibrating symbols to show earthquake magnitude—254 participants completed the tasks. This group included 108 men, 135 women, 4 people who identified as "other," and 7 participants who chose not to disclose their gender.

Participants' ages ranged from 18 to 77 years, with an average age of 41.7 years, giving the sample a broad demographic spread. When asked about their everyday use of maps, the group showed a wide range of habits:

- 74 participants reported using maps *daily*,
- 52 used them *several times a week*,
- 47 used maps *several times a month*,
- and the largest subgroup—81 people—used maps *less than once a month.*

Self-evaluations of map-reading skills also varied:

- 30 people described their skills as *very good*,
- 61 as *good*,
- 106 as *average* (this was the largest group),
- 35 as *low*,
- and 22 as *very low.*

In terms of education,

- 94 participants had completed *higher education*,
- 138 had *secondary education*,
- 23 had *elementary education*,
- and 2 people did not specify their education level.

Overall, map-usage frequency differed widely across participants. Some relied on maps almost every day, while others used them only occasionally. This diversity is valuable—it reflects a range of real-world map users, from confident daily navigators to those who consult maps only when necessary.

### 3.4.3.2 Map Variation "B"—Pictorial Symbol Motion for Earthquake Magnitude

For map variant B—the version that used a changing pictorial symbol to show earthquake magnitude—305 participants completed the survey. This group consisted of 136 men, 167 women, and 2 participants who preferred not to disclose their gender.

Participants were generally younger than in the previous group: the average age was 35.4 years, with ages ranging from 19 to 66. As in variant A, participants showed diverse habits when it came to using maps in everyday life:

- 81 people reported using maps *every day*,
- 58 used them *several times a week*,
- 51 used them *several times a month*,
- and the largest subgroup—115 participants—used maps *less than once a month*.

Self-assessed map-reading skills covered the full range:

- 48 participants rated their skills as *very good*,
- 82 as *good*,
- 119 as *average* (the most common answer),
- 35 as *low*,
- and 21 as *very low*.

Regarding education levels,

- 114 participants had completed *higher education,*
- 174 had *secondary education,*
- and 17 had *elementary education.*

Overall, the map-usage habits in this group were similar to those observed in map variant A. As before, we see a wide spectrum of behaviors: some participants used maps regularly—often as part of their daily routine—while others turned to maps only occasionally. Just like in the first group, this diversity of experience levels created a balanced sample, making it possible to compare how different types of users responded to this particular symbol design.

#### 3.4.3.3 Map Variation "C"—Pulsation for Earthquake Magnitude

For map variant C—the version that used pulsation to represent earthquake magnitude—303 participants completed the survey. This group consisted of 156 men, 143 women, one participant who identified as "other," and three who preferred not to disclose their sex.

The average age in this group was 35.4 years, with participants ranging from 18 to 62 years old, making this sample similar in age profile to variant B. As with the previous variants, participants reported very different habits when it came to using maps in everyday life:

- 102 people reported using maps *every day*,
- 120 used them *several times a week*,
- 22 used them *several times a month*,
- and 59 participants used maps *less than once a month.*

Self-assessed map-reading skills again covered the full ability range:

- 51 participants rated their skills as *very good,*
- 108 as *good* (the most common answer),
- 100 as *average,*
- 30 as *low,*
- and 14 as *very low.*

Education levels were also diverse:

- 146 participants had *higher education,*
- 144 had *secondary education,*
- and 13 had *elementary education.*

Overall, this group resembled the previous ones in its mix of frequent and occasional map users, as well as a broad range of skill and education levels.

#### 3.4.3.4 Map Variation "D"—Static Pictorial Symbols for Earthquake Magnitude

For map variant D—the version that used static pictorial symbols as a nonanimated reference—337 participants completed the survey, making this the largest group in the online study. The sample consisted of 141 men, 181 women, nine participants who identified as "other," and six who preferred not to disclose their sex. Participants in this group were, on average, slightly younger than those in variants A and C, with a mean age of 34.9 years and an age range spanning from 20 to 71 years.

Their everyday map-use habits showed a wide spread, similar to the previous variants:

- 81 people reported using maps *every day,*
- 53 used them *several times a week,*
- 67 used them *several times a month,*
- and the largest subgroup—136 participants—used maps *less than once a month.*

Self-rated map-reading skills also varied across the full spectrum:

- 34 participants described their skills as *very good,*
- 97 as *good,*
- 133 as *average* (again the most common response),
- 48 as *low,*
- and 25 as *very low.*

Educational backgrounds reflected a similar level of diversity:

- 139 participants reported *higher education,*
- 173 had *secondary education,*

- 25 had *elementary education*,
- and 3 did not specify their education level.

Overall, this group closely resembled the previous samples in diversity and balance. With a mix of frequent and infrequent map users, different skill levels, and varied educational backgrounds, the variant D group provides a strong baseline for comparing static symbols with the three dynamic symbol techniques.

A summary of all participants from the online survey, along with a crosstabulation of their characteristics, is presented in Table 3.6. While the survey provides useful demographic information, it also comes with certain limitations. Online questionnaires typically allow only a basic level of insight into who the respondents are, and this study was no exception. The participant pool included both very young and much older individuals; however, these age patterns should be interpreted cautiously. Some response times in the dataset were extremely short or unusually long, suggesting that a portion of the data may need to be excluded during performance-based

**TABLE 3.6**

**Crosstabulation of Online Survey Questionnaire Participants Characteristics**

| Category | Map Variant A (*Vibration*) | Map Variant B (*Pictorial Symbol Motion*) | Map Variant C (*Pulsation*) | Map Variant D (*Static Pictorial Symbols*) |
|---|---|---|---|---|
| **Total number of participants** | 254 | 305 | 303 | 337 |
| **Sex** | M: 108<br>W: 135<br>O: 4<br>N: 7 | M: 136<br>W: 167<br>O: 0<br>N: 2 | M: 156<br>W: 143<br>O: 1<br>N: 3 | M: 141<br>W: 181<br>O: 9<br>N: 6 |
| **Age range** | 18–77 (41.7 on average) | 19–66 (35.4 on average) | 18–62 (35.4 on average) | 20–71 (34.9 on average) |
| **Map-usage frequency** | Daily—74<br>Several times a week—52<br>Several times a month—47<br>Less than once a month—81 | Daily—65<br>Several times a week—62<br>Several times a month—78<br>Less than once a month—100 | Daily—102<br>Several times a week—120<br>Several times a month—22<br>Less than once a month—59 | Daily—81<br>Several times a week—53<br>Several times a month—67<br>Less than once a month—136 |
| **Map-reading skill** | Very good—30<br>Good—61<br>Average—106<br>Low—35<br>Very low—22 | Very good—48<br>Good—82<br>Average—119<br>Low—35<br>Very low—21 | Very good—51<br>Good—108<br>Average—100<br>Low—30<br>Very low—14 | Very good—34<br>Good—97<br>Average—133<br>Low—48<br>Very low—25 |
| **Education level** | Higher—93<br>Secondary—138<br>Elementary—23 | Higher—114<br>Secondary—174<br>Elementary—17 | Higher—146<br>Secondary—144<br>Elementary—13 | Higher—139<br>Secondary—173<br>Elementary—25 |

*M—men; W—woman; O—other; N—not specified.*

analysis. For this reason, greater emphasis was placed on participants' self-assessed characteristics, such as their map-reading skills and frequency of map use, as these measures offered a more reliable basis for interpreting user performance.

### 3.4.4 Participants in the Eye-Tracking Study

The eye-tracking experiment took place in the laboratory at the Faculty of Geographical and Geological Sciences at Adam Mickiewicz University in Poznań. Because the study was conducted on-site and required the use of specialized equipment, all participants were recruited from among the faculty's students. As a result, none of the individuals taking part had only elementary education; instead, every participant had at least a secondary-school background and was currently pursuing higher education. This created a more academically homogeneous group compared with the online survey sample.

#### 3.4.4.1 Map Variation "A"—Vibration for Earthquake Magnitude

For map variant A—the version using vibration to show earthquake magnitude—27 students took part in the eye-tracking experiment. The group included 12 men and 15 women, with an average age of 24.1 years (the youngest participant was 19, the oldest 36). Because all participants were enrolled at Adam Mickiewicz University in Poznań, none had only elementary education: 23 reported having completed secondary school, while 4 already held a higher education degree.

Participants also differed in how often they used maps in daily life:

- 10 participants said they used maps *every day*,
- 14 used them *several times a week*,
- and 3 used them *several times a month*.

Notably, no one indicated using maps less than once a month.

Self-evaluations of map-reading skills varied as well:

- 6 participants described their skills as *very good*,
- 15 as *good*,
- 5 as *average*,
- and 1 as *low*.

No one rated their map-reading skills as very low, which is typical for a university-level sample familiar with geographic content.

#### 3.4.4.2 Map Variation "B"—Pictorial Symbol Motion for Earthquake Magnitude

For map variant B—the version that used a changing pictorial symbol to represent earthquake magnitude—24 students took part in the eye-tracking experiment. The group had an even gender distribution, with 12 men and 12 women. The average age was 24 years, though the full range extended from 19 to 45, making this group slightly more diverse in age than the previous one. As expected for a university-based sample, nearly all participants had finished secondary education (23 people), and 1 participant had already completed higher education.

Participants also differed in how often they used maps in everyday situations:

- 5 people reported using maps *every day,*
- 10 used them *several times a week,*
- 8 used them *several times a month,*
- and 1 participant used maps *less than once a month.*

Their confidence in map reading also varied:

- 6 participants rated their skills as *very good,*
- 9 as *good,*
- 6 as *average,*
- and 3 as *low.*

#### 3.4.4.3 Map Variation "C"—Pulsation for Earthquake Magnitude

For map variant C—the version that used a pulsating symbol to show earthquake magnitude—20 students participated in the eye-tracking experiment. The group consisted of 12 men and 8 women, with an average age of 22.5 years. Compared to the previous variants, this was the youngest group overall, with ages ranging from 18 to 28. As expected in a university setting, the majority had completed secondary education (18 people), and 2 participants had already earned a higher education degree.

Participants also varied in how frequently they used maps in their daily lives:

- 5 people used maps *every day,*
- 12 used them *several times a week,*
- 2 used them *several times a month,*
- and 1 participant used maps *less than once a month.*

Their self-assessed map-reading skills were generally positive:

- 3 participants rated their skills as *very good,*
- 14 as *good,*
- 2 as *average,*
- and 1 as *low.*

None of the participants judged their skills as "*very low,*" making this group one of the more confident ones in terms of map literacy.

#### 3.4.4.4 Map Variation "D"—Static Pictorial Symbols for Earthquake Magnitude

For map variant D—the version that used traditional static pictorial symbols—30 participants took part in the eye-tracking study. The group consisted of 16 men and 14 women, with an average age of 22.7 years. Although the group was generally young, there was still some age diversity: the youngest participant was 20, and the oldest was 41. Most participants had completed secondary education (26 people), while 4 already held higher education degrees.

Participants also differed in how often they used maps in everyday life:

- 12 used maps *daily*,
- 15 used them *several times a week*,
- 3 used them *several times a month*,
- and none reported using maps *less than once per month.*

When asked to rate their map-reading skills, the group showed a generally high level of confidence:

- 4 participants described their skills as *very good*,
- 18 as *good*,
- 8 as *average*,
- and no one rated their skills as *low* or *very low*.

This made variant D's group one of the most confident and map-experienced participant samples across the four symbol types.

A complete overview of the eye-tracking participants and their characteristics can be found in Table 3.7.

When we compare the participant groups from the controlled eye-tracking studies with those from the online survey, clear differences in group composition

**TABLE 3.7**
**Crosstabulation of Online Survey Questionnaire Participants Characteristics**

| Category | Map Variant A (*Vibration*) | Map Variant B (*Pictorial Symbol Motion*) | Map Variant C (*Pulsation*) | Map Variant D (*Static Pictorial Symbols*) |
|---|---|---|---|---|
| **Total number of participants** | 27 | 24 | 20 | 30 |
| **Sex** | M: 12<br>W: 15<br>O: 0<br>N: 0 | M: 12<br>W: 12<br>O: 0<br>N: 0 | M: 12<br>W: 8<br>O: 0<br>N: 0 | M: 16<br>W: 14<br>O: 0<br>N: 0 |
| **Age range** | 19–36 (24.1 on average) | 19–45 (24.0 on average) | 18–28 (22.5 on average) | 20–41 (22.7 on average) |
| **Map-usage frequency** | Daily—10<br>Several times a week—14<br>Several times a month—3<br>Less than once a month—0 | Daily—5<br>Several times a week—10<br>Several times a month—8<br>Less than once a month—1 | Daily—5<br>Several times a week—12<br>Several times a month—2<br>Less than once a month—1 | Daily—12<br>Several times a week—15<br>Several times a month—3<br>Less than once a month—0 |

*(Continued)*

**TABLE 3.7 (*Continued*)**

**Crosstabulation of Online Survey Questionnaire Participants Characteristics**

| Category | Map Variant A (*Vibration*) | Map Variant B (*Pictorial Symbol Motion*) | Map Variant C (*Pulsation*) | Map Variant D (*Static Pictorial Symbols*) |
|---|---|---|---|---|
| **Map-reading skill** | Very good—6 | Very good—6 | Very good—3 | Very good—4 |
| | Good—15 | Good—9 | Good—14 | Good—18 |
| | Average—5 | Average—6 | Average—2 | Average—8 |
| | Low—1 | Low—3 | Low—1 | Low—0 |
| | Very low—0 | Very low—0 | Very low—0 | Very low—0 |
| **Education level** | Higher—4 | Higher—3 | Higher—2 | Higher—4 |
| | Secondary—23 | Secondary—21 | Secondary—18 | Secondary—26 |
| | Elementary—0 | Elementary—0 | Elementary—0 | Elementary—0 |

*M—men; W—woman; O—other; N—not specified.*

emerge. The eye-tracking sessions involved relatively small and highly homogeneous groups—mostly university students with similar educational backgrounds and ages. This kind of uniformity is typical for laboratory-based research, where the setup, timing, and equipment requirements limit who can realistically participate. In contrast, the online survey attracted a much larger and far more diverse group of respondents. Participants varied widely in age, education, and map-use habits, giving this part of the study a broader demographic profile. This diversity is one of the strengths of online research, as it allows us to gather feedback from people with many different levels of experience and backgrounds.

This greater diversity in the online group also means that participants brought a much wider range of map-use experience and familiarity to the study. In the laboratory sessions, most participants reported using maps frequently and rated their map-reading skills relatively highly. This suggests that the eye-tracking participants were generally more confident and experienced in working with maps. In contrast, the online survey included many individuals who used maps only occasionally and who often rated their skills as average or even below average. As a result, the online sample provided insight into how a broader, more heterogeneous audience interprets the maps—an important perspective when evaluating visualizations intended for the general public.

# 4 Understanding How Well the Maps Worked

## *Results from the Online Study*

## 4.1 HOW ACCURATE WERE THE PARTICIPANTS

To compare how well participants performed across the different map tasks, there first needs to be a clear way to measure accuracy (effectiveness). In this study, accuracy is determined by measuring the distance between each response and the correct answer.

A simple scoring system was applied:

- A completely correct answer receives a score of 0, meaning no deviation.
- Answers that are one class above or below the correct one receive a score of ±1.
- Answers that fall two classes away receive ±2, and so on.

In this way, larger deviation scores reflect answers that are further from the correct value. This scoring method enables consistent accuracy comparisons across all four map-based tasks, despite each task focusing on a distinct aspect of the earthquake data.

However, when presenting the results, two different perspectives are applied. The first is a strict approach, where only completely correct answers—those with zero deviation—are counted as accurate. The second is a more inclusive approach, recognizing that the tasks were genuinely demanding, especially given the spatiotemporal scale of more than 20 years of data. Under this approach, answers that miss the correct class by just one step (±1) are also treated as effectively accurate.

### 4.1.1 Finding the Strongest Earthquakes (Using Motion Cues)

This task focused on whether participants could correctly identify the strongest earthquakes by reading the motion-based symbols. Since this is a new and less familiar way of showing magnitude, it was important to compare the results with map variant D, which used traditional static symbols. Variant D serves as a baseline—because static symbols are well-established in earthquake mapping, any improvement (or drop) in accuracy with the dynamic symbols in variants A, B, and C becomes easier to interpret.

For the task involving earthquake magnitude shown with the vibrating symbol in map variant A, only a minority of participants were able to identify the correct

DOI: 10.1201/9781003728535-4

answer with perfect accuracy. A total of 48 participants (18.8%) selected the correct class (0 deviation). Another 40 participants (15.7%) were close, giving answers that were off by just one class (±1). A larger group—56 participants (22.0%)—missed the correct value by two classes (±2). The biggest portion of respondents, however, struggled with this task. 109 participants (42.9%) selected answers that were three or more classes away from the correct magnitude. One participant skipped the question entirely.

For the task in which earthquake magnitude was shown using pictorial symbol motion in map variant B, the accuracy pattern looked quite similar to variant A. A total of 55 participants (18.0%) selected the correct magnitude class with no deviation (0). Another 53 participants (17.3%) were close, giving answers that were just one class off (±1). A larger group—73 participants (23.9%)—missed the correct answer by two classes (±2). As in the previous variant, the largest share of responses fell into the least accurate group. 124 participants (40.6%) selected answers that were three or more classes away from the correct magnitude. Two participants did not provide an answer.

For the task where earthquake magnitude was shown using a pulsating symbol in map variant C, the accuracy results again followed a similar pattern. A total of 54 participants (17.8%) identified the correct magnitude class with full accuracy (0 deviation). A slightly larger group—73 participants (24.1%)—gave answers just one class away from the correct value (±1). Another 79 participants (26.1%) deviated by two classes (±2), placing their answers in the midrange of accuracy. Finally, 97 participants (32.0%) selected answers three or more classes away from the correct magnitude (±3), forming the largest response category for this question.

The results for map variant D, which used static pictorial symbols as a baseline for comparison, followed the same general trend but with slightly higher accuracy overall. In this variant, 71 participants (21%) selected the correct magnitude class with full accuracy (0 deviation)—the highest proportion among all four map versions. Another 70 participants (20.8%) were close, choosing answers just one class away from the correct value (±1). A larger group—95 participants (28.2%)—gave answers deviating by two classes (±2). Finally, 99 participants (29.4%) selected answers three or more classes away from the correct one (±3). Two people did not provide any response.

A summary of how well participants identified the number of strongest earthquakes is shown in Figure 4.1. Overall, the proportion of fully correct answers was fairly similar across all four map versions. While the number of perfect responses was not very high, it remained remarkably consistent, regardless of whether the symbols were static or dynamic. This pattern suggests that, for a task requiring users to count events on a map, dynamic motion does not necessarily improve accuracy—and it does not seem to make performance worse either. Statistical tests support this interpretation: both the Pearson chi-square test and Fisher's exact test showed no statistically significant differences in accuracy between the map variants. Interestingly, the results for map variant D (static pictorial symbols) were most similar to those for variant C (pulsating symbols). This hints that, for this particular type of magnitude-identification task, symbol designs that rely on pictorial motion may be more closely aligned with how users process static pictorial cues.

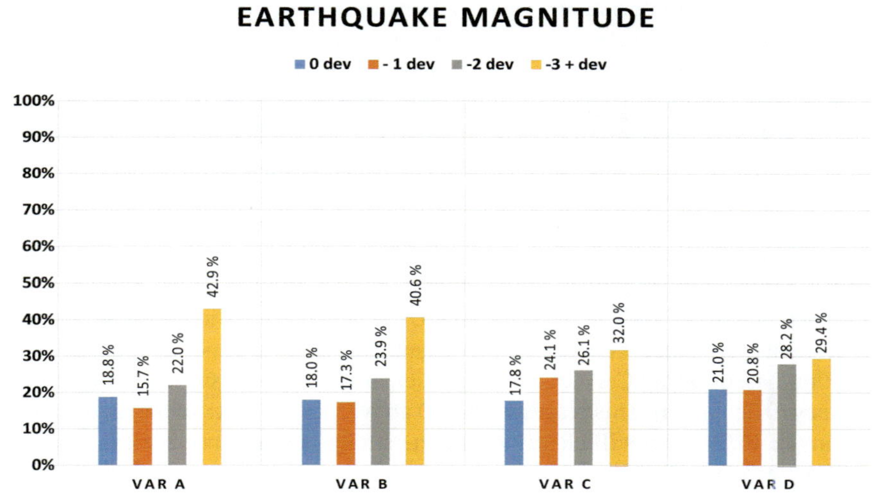


**FIGURE 4.1** Summary of the effectiveness in communicating earthquake magnitudes using dynamic point symbols compared to static symbols on a spatiotemporal map.

### 4.1.2 Finding the Most Fatal Earthquakes (Using Symbol Size)

For the task where participants had to identify how many times the most fatal earthquakes occurred—using symbol size as the visual cue—in map variant A, only 27 participants gave a fully accurate answer (0 deviation). This represents 10.6% of everyone who completed this version of the survey. A larger portion of participants—69 people (27.2%)—selected an answer that was only one class away from the truth (±1). Another 26 participants (10.2%) were two classes away (±2). The majority, 132 participants (52.0%), gave answers that deviated by three or more classes (±3), indicating that they struggled with interpreting the fatalities using symbol size in this map variant.

For the fatalities task in map variant B, where magnitude was represented using pictorial symbol motion, the pattern of results was similar to what was observed in variant A, though with slightly higher accuracy. A total of 49 participants (16.0%) answered perfectly (0 deviation), which is a modest improvement compared to the 10.6% accuracy in variant A. As with variant A, the largest share of participants fell into the "near-miss" category. Here, 99 participants (32.4%) selected an option just one class away from the correct answer (±1). A smaller group—22 participants (7.2%)—provided answers that deviated by two classes (±2), which is comparable to the proportion seen in variant A. Finally, 135 participants (44.3%) gave responses that were three or more classes off (±3). Although still a substantial portion, this is slightly lower than the 52% observed in variant A, again suggesting a small improvement when switching from vibration to pictorial motion.

For the fatalities task in map variant C, where pulsation was used as the dynamic cue for magnitude, participants performed noticeably better than in the two previous variants. A total of 93 participants—30.7% of the group—answered the question with complete accuracy (0 deviation). This is almost three times higher than the accuracy in variant A (10.6%) and about double the accuracy in variant B (16.0%), making variant C the strongest performer so far for this particular task. As in the earlier variants, a substantial portion of responses fell into the "close but not correct" category. Here, 86 participants (28.3%) chose an answer just one class away from the truth (±1), a percentage similar to that observed in both A and B. The share of participants whose answers were two classes off (±2) was relatively small—19 participants (6.2%)—making it the lowest among the three variants. Meanwhile, 105 participants (34.6%) gave responses three or more classes away (±3), which is noticeably lower than in variant A (52.0%) and somewhat lower than in variant B (44.3%).

For the fatalities task in map variant D, which used static pictorial symbols, participants performed better than in variants A and B, but not as well as in the pulsation-based variant C. A total of 77 participants (22.8%) answered with complete accuracy (0 deviation). This places variant D above vibration (A: 10.6%) and pictorial motion (B: 16.0%), but still clearly below the strong performance seen in variant C (30.7%). Responses that were close to the correct answer—just one class away (±1)—came from 66 participants (19.6%). This share is lower than in A (27.2%), B (32.4%), and C (28.3%), suggesting that fewer people landed in the "almost correct" category in this variant. When looking at larger deviations, 45 participants (13.3%) were off by two classes (±2), which is higher than in variant C (6.2%) but still roughly comparable to variants A (10.2%) and B (7.2%). The largest group consisted of those whose answers were three or more classes off (±3): 144 participants (42.7%). This places variant D in a similar range to A (52.0%) and B (44.3%), though still worse than the pulsation-based variant C (34.6%). Five participants did not answer this question.

A summary of the results is presented in Figure 4.2, and the overall pattern is quite clear. The most accurate answers—those with no deviation at all—were given when

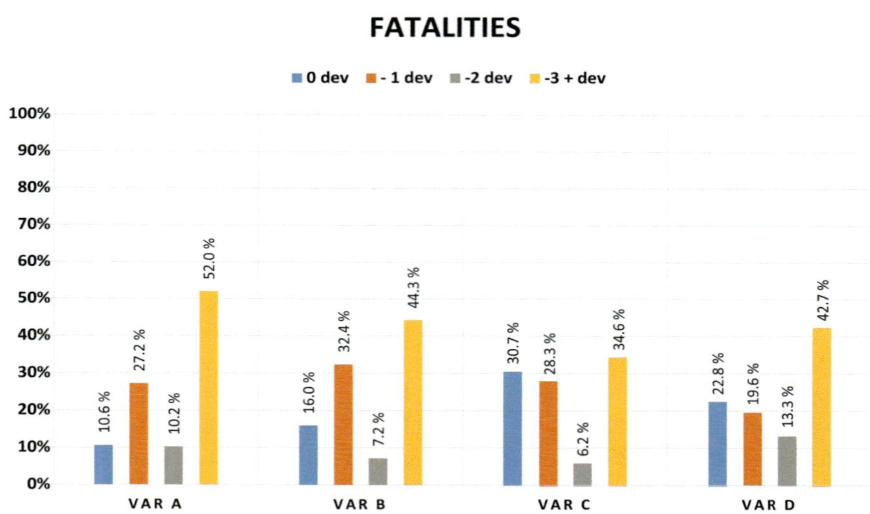


**FIGURE 4.2** Summary of the effectiveness in communicating the number of fatalities using symbol size, in comparison to static symbols on a spatiotemporal map.

participants used map variant C, the pulsation-based design. In this task, pulsation seemed to help users focus more effectively on symbol size, which was the variable encoding the number of fatalities. The next best performer was map variant D, which used static pictorial symbols. However, when answers those were close to correct (off by only one class, ±1 deviation) are also taken into account, the ranking shifts slightly: map variant B, which used pictorial motion, becomes the second most effective. This suggests that although pictorial motion did not produce the highest number of perfectly correct answers, it often helped users get near the correct value. On the other hand, map variant A, which used vibration to show magnitude, resulted in the highest number of incorrect responses. This does not mean that vibration is inherently unsuitable—but in this task, where users had to judge symbol size (fatalities) rather than motion, the strong vibrational effect likely interfered with their ability to make stable size comparisons. Finally, the statistical analyses (both the Pearson chi-square test and Fisher's exact test) showed no statistically significant differences between the map variants. In practical terms, this means that even though some symbol types appear more effective than others, the observed differences were not strong enough to be considered statistically meaningful.

### 4.1.3 Identifying the Most Common Earthquake Depth (Using Color)

For the task where participants had to identify the most common earthquake depth—shown on the map through color—map variant A (the version that used vibration for magnitude) produced fairly strong results. In this group, 110 participants (43.3%) gave a fully accurate answer with no deviation. A similarly large group, 106 participants (41.7%), selected an answer that was just one class away from the correct depth (±1 deviation). Far fewer responses fell into the less accurate categories: 24 participants (9.4%) were off by two classes (±2), and only 10 participants (3.9%) selected values that deviated by three or more classes. Four participants did not submit an answer.

In map variant B—the version that used pictorial motion to show magnitude—the results for identifying the most common earthquake depth were very similar to those from variant A. Here, 124 participants (40.6%) selected the fully correct depth class, only slightly fewer in proportion than in variant A. As before, a large share of responses fell into the "almost correct" category: 127 participants (41.6%) chose an answer just one class away from the correct one (±1 deviation), a proportion nearly identical to that observed in variant A. Less accurate responses were again relatively uncommon. 34 participants (11.1%) selected answers with a two-class deviation (±2), while only 11 participants (3.6%) chose options three or more classes away from the correct depth. Nine participants chose not to answer the question.

In map variant C, where pulsation was used to represent earthquake magnitude, participants performed strikingly better in identifying the most common earthquake depth than in the two previous variants. An impressive 238 participants—that is 78.5% of everyone in this group—selected the correct depth class with no deviation at all. This is nearly double the accuracy observed in variants A and B, both of which hovered around 40%. Only 64 participants (21.1%) chose an option just one class away from the correct answer (±1 deviation), and this group was much smaller than in the earlier variants, where "almost correct" answers made up more than 40% of all responses. Remarkably, only one participant (0.3%) deviated by two classes (±2), and no one selected an answer three or more classes away.

In map variant D, which used static symbols, participants performed better than in variants A and B but still far below the exceptionally high accuracy observed in variant C. In this group, 154 participants—or 45.7%—identified the most common earthquake depth perfectly (0 deviation). This places variant D slightly above both A (43.3%) and B (40.6%) in terms of full accuracy, but still dramatically behind variant C's standout 78.5%. A further 109 participants (32.3%) missed the correct class by only one step (±1 deviation), a proportion similar to variants A and B, where near-correct answers also made up large shares of the responses. Meanwhile, 44 participants (13%) deviated by two classes (±2), which is higher than in the previous variants, especially when compared to variant C, where only a single person selected an answer this far off. Finally, 26 participants (7.7%) selected answers three or more classes away (±3), again notably higher than in any of the dynamic-symbol variants.

In summary, when it came to identifying the most common earthquake depth (Figure 4.3), map variant C—using pulsation—stood out clearly as the most effective. No other map variation came close; its accuracy rate was dramatically higher than the rest. The remaining variants (A, B, and D), while showing differences in percentages, performed at broadly similar levels and did not approach the success seen in the pulsation condition. Despite these visible differences, the statistical tests (Pearson chi-square and Fisher's exact test) showed no statistically significant differences between the map variants. This means that although variant C appears to outperform the others, the variation in results is not strong enough, from a statistical standpoint, to confirm a meaningful difference in effectiveness across the four mapping methods.

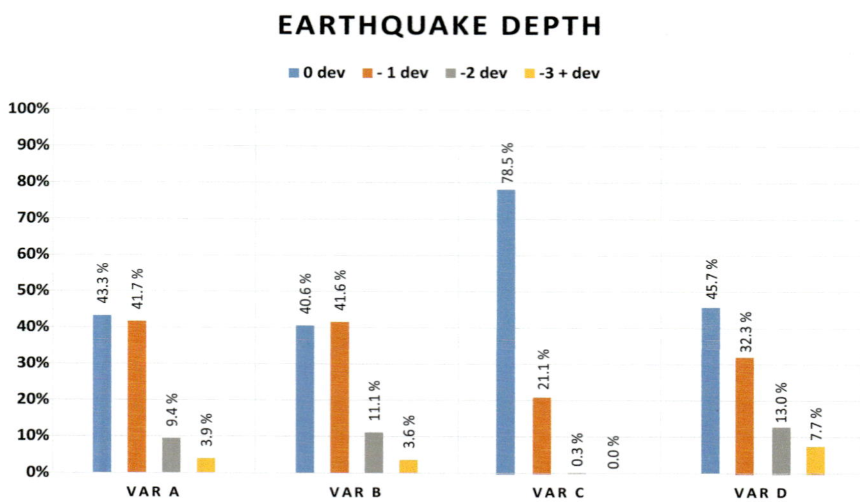


**FIGURE 4.3** Summary of the effectiveness in communicating the most frequent earthquake depths in Europe using symbol color hue, compared to static symbols on a spatiotemporal map.

### 4.1.4 Finding the Year with the Most Earthquakes (Timeline Task)

For the task that required identifying the year with the highest number of earthquakes using the timeline in map variant A, 77 participants selected the correct year (0 deviation), which represents 30.3% of all respondents in this group.

Only 11 participants (4.3%) chose a year that was off by one position on the timeline (±1 deviation), and 7 participants (2.7%) were off by 2 years (±2 deviation). However, the majority of respondents—159 participants, or 62.6%—gave answers that were 3 or more years away from the correct value (±3 deviation), indicating that most participants struggled with this timeline-based task when using map variant A.

For the same timeline-based task in map variant B, the pattern looked very similar to what was observed in variant A. A total of 91 participants (29.8%) identified the correct year (0 deviation), which is nearly identical to the accuracy level seen in variant A. Only 8 participants (2.6%) were off by a single year (±1 deviation), and 9 participants (2.9%) were off by 2 years (±2 deviation). As in variant A, most respondents struggled with this task: 197 participants—a substantial 64.6% of the group—gave answers that were 3 or more years away from the correct one (±3 deviation). This confirms that switching from vibrating symbols (variant A) to pictorial motion (variant B) did not noticeably improve performance on this timeline-based question.

In map variant C, participants performed much better on the timeline task than in the previous two variants. A total of 190 participants—a noticeably higher 62.7%—identified the correct year (0 deviation). This is more than double the accuracy seen in variants A and B, where only about 30% of respondents answered correctly. Interestingly, this group produced no responses that were just 1 year off (±1 deviation). Instead, the remaining errors were distributed across larger deviations: 39 participants (12.8%) were 2 years off (±2 deviation), and 74 participants (24.4%) selected a year that differed from the correct one by 3 or more years (±3 deviation). Compared to the earlier variants, this represents a clear improvement in accuracy.

In map variant D, the static-symbol version, 112 participants—or 33.2%—correctly identified the year with the most earthquakes (0 deviation). This accuracy level is very similar to what was observed in variants A and B, both of which also hovered around the 30% mark. However, it remains far below the much stronger performance seen in variant C, where nearly two-thirds of participants answered correctly. Responses that were close but not exact were relatively rare—21 participants (6.2%) were off by just 1 year (±1 deviation), and only four participants (1.1%) missed the correct year by 2 years (±2 deviation). As in the earlier variants (except for the standout case of C), most errors were considerable: 199 participants—a substantial 59.0%—selected a year that differed by 3 or more years (±3 deviation). One participant did not provide an answer.

Taken together, the results for this timeline task (Figure 4.4) show a pattern similar to what appeared in earlier sections. Map variant C, which used pulsation, again produced the highest proportion of correct answers by a wide margin. The other three variants—A, B, and D—clustered much more closely together, with all of them showing noticeably lower accuracy. Among these remaining variants, the

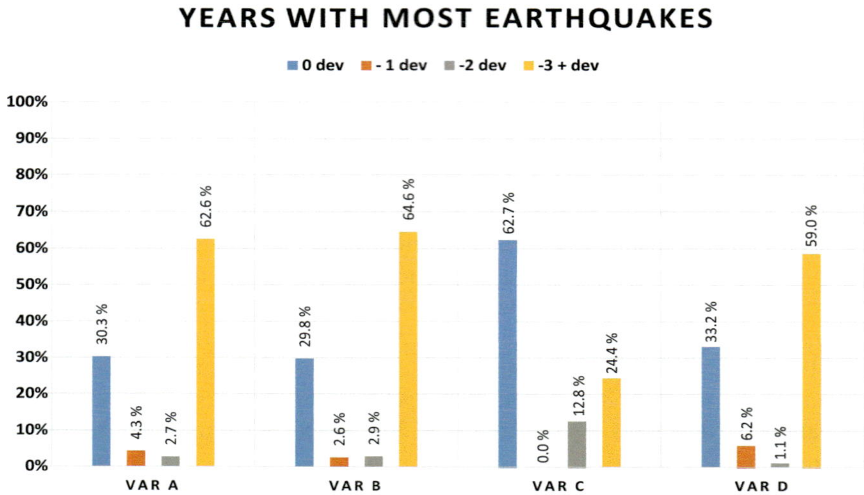


**FIGURE 4.4** Summary of the effectiveness in communicating the years with the highest number of earthquakes in Europe.

static version (D) performed slightly better than A and B, but the differences were modest. More importantly, statistical testing (Pearson chi-square and Fisher's exact test) showed that none of these differences reached statistical significance. In practical terms, this means that although pulsation stood out descriptively, the variations in accuracy between the four symbol types cannot be considered statistically meaningful for this particular task.

## 4.2 HOW FAST DID PARTICIPANTS COMPLETE THE TASKS

In addition to accuracy, it is also important to understand how quickly participants were able to complete each map-based task. This measure is called efficiency. In simple terms, efficiency tells us how much time someone needed to look at the map, search for the required information, and choose an answer. It is important to emphasize that this measure includes all response times, not just those that produced correct answers. A fast response is still counted as fast—even if the answer was wrong. However, to better understand how symbol types might influence performance, the analysis also compares only those participants who answered correctly, allowing a fairer look at how efficient successful map readers were. Before comparing groups, the response-time data were checked to see whether they followed a normal distribution (a bell-shaped curve). They did not, as confirmed by the Shapiro–Wilk normality test, a standard statistical test for this purpose. Because of this, a nonparametric test—the Mann–Whitney U test—was used to compare response times between participants who answered correctly and those who did not. This approach makes it possible to see not only who was correct but also how quickly correct answers were obtained

### 4.2.1 Speed of Finding the Strongest Earthquakes

When looking at how quickly participants identified the strongest earthquakes, the four map variants showed noticeable differences. Variant "A," which used vibrating symbols to show earthquake magnitude, clearly stood out—mainly because participants needed much more time to complete the task. For this variant, the average response time was 202 seconds, with a median time of 150 seconds. When focusing only on participants who answered correctly, the task became even slower: the average time increased to 224 seconds, and the median rose to 175 seconds. This suggests that even participants who understood the vibrating symbols well enough to answer correctly still needed considerable time to interpret them.

To check whether correct answers were consistently faster (or slower) than incorrect ones, the Mann–Whitney U test was applied. The result showed no statistically significant difference between the two groups ($p = 0.09$, $z = 1.65$). In other words, for variant "A," taking more or less time did not meaningfully predict whether someone would answer correctly. This indicates that vibration—at least in this task—may slow users down overall, regardless of whether they ultimately succeed.

For variant "B," which used pictorial symbol motion to show earthquake magnitude, participants completed the task noticeably faster than in variant A. The average response time was 142 seconds, with a median of 92 seconds—a substantial improvement compared to the much longer times recorded for the vibrating symbols. Participants who answered correctly were even quicker: their responses averaged 127 seconds, with a median of 86 seconds. This contrasts with variant A, where correct responders actually took slightly longer than the overall group. In variant B, by comparison, correctly solving the task generally required less time.

To determine whether this difference in speed between correct and incorrect answers was meaningful, the Mann–Whitney U test was applied. The test showed no statistically significant difference ($p = 0.80$; $z = 0.25$). In practical terms, this means that—just like in variant A—the amount of time participants spent on the task did not strongly relate to whether they answered correctly. However, participants overall needed much less time with pictorial motion than with vibration, suggesting that this form of motion was easier and quicker for users to interpret.

For variant "C," which used pulsation to represent earthquake magnitude, task completion times fell between those observed for variants A and B. The average search time was 169 seconds, with a median of 156 seconds—faster than the vibrating symbols in variant A, but slower overall than the pictorial-motion symbols in variant B. Interestingly, participants who answered correctly in this variant took slightly longer than the overall group, with an average of 179 seconds (median = 158 seconds). This pattern more closely resembles what was observed in variant A (where correct responders were also slower), rather than variant B (where correct responders were faster). Because the median times for correct and incorrect answers in variant C were almost identical, the Mann–Whitney U test again showed no statistically significant difference between the two groups ($p = 0.95$; $z = 0.05$).

The most efficient performance was observed in variant "D," the version that used static symbols without any dynamic effects. Here, the average task completion time dropped sharply to 61 seconds, with a median of 48 seconds—making it dramatically faster than any of the dynamic-symbol variants. This clearly shows that removing motion entirely made it much easier and quicker for participants to find the strongest earthquakes. Participants who answered correctly in variant D showed no slowdown at all—their response times matched the overall group almost exactly (median = 58 seconds), unlike variants A and C where correct responders tended to take longer, or variant B where they were faster. As with all previous variants, the Mann–Whitney U test did not detect any statistically significant difference between the response times of correct and incorrect responders ($p = 0.41$; $z = 0.82$).

A summary of the task efficiency results is presented in Figure 4.5. Overall, the static-symbol map (variant D) proved to be the fastest and most efficient option for identifying and counting the strongest earthquakes. Among the dynamic-symbol techniques, variant B, which used pictorial symbol motion, stood out as the most efficient—performing noticeably better than both the vibrating and pulsating-symbol variants.

When these efficiency results are compared with the earlier findings on accuracy, a clear pattern emerges. Searching through static symbols (variant D) is consistently faster than searching through dynamic ones, even though the map variants performed similarly in terms of accuracy. In other words, participants could usually find the correct answer regardless of symbol type, but they did so much more quickly when the map contained no animation. Among the dynamic-symbol variants, variant B—the one that used pictorial symbol motion—again stood out as the most efficient option. This is especially evident when considering the median response time, which provides a more stable picture of performance than the average in cases with large variability. The statistical tests further support these observations.

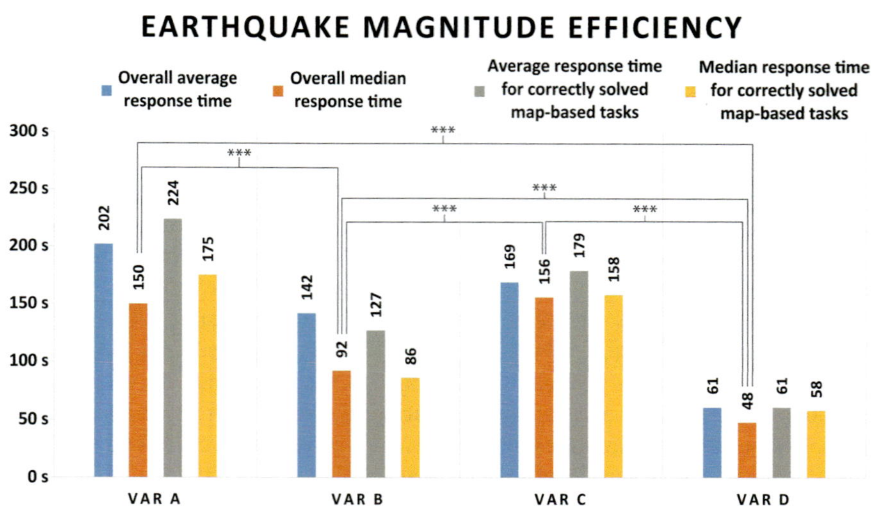


**FIGURE 4.5** Participants' response times (efficiency) for map variations in tasks involving finding and counting the highest magnitude earthquakes. (*** denotes statistical significance at a level of $p < 0.0005$.)

The Mann–Whitney U test confirmed significant differences in efficiency between several pairs of map variants:

- Variant A vs. B: $p=0.0001$, $z=6.77$
- Variant A vs. D: $p=0.0001$, $z=11.95$
- Variant B vs. C: $p=0.0001$, $z=5.98$
- Variant B vs. D: $p=0.0001$, $z=5.59$
- Variant C vs. D: $p=0.0001$, $z=11.52$

These results show that the differences in task speed between symbol types were not random—they were statistically meaningful. Finally, the efficiency of only the correct responses was not tested separately. The assumption was that if significant differences appear in the overall response time for each map variant, the same differences would naturally carry over to the group of participants who answered correctly.

### 4.2.2 Speed of Identifying the Highest Fatalities

When looking at how quickly participants were able to identify the earthquakes with the highest number of fatalities, the results again varied across the different map variants. For map variant A, which used vibration as the dynamic element, the average time required to complete this task was 202 seconds (median = 148 seconds). Interestingly, participants who answered correctly needed even more time: their average response time rose to 250 seconds (median = 232 seconds). This suggests that, in this variant, reaching the correct answer required more careful searching rather than quick guessing. To test whether participants who answered correctly tended to take systematically more (or less) time than those who answered incorrectly, the Mann–Whitney U test was applied. As in the previous tasks, the test did not show statistically significant differences between the two groups ($p=0.12$, $z=1.53$). This means that, for variant A, correct and incorrect responses were completed in similar amounts of time, and accuracy did not depend on speed.

For map variant B, the symbols used pictorial motion to represent earthquake magnitude, while symbol size—as in all variants—indicated the number of fatalities. The overall average response time for this task was 142 seconds (median = 76 seconds), which is noticeably faster than the 202-second average recorded for variant A. Interestingly, participants who answered the fatalities task correctly were slower than the group as a whole, taking 182 seconds on average (median = 97 seconds). This pattern mirrors what was observed in variant A: participants who arrived at the correct answer tended to spend more time examining the map. Despite these differences, the Mann–Whitney U test showed no statistically significant difference in median response time between correct and incorrect responses ($p=0.10$, $z=1.62$), indicating that accuracy was not strongly tied to how quickly participants completed the task.

For map variant C, where the point symbols included a pulsing animation (while symbol size again represented the number of fatalities), participants completed the task more quickly than in variants A and B. The average response time was 123 seconds (median = 89 seconds), making it the fastest among all dynamic-symbol variants studied so far. Participants who answered correctly showed almost the same level of efficiency: their average response time was 128 seconds, with a median of 83 seconds. This stands in contrast to variants A and B, where correct responses tended to take noticeably longer than the overall average. As with the previous variants, the Mann–Whitney U test showed no statistically significant difference in median response times between correct and incorrect responses ($p=0.72$, $z=0.35$). This again suggests that spending more or less time on the task did not reliably predict the correctness of the final answer.

Finally, map variant D, which used static point symbols, proved to be the most efficient of all variants—not only compared with the dynamic approaches (A, B, and C), but also overall. The average response time for this task was just 53 seconds (median = 45 seconds), making it substantially faster than any of the dynamic-symbol variants. Participants who answered correctly were nearly just as fast, with an average response time of 56 seconds (median = 47 seconds). This pattern contrasts sharply with variants A and B, where correct responses required noticeably more time than the overall average, and even outperforms variant C, where correct and incorrect response times were similar but still slower than in the static case. As observed in all previous variants, the Mann–Whitney U test showed no statistically significant difference between the response times of participants who answered correctly and those who did not ($p=0.28$, $z=1.07$). Thus, even though variant D was the most efficient, greater speed did not guarantee a correct answer—mirroring the trend seen in the other mapping approaches.

The efficiency results for identifying and counting all earthquake locations with the highest number of fatalities are summarized in Figure 4.6. Overall, the fastest map to work with was variant D, which used static pictorial symbols—a result consistent with earlier findings that static symbols tend to support quicker visual search. Among the dynamic-symbol variants, variant B, which incorporated pictorial motion, emerged as the most efficient. This conclusion is based on the median response time, which is the more appropriate measure here, given the non-normal distribution of the data.

However, variant C, which used pulsation, performed nearly as fast as variant B. This places pulsation among the most promising dynamic methods, especially when its effectiveness is taken into account. In fact, variant C produced the highest accuracy of all mapping methods when participants were asked to identify and count the earthquakes with the most fatalities. This combination of strong accuracy and solid efficiency makes pulsation a noteworthy alternative to the more established approaches. The differences in efficiency between the map variants were supported by the Mann–Whitney U test, and the degree of similarity or difference between variants is shown in Figure 4.6. As before, statistical comparisons were also made for participants who gave correct answers, based on the assumption that any difference visible in the overall response time would likely appear in the median response time of successful participants as well.

### 4.2.3 Speed of Determining the Most Frequent Depth

The next task asked participants to determine which earthquake depth range occurred most often across the 23-year dataset. This task relied on interpreting color, not motion or size. For map variant A, which used vibrating symbols (even though depth itself was shown by color), the average response time was 135 seconds

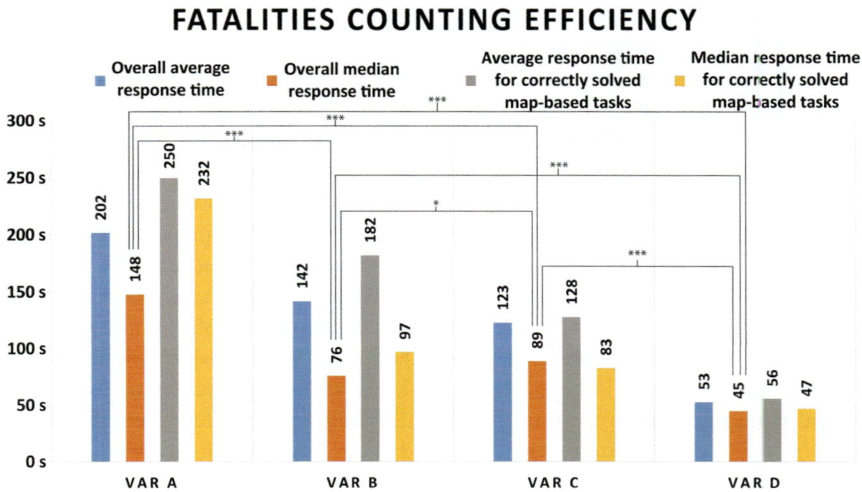


**FIGURE 4.6** Participants' response times (efficiency) for map variations in tasks involving identifying and counting the earthquake location with the highest number of fatalities. (* denotes statistical significance at a level of $p < 0.05$; *** denotes statistical significance at a level of $p < 0.0005$.)

(median = 100 seconds). Interestingly, participants who answered correctly tended to take more time: their average was 146 seconds (median = 118 seconds). A statistically significant difference emerged between correct and incorrect responders, confirmed by the Mann–Whitney U test ($p = 0.01$, $z = 2.51$). Participants who gave incorrect answers completed the task much faster, with a median of 86 seconds, suggesting that quick responses often came at the cost of accuracy for this particular task.

For map variant B, which used motion in pictorial symbols (while depth was still encoded by color), participants were generally faster than in variant A. The average response time was 128 seconds (median = 65 seconds), noticeably lower than the vibrating-symbol variant. Participants who answered correctly were just as quick, with an average time of 123 seconds (median = 64 seconds), almost identical to the overall group. In contrast to variant A, the Mann–Whitney U test showed no statistically significant difference between correct and incorrect responders ($p = 0.54$, $z = 0.60$). This suggests that, in variant B, speed did not strongly influence accuracy, and participants who took their time were not necessarily more likely to answer correctly.

For map variant C, which used pulsating-point symbols, participants were even faster than in the previous two dynamic-symbol variants. This variant recorded the lowest response time of all dynamic methods, with an average of 87 seconds (median = 56 seconds). Those who answered correctly performed almost identically: their average response time was 88 seconds, with the same median of 56 seconds.

Compared to variants A and B, this shows a clear pattern:

- Faster than A, which required much longer response times.
- Faster than B, which still required about twice as much time (median 65 seconds).
- And importantly, efficiency did not come at the cost of accuracy.

As with variant B, the Mann–Whitney U test showed no statistically significant difference between correct and incorrect responders ($p = 0.77$, $z = 0.29$). This suggests that, for pulsation, participants tended to work at the same pace regardless of whether they reached the correct answer.

Finally, map variant D, which relied on static symbols, was by far the most efficient of all four variants. Participants completed the task in an average of 36 seconds (median = 24 seconds)—less than half the time needed for any dynamic-symbol map. Those who answered correctly performed almost identically, with an average response time of 38 seconds (median = 24 seconds). Compared with the dynamic versions, it was dramatically faster than variant A (median 100 seconds), much quicker than variant B (median 65 seconds), and still substantially faster than variant C, even though C was the quickest among the dynamic maps (median 56 seconds). Just like in variant C, the Mann–Whitney U test confirmed that there were no statistically significant differences in response times between participants who answered correctly and those who answered incorrectly ($p = 0.77$, $z = 0.29$).

The combined efficiency results—supported by the statistical comparisons shown in Figure 4.7—paint a consistent picture. The static-map variant (D) was the fastest, making it the most efficient solution in terms of response time. However, fastest does not mean best: variant D was not the most effective in terms of accuracy. This creates an important contrast. Participants worked through the static map very quickly—likely because static symbols are familiar and easy to scan—but this speed often came at the expense of precision. In other words, people may answer faster on static maps, but not necessarily more accurately.

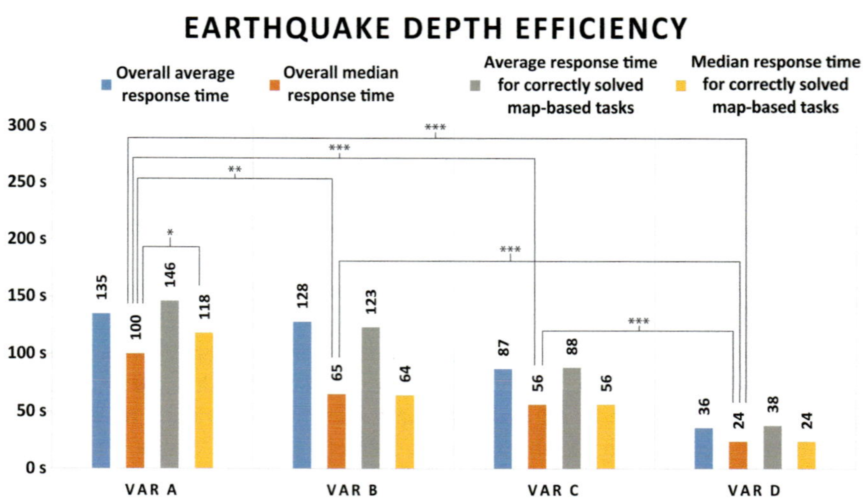


**FIGURE 4.7** Participants' response times (efficiency) for map variations in tasks involving identifying and counting earthquake locations with the most frequent depth. (* denotes statistical significance at a level of $p < 0.05$; ** denotes statistical significance at a level of $p < 0.005$; *** denotes statistical significance at a level of $p < 0.0005$.)

The dynamic variants tell a different story. Among them, map variant C (pulsation) stands out. It was slightly more efficient than variant B (pictorial motion), the most accurate of all variants, and, when focusing on correct responses, the top performer overall. This pattern highlights a key lesson about working with spatiotemporal map tasks:

- Static maps feel intuitive and quick to use, especially when scanning for patterns.
- Dynamic symbols require a little more processing time—but they often lead to more precise and reliable answers.

In short, dynamic does not always mean slower, and static does not always mean clearer. The results suggest a trade-off: speed favors static symbols, but accuracy favors carefully designed dynamic ones—especially pulsation (variant C).

### 4.2.4 Speed of Finding the Year with the Most Earthquakes

For the final task—identifying the year with the highest number of earthquakes—map variant A (the vibrating-symbol version) produced relatively slow response times. Participants needed, on average, 187 seconds to complete the task (median = 131 seconds). Those who answered correctly took even longer: their average time increased to 234 seconds (median = 169 seconds). A statistically significant difference emerged between correct and incorrect respondents, as shown by the Mann–Whitney U test ($p = 0.008$, $z = 2.63$). Interestingly, participants who provided incorrect answers were much faster, with a median response time of 123 seconds. This pattern is similar to what we observed in earlier tasks: variant A tends to encourage quicker but less accurate responses, while participants who take the time to examine the vibration-based symbols more carefully are more likely to answer correctly—but at the cost of additional time.

In map variant B, which used pictorial motion, participants completed the task more quickly than those using variant A. The overall average response time was 134 seconds (median = 77 seconds). Those who identified the correct year performed almost identically, with an average of 134 seconds and a median of 78 seconds. Unlike variant A—where incorrect answers tended to be faster—variant B showed no meaningful difference between correct and incorrect responders. The Mann–Whitney U test confirmed that median response times did not significantly differ ($p = 0.73$, $z = 0.34$). Compared with variant A, variant B appears to support more consistent performance, suggesting that the dynamic pictorial motion may place a smaller cognitive load on users when interpreting temporal patterns.

In map variant C, which used pulsing symbols, participants performed at a speed similar to those using variant B, though still noticeably faster than in variant A. The average response time was 137 seconds (median = 102 seconds). Participants who correctly identified the year with the most earthquakes showed nearly the same performance, averaging 133 seconds with a median of 106 seconds. Just like in variant B, the response times for correct and incorrect answers were statistically indistinguishable. The Mann–Whitney U test confirmed that there was no significant difference between the two groups ($p=0.62$, $z=0.49$). Compared with the earlier variants, variant C sits comfortably between them, being much faster and more consistent than variant A, where incorrect answers tended to be rushed. Slightly slower than variant B, but still showing the same pattern of balanced performance across participants.

In map variant D, which used only static point symbols to show the spatiotemporal earthquake data, participants were once again the fastest overall. The average response time was just 65 seconds (median = 52 seconds), making it by far the most efficient option compared with the dynamic-symbol variants. Participants who answered the task correctly took a bit longer—as was also the case in variants A and D—with an average response time of 74 seconds (median = 65 seconds). Even so, their performance remained substantially faster than in all three dynamic variants. As in variants B and C, the difference between correct and incorrect responses was not statistically significant. The Mann–Whitney U test showed no meaningful difference in median response times ($p=0.06$, $z=1.84$). Overall, static symbols enabled the fastest task completion, even if they did not always lead to the highest accuracy.

Taken together (Figure 4.8), these findings highlight an important pattern in how people interact with different types of map symbols. Static symbols (variant D) make it easier for participants to work quickly—so if speed is the main goal, this method performs best. However, this speed often comes at the cost of accuracy, especially in tasks that require scanning across many years of data on the timeline. Variant B, which uses pictorial motion, emerges as a strong middle-ground solution. It is nearly as fast as the static map and maintains solid accuracy, suggesting that subtle dynamic cues can help guide attention without overwhelming the viewer. Finally, variant C, based on pulsation, demonstrates a different strength: although participants spend

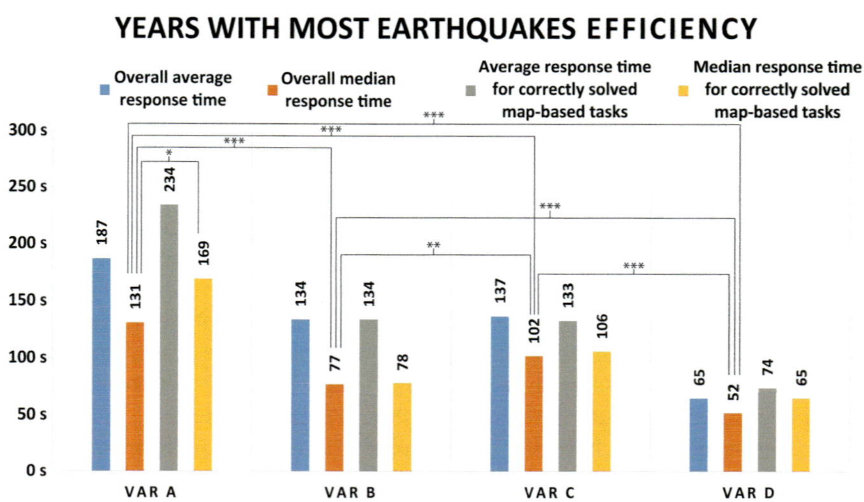


**FIGURE 4.8** Participants' response times (efficiency) for map variations in tasks involving the identification of the years with the highest number of earthquakes. (* denotes statistical significance at a level of $p<0.05$; ** denotes statistical significance at a level of $p<0.005$; *** denotes statistical significance at a level of $p<0.0005$.)

more time examining the map, they tend to answer more accurately. This makes pulsation particularly promising for tasks where precision matters more than raw speed—an important consideration for maps dealing with risk, public safety, or scientific analysis. In short, the results show that there is no single "best" dynamic symbol, but rather a set of trade-offs between speed, accuracy, and visual complexity. Understanding these trade-offs is essential for choosing the right mapping technique for a given task or audience.

## 4.3 HOW PARTICIPANT CHARACTERISTICS SHAPED THE RESULTS

As described in the Introduction, map-reading performance can be evaluated using two objective measures: effectiveness (how accurate the responses were) and efficiency (how quickly participants completed the tasks). Now that the main results from the online survey have been presented, it is important to look one step deeper and examine whether different groups of participants—defined by characteristics such as gender/sex or self-assessed map-reading ability—performed differently on these tasks.

This section explores whether these personal characteristics are linked to actual performance. In other words:

- Do men and women perform differently on any of the tasks?
- Do people with higher education perform better or faster?
- Do people who think they are good at reading maps actually perform better?
- Does frequent map use lead to better or faster results?

By examining these questions, this analysis helps reveal whether dynamic and static-map symbols work equally well for all users—or whether certain groups benefit more than others.

### 4.3.1 Comparing Responses by Gender

#### 4.3.1.1 Map Variant A

**For map variant A**, participants who identified as men and women performed very similarly in the task of identifying and counting the **earthquakes with the highest magnitude**.

- 20.4% of men answered the task correctly,
- compared with 18.0% of women.

This difference is very small, and the statistical tests confirmed that it is not significant. Both the Pearson chi-square test ($\chi^2 = 1.53$, $p = 0.21$) and the Fisher's exact test (two-tailed $p = 0.74$) showed that gender was not related to whether a participant answered correctly or incorrectly. The same pattern appeared when looking at efficiency (the time needed to complete the task).

- Men had a median response time of 157 seconds,
- while women had a median of 149 seconds.

The Mann–Whitney U test showed no significant difference between these times ($p=0.76$, $z=0.30$). Even when only participants with correct answers were considered, the difference remained statistically insignificant ($p=0.40$, $z=0.84$). For map variant A, gender did not influence accuracy or speed. Men and women performed at nearly the same level, and none of the statistical tests suggested that gender played any meaningful role in task performance.

For the task involving the identification of **earthquakes with the highest number of fatalities**, the results again showed only small differences between men and women.

- 7.4% of men answered correctly,
- compared with 11.8% of women, indicating that women performed slightly better on this task.

However, just as in the previous analysis, these differences were not statistically significant. Both the Pearson chi-square test ($\chi^2=0.19$, $p=0.66$) and the Fisher's exact test (two-tailed $p=0.23$) confirmed that gender was not meaningfully related to answer correctness. Efficiency results followed the same pattern.

- Men had a median response time of 150 seconds,
- Women had a nearly identical median of 149 seconds.

The Mann–Whitney U test found no statistically significant difference between these groups ($p=0.59$, $z=0.53$). Among participants who actually solved the task correctly, response times were also similar, with no significant difference ($p=0.70$, $z=0.40$). Although women showed slightly higher accuracy in this particular task, the differences were small and not statistically meaningful. As with the previous map variant, gender did not influence either how accurately or how quickly participants completed the task.

For the task that required **identifying the most common earthquake depth** across the 23-year period, the pattern shifted slightly compared with the previous tasks.

- 51% of men answered this question correctly,
- While 38% of women selected the correct depth.

This suggests that men performed better on this specific task. However—just as in the earlier analyses—these differences did not reach statistical significance. Both the Pearson chi-square test ($\chi^2=0.05$, $p=0.83$) and the Fisher's exact test (two-tailed $p=0.84$) showed that gender was not meaningfully related to the correctness of responses. Efficiency results again followed the same general pattern observed in earlier tasks.

- Men had a median response time of 108 seconds,
- while women responded slightly faster, at 98 seconds.

Despite this small difference, the Mann–Whitney U test confirmed that the gap was not statistically significant ($p = 0.84$, $z = 0.19$). Among those who actually answered correctly, the results remained consistent—no significant difference was found between men and women ($p = 0.38$, $z = 0.88$). Even though men achieved higher accuracy in this particular task, the statistical tests show that these differences could easily be due to chance. Just like in the previous tasks, gender did not influence how accurately or how efficiently participants solved the problem.

For the final task—identifying the years with the **highest number of earthquakes on a timeline**—the results again showed differences between men and women, but these differences were not statistically meaningful. In terms of effectiveness:

- 38% of men selected the correct year,
- compared to 23.7% of women.

This suggests that men performed better on this task. However, neither the Pearson chi-square test ($\chi^2 = 0.73$, $p = 0.39$) nor the Fisher's exact test (two-tailed $p = 0.49$) found this difference to be statistically significant. In other words, the variation in success rates between men and women could easily be due to chance—just like in the previous tasks. When looking at efficiency, measured through median response time, the pattern was again consistent with earlier findings. There were no statistically significant differences between men and women, whether considering the full participant group or only those who answered correctly ($p = 0.91$, $z = 0.11$).

However, one interesting exception appeared within the group of women. Women who answered correctly took much longer to do so:

- Median time for correct responses: 205 seconds
- Median time for all women: 129 seconds

This difference was statistically significant ($p = 0.03$, $z = 2.20$). This suggests that among women, those who arrived at the correct answer tended to invest more time and effort in examining the timeline before making a decision.

#### 4.3.1.2 Map Variant B

For **map variant B**, which used pictorial motion for magnitude, the pattern of results was similar to what was observed in variant A, although with a slight shift in which group performed better. In terms of effectiveness,

- 19.8% of women answered **the magnitude task** correctly,
- compared to 14.7% of men.

This gives women a small advantage for this particular variant. However, statistical testing showed that this difference was not significant. Both the Pearson chi-square test ($\chi^2 = 1.37$, $p = 0.24$) and Fisher's exact test (two-tailed $p = 0.24$) indicated that men and women performed similarly from a statistical standpoint. When considering efficiency, men and women also behaved much the same:

- Median response time for men: 97 seconds
- Median response time for women: 75 seconds

Although the numbers differ slightly, the Mann–Whitney U test showed no statistically significant difference between the two groups ($p=0.34$, $z=0.95$). The same was true when looking only at participants who answered correctly—again, no meaningful differences emerged. Overall, just like in map variant A, both effectiveness and efficiency were comparable for men and women in variant B. Women were slightly more accurate here, but the difference was not strong enough to be statistically meaningful.

For **the fatalities task** in map variant B, the performance of men and women was again very similar—much like in the previous magnitude task. In terms of effectiveness, the accuracy was almost identical between groups:

- Men: 15.4% correct
- Women: 15.0% correct

This tiny difference was not statistically meaningful. Both the Pearson chi-square test ($\chi^2=1.90$, $p=0.17$) and Fisher's exact test (two-tailed $p=0.18$) confirmed that men and women performed equally well in identifying the earthquakes with the highest number of fatalities. The same pattern appeared in efficiency. Among participants who answered correctly:

- Median time for men: 80 seconds
- Median time for women: 77 seconds

The Mann–Whitney U test indicated no statistically significant difference ($p=0.45$, $z=0.76$). Even when considering all participants—regardless of whether they answered correctly or not—the response times showed no meaningful difference between men and women ($p=0.84$, $z=0.20$). Overall, for this task as well as the previous one, men and women performed almost identically in both speed and accuracy when using map variant B.

For the **depth-identification task** in map variant B, the performance of men and women was once again almost identical—continuing the pattern observed in the previous tasks for this variant. In terms of effectiveness, both groups performed equally well:

- Men: 41.2% correct
- Women: 41.3% correct

This near-perfect match was confirmed statistically. Neither the Pearson chi-square test ($\chi^2=0.90$, $p=0.34$) nor Fisher's exact test (two-tailed $p=0.37$) indicated any meaningful relationship between gender and answer correctness. Efficiency results showed the same pattern of similarity. For all participants:

- Median time for men: 66 seconds
- Median time for women: 65 seconds (clarified value)

The Mann–Whitney U test again showed no significant difference between men and women in overall response time ($p=0.73$, $z=0.35$). Among those who answered correctly, response times were also statistically indistinguishable ($p=0.46$, $z=0.74$). In summary, for map variant B and this depth-related task, men and women displayed almost identical levels of both accuracy and speed—consistent with the previous findings for this variant.

For **the timeline task** in map variant B—identifying the year with the highest number of earthquakes—men performed slightly better than women, but the difference was small and not statistically meaningful. In terms of effectiveness:

- Men: 32.4% correct
- Women: 27.5% correct

Although men showed a modest advantage, neither the Pearson chi-square test ($\chi^2=1.54$, $p=0.21$) nor Fisher's exact test (two-tailed $p=0.29$) indicated that this difference was statistically significant. Efficiency results were also aligned with what was observed in the previous tasks for variant B:

- Median response time for men: 82 seconds
- Median response time for women: 73 seconds

The Mann–Whitney U test showed no statistically significant difference in overall median response times ($p=0.61$, $z=0.51$). This lack of difference also held true when considering only those who answered correctly ($p=0.90$, $z=0.12$). Overall, for this final task in map variant B, men and women again showed comparable performance in both accuracy and speed, consistent with the pattern seen across the other tasks in this variant.

#### 4.3.1.3 Map Variant C

For map variant C, which used pulsating-point symbols to represent magnitude, men and women performed almost identically on the task of **identifying the strongest earthquakes**. In terms of effectiveness:

- Men: 16.7% correct
- Women: 18.2% correct

This 1.5% difference is minimal, and both the Pearson chi-square test ($\chi^2=0.04$, $p=0.83$) and Fisher's exact test (two-tailed $p=0.99$) confirmed that there is no statistically significant difference between the two groups. Efficiency results tell the same story.

- Median response time for men: 156 seconds
- Median response time for women: 159 seconds

The Mann–Whitney U test showed no statistically significant difference in response times ($p = 0.34$, $z = 0.96$). Among those who answered correctly, response times were again similar, with no significant differences ($p = 0.15$, $z = 1.46$). In summary, for this first task in variant C, men and women showed essentially the same performance—both in accuracy and in speed—mirroring the balanced results observed in variants A and B.

For the second task in map variant C—**identifying and counting the earthquake locations with the highest number of fatalities**—the results showed a clearer distinction between men and women than in the previous task. In terms of effectiveness:

- Men: 32.7% correct
- Women: 27.8% correct

This difference of nearly 5% points favored men, and unlike in earlier comparisons, it was statistically significant. Both the Pearson chi-square test ($\chi^2 = 4.17$, $p = 0.04$) and Fisher's exact test (two-tailed $p = 0.04$) confirmed that men outperformed women in this particular task. However, when looking at efficiency, the pattern levels out again:

- Median response time for men: 89 seconds
- Median response time for women: 111 seconds

Despite women taking slightly longer on average, the Mann–Whitney U test showed that this was not a statistically significant difference ($p = 0.87$, $z = 0.16$). The same was true when considering only participants who answered correctly ($p = 0.32$, $z = 0.99$). In summary, while men were significantly more accurate than women on this specific task, both groups worked at similar speeds. The efficiency pattern aligns with earlier findings in variants A and B: even when effectiveness differs, response times between men and women typically do not.

For the third task in map variant C—**identifying the most common earthquake depth** over the 23-year period—the results again showed only small differences between men and women. In terms of effectiveness:

- Men: 25.6% correct
- Women: 22.9% correct

Although men performed slightly better, the difference was small, and neither the Pearson chi-square test ($\chi^2 = 1.79$, $p = 0.18$) nor Fisher's exact test (two-tailed $p = 0.22$) indicated a statistically significant relationship between gender and accuracy. The pattern was similar for efficiency. Median response times between men and women were nearly identical, and the Mann–Whitney U test confirmed that the difference was not statistically significant ($p = 0.93$, $z = 0.08$). Even when the analysis focused only on participants who answered correctly, the result remained nonsignificant ($p = 0.78$, $z = 0.28$). In short, for this depth-related task—unlike the previous fatalities task—men and women showed very similar levels of both accuracy and speed.

For the final task in map variant C—**identifying the year with the highest number of earthquakes**—the performance of men and women was again very similar. In terms of effectiveness:

- 66.6% of men answered correctly,
- 61.8% of women did so.

Although men performed slightly better, the difference was small, and neither the Pearson chi-square test ($\chi^2=2.25, p=0.13$) nor Fisher's exact test (two-tailed $p=0.14$) indicated statistical significance. In other words, gender did not meaningfully influence accuracy for this task. The same pattern appeared in the efficiency results. Median response times were nearly identical—107 seconds for men and 100 seconds for women—and the Mann–Whitney U test confirmed that this difference was not statistically significant ($p=0.91$, $z=0.13$). When the analysis was limited only to participants who answered correctly, the results remained nonsignificant ($p=0.35$, $z=0.94$). Overall, for this time-based task, men and women performed almost identically, both in accuracy and in the time required to find the correct year.

#### 4.3.1.4 Map Variant D:

For map variant D, which used static point symbols, the task of **identifying and counting earthquake locations with the highest magnitude** produced nearly identical results for men and women. In terms of effectiveness:

- 19.9% of women answered correctly,
- 19.1% of men did so.

The difference is very small, and both the Pearson chi-square test ($\chi^2=0.15$, $p=0.70$) and Fisher's exact test (two-tailed $p=0.79$) confirmed that it is not statistically significant. In other words, gender did not influence accuracy for this task. The efficiency results told the same story. Median response times were almost identical—46 seconds for men and 47 seconds for women—and the Mann–Whitney U test showed no statistically significant difference ($p=0.73$, $z=0.34$). The same result held when considering only participants who answered correctly ($p=0.93$, $z=0.09$). Overall, both men and women performed similarly in this static-symbol variant, with no meaningful differences in accuracy or response time.

For the task involving the **identification and counting of earthquakes with the highest number of fatalities**, the results for map variant D again showed only small differences between men and women—similar to the previous task in this variant. In terms of effectiveness:

- 19.1% of men answered correctly,
- 24.9% of women responded correctly.

Although this appears to give women a slight advantage, the difference was not statistically significant, as confirmed by both the Pearson chi-square test ($\chi^2=0.79$, $p=0.37$) and Fisher's exact test (two-tailed $p=0.46$). Thus, just like in the strongest-earthquake task, gender did not play a meaningful role in accuracy. The efficiency results mirror this pattern. Median response times were nearly the same—50 seconds for men and 48 seconds for women—and the Mann–Whitney U test indicated no statistically significant difference ($p=0.89$, $z=0.14$). The same held true even when analyzing only participants who completed the task correctly ($p=0.72$, $z=0.37$). Overall, as in the previous task for map variant D, men and women performed similarly in both accuracy and speed.

In the next task for map variant D, which required participants to **identify the most common earthquake depth** across the full spatiotemporal range, performance patterns again closely mirrored those observed in the earlier tasks for this variant. In terms of effectiveness,

- 41.1% of men answered correctly,
- 47.5% of women responded correctly.

Although women showed a modest advantage, the difference was not statistically significant, as confirmed by both the Pearson chi-square test ($\chi^2=0.15$, $p=0.69$) and Fisher's exact test (two-tailed $p=0.73$). As in the previous two tasks, gender did not meaningfully influence accuracy. The efficiency results were similarly close between groups. Median response times differed only slightly—27 seconds for men and 23 seconds for women—and the Mann–Whitney U test indicated no statistically significant difference ($p=0.71$, $z=0.37$). This pattern also remained consistent when analyzing only those participants who provided correct answers ($p=0.41$, $z=0.83$). Once again, as in the earlier analyses for map variant D, men and women performed nearly identically, both in terms of accuracy and response speed.

In the final task for map variant D, participants were asked to **identify the year with the most recorded earthquakes**. As with the earlier tasks in this variant, the performance of men and women was very similar. In terms of effectiveness,

- 33.3% of men selected the correct year,
- 29.3% of women did so.

Although men scored slightly higher, the difference was minimal, and both the Pearson chi-square test ($\chi^2=0.09$, $p=0.75$) and Fisher's exact test (two-tailed $p=0.84$) confirmed that this gap was not statistically significant. The efficiency results again followed the same pattern observed in the previous D-variant tasks. Median response times were nearly identical—48 seconds for men and 55 seconds for women—and the Mann–Whitney U test showed no statistically significant difference between the two groups ($p=0.71$, $z=0.37$). This was also true when examining only participants who answered correctly ($p=0.99$, $z=0.01$). Overall, this final task reinforces a consistent finding for map variant D: men and women performed equally well, both in how accurately and how quickly they were able to read and interpret the static-symbol map.

Across all the map types and all four tasks, men and women performed almost the same. In most cases, the differences in accuracy were so small that they were not meaningful from a statistical point of view. In simple terms, being male or female did not noticeably affect how well participants answered the map-based questions. There was one exception. In map variant C (the pulsating symbols), men did better than women on the task that involved finding and counting the earthquakes with the highest number of fatalities. This was the only situation where the difference was large enough to be considered statistically significant. Apart from this one task, however, the results clearly showed that men and women understood and interpreted the maps equally well, whether the symbols were static or dynamic. A visual overview of these comparisons—and which differences were statistically meaningful—can be found in Figure 4.9.

When looking at efficiency—that is, how quickly participants completed the tasks—there were no meaningful differences between men and women. In every map variant and for every task, both groups responded at very similar speeds, and statistical tests confirmed that these small differences were not significant. In other words, men and women were equally fast at solving the map-based questions, regardless of whether the symbols were static or dynamic. A visual summary of all efficiency comparisons is provided in Figure 4.10.

## 4.3.2 Does Education Level Matter?

### 4.3.2.1 Map Variant A

To explore whether educational background influenced how people interpreted the maps, participant performance was compared across three groups: those with higher, secondary, and elementary education. Map variant A (the vibrating-symbol map)

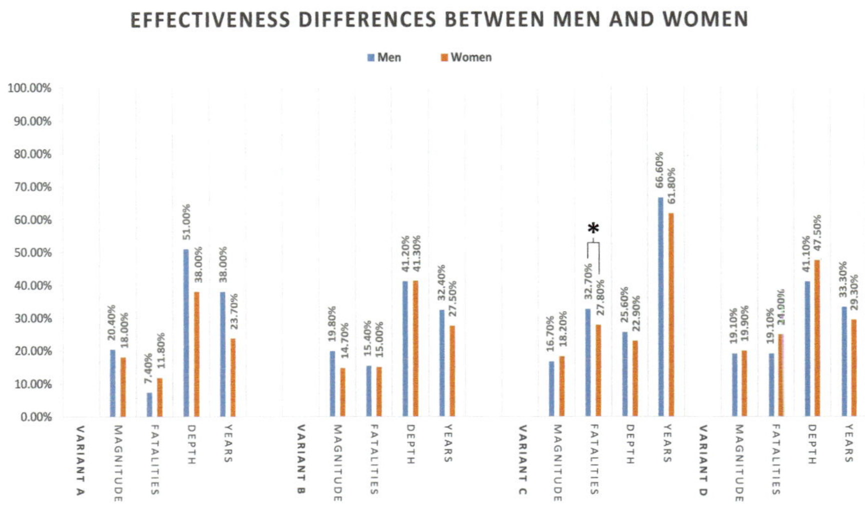


**FIGURE 4.9** Bar chart illustrating the differences in effectiveness between men and women across map variants in specific map-based tasks. (* denotes statistical significance at $p < 0.05$.)

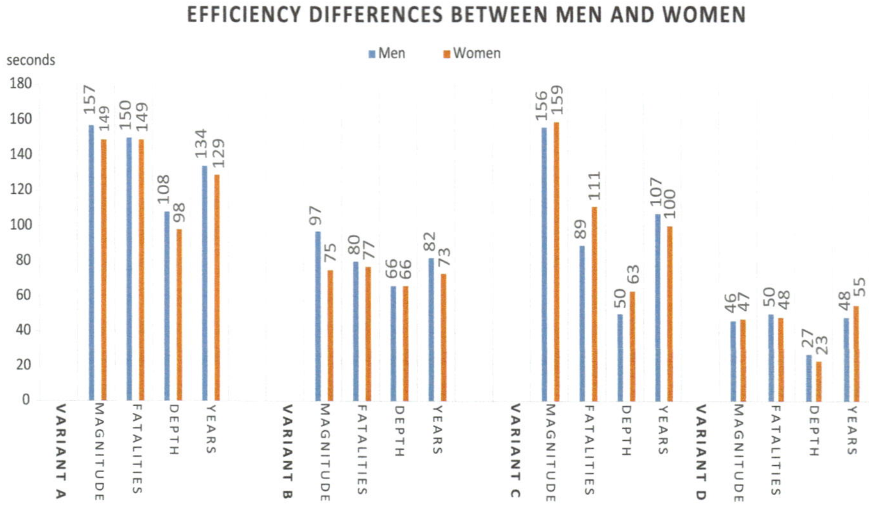


**FIGURE 4.10** Differences in response times between male and female participants across specific map variants, with no statistically significant differences observed.

provides a good starting point for this comparison. For the task of **identifying and counting earthquake locations with the highest magnitude**, the effectiveness scores were very similar across all education levels:

- 20.4% correct among participants with *higher education*,
- 17.4% correct among those with *secondary education*,
- 21.7% correct among those with *elementary education.*

Statistical tests (Pearson chi-square and Fisher's exact test) confirmed that none of these differences were statistically significant, meaning that education level did not meaningfully influence accuracy in this task. A similar picture emerged when examining efficiency—the time needed to complete the task. Median response times were:

- 149 seconds for participants with *higher education*,
- 151 seconds for those with *secondary education*,
- 137 seconds for those with *elementary education.*

Once again, according to the Mann–Whitney U test, these differences were not statistically significant. In short, for this task in map variant A, education level did not appear to affect either accuracy or response time, suggesting that the vibrating-symbol map was interpreted similarly across educational backgrounds.

In the next task within map variant A—**identifying the earthquake locations with the highest number of fatalities**—the results again showed only small differences between education groups. Participants with higher education answered correctly 12.9% of the time, those with secondary education scored 8.0%, and participants with elementary education achieved 13.0% correct responses. Although these percentages vary slightly, statistical tests (Pearson chi-square and Fisher's exact test) confirmed that none of these differences were significant. When looking at efficiency, the pattern was similar. Median response times were:

- 156 seconds for participants with *higher education*,
- 148 seconds for those with *secondary education*,
- 105 seconds for those with *elementary education*.

Despite these numerical differences, the Mann–Whitney U test showed that none of the groups were significantly faster or slower than the others. Overall, in this task as well, education level did not have a meaningful impact on how accurately or how quickly participants solved the problem.

In the third task of map variant A, which focused on **identifying the most common earthquake depth**, participants achieved their highest overall effectiveness within this variant. Those with higher education performed best, with 45.2% correct responses. Participants with secondary education followed at 42.0%, while those with elementary education scored 39.1%. Statistical testing revealed only one meaningful difference: according to the Pearson chi-square test ($\chi^2=4.13, p=0.042$) and Fisher's exact test ($p=0.049$), **participants with higher education performed significantly better than those with secondary education**. However, no significant differences were found between:

- higher vs. elementary education, or
- secondary vs. elementary education.

When considering efficiency, participants with higher education had a median response time of 96 seconds, those with secondary education 104 seconds, and those with elementary education 67 seconds. Despite these numerical differences, the Mann–Whitney U test showed that none of the groups were significantly faster or slower. Overall, while this task did reveal a small, statistically significant advantage for the higher education group over the secondary-education group, the broader pattern still suggests that education level had only a limited influence on performance.

In the final task of map variant A, participants were asked to **identify the year with the highest number of earthquakes**. Effectiveness varied across education levels:

- 30.1% of participants with *higher education* answered correctly,
- 28.3% of those with *secondary education*, and
- 39.1% of participants with *elementary education*.

Although these percentages suggest some variation—particularly the higher score among participants with only elementary education—the statistical tests did not confirm any meaningful differences between the groups. Both the Pearson chi-square test ($\chi^2=3.47, p=0.062$) and Fisher's exact test ($p=0.075$) indicated that the observed differences were not statistically significant, even in the comparison between elementary and secondary education, where the gap appeared the most pronounced.

The efficiency results followed a similar pattern. Participants with higher education had a median response time of 147 seconds, those with secondary education 131 seconds, and those with elementary education 93 seconds. Once again, despite the numerical differences, statistical testing showed no significant differences in the time required to complete the task. Overall, for this final task in variant A, education level did not play a statistically meaningful role in either accuracy or speed.

#### 4.3.2.2 Map Variant B

In **map variant B**, which used pictorial-motion symbols, the first task focused on **identifying and counting the earthquake locations with the highest magnitude**. Effectiveness varied slightly across education levels:

- Participants with *higher education* achieved 21.0% correct answers,
- Those with *secondary education* scored 14.9%,
- Participants with *elementary education* reached 17.6%.

Although these numbers suggest modest differences, statistical tests (Pearson chi-square and Fisher's exact test) showed that none of these differences were significant. In other words, education level did not meaningfully affect accuracy for this task. Efficiency tells a similar story:

- Participants with *higher education* completed the task with a median time of 79 seconds,
- Those with *secondary education* took 93 seconds,

Participants with *elementary education* needed 129 seconds.

Despite these time differences, the Mann–Whitney U test indicated no statistically significant differences between education groups. Overall, for this magnitude-identification task in variant B, education level did not significantly influence either accuracy or response time, even though small numerical differences were present.

In the next task, which required participants to **identify and count earthquake locations with the highest number of fatalities**, effectiveness again differed slightly across education groups. Participants with higher education correctly completed the task in 14.9% of cases, while those with secondary education achieved a slightly higher score of 16.7%. Notably, none of the participants with only an elementary education provided a correct response. Despite this variation, statistical testing showed that the difference between higher educated and secondary-educated participants was not significant. In practical terms, education level did not show a meaningful impact on accuracy for this task. Response times showed a different pattern:

- Participants with *higher education* had a median response time of 75 seconds,
- Those with *secondary education* took 82 seconds,

Participants with *elementary education* responded fastest, with a median of 64 seconds.

Even though these numbers vary, the Mann–Whitney U test indicated that none of the differences in response time were statistically significant. Overall, for this fatalities-identification task in variant B, education level did not significantly affect either accuracy or efficiency, even though the group with elementary education appeared numerically faster.

In the task focused on **identifying the most common earthquake depth**, the differences between education groups became more noticeable—at least numerically. Participants with higher education achieved the highest effectiveness rate at 46.5%, followed by those with secondary education at 39.6%. The group with elementary education performed considerably lower, with only 11.7% answering correctly. Despite these sizable numerical gaps, statistical testing tells a different story. Both the Pearson chi-square test and Fisher's exact test showed that none of these differences were statistically significant. This means that while performance appeared different, the variation could just as easily be due to chance. When it comes to efficiency, all three groups performed similarly. Median response times were:

- 62 seconds for participants with *higher education*,
- 69 seconds for those with *secondary education*,
- 65 seconds for the *elementary-education group.*

Once again, the Mann–Whitney U test confirmed that these differences were not statistically significant. Overall, for this depth-identification task in variant B, education level did not meaningfully influence either accuracy or speed, even if the numerical trends suggest some separation between the groups.

The final task in map variant B asked participants to **identify the year with the highest number of earthquake events**. As with earlier tasks, some differences appeared between education groups, but none were strong enough to be considered meaningful. Participants with higher education achieved the best accuracy, with 33.3% answering correctly. Those with secondary education followed at 27.0%, while participants with elementary education scored 23.5%. Although this pattern suggests that accuracy tended to increase with education level, the Pearson chi-square test and Fisher's exact test showed no statistically significant differences. In other words, the apparent differences were not strong enough to conclude that education level influenced effectiveness in this task. The efficiency results followed a similar pattern. Median response times were:

- 72 seconds for participants with *higher education*,
- 80 seconds for those with *secondary education*,
- 93 seconds for participants with *elementary education.*

Despite these gradual increases in time, the Mann–Whitney U test confirmed that none of the differences were statistically significant. Taken together, the results indicate that for this temporal-identification task in variant B, education level did not have a meaningful impact on either accuracy or speed, even if the numerical values show a modest trend.

#### 4.3.2.3 Map Variant C:

Map variant C used pulsating-point symbols to represent earthquake magnitude. In the first task—**identifying and counting the locations with the highest magnitudes**—the differences between education groups were small and showed no consistent pattern. Participants with higher education achieved an effectiveness rate of 19.2%, while those with secondary education scored slightly lower at 16.1%. Surprisingly, the group with elementary education performed the best, with 23.1% correct responses. Even so, these differences were not statistically significant, as confirmed by both the Pearson chi-square test and Fisher's exact test. Efficiency results also showed no meaningful differences. Median response times were nearly identical across groups:

- *Higher education*: 152 seconds
- *Secondary education*: 159 seconds
- *Elementary education*: 153 seconds

The Mann–Whitney U test confirmed that these differences were not statistically significant. Overall, for this first task in variant C, education level did not have a measurable impact on either how accurately or how quickly participants completed the task.

In the second task for map variant C—**identifying and counting earthquake locations with the highest number of fatalities**—participants performed noticeably better overall than in the previous magnitude-based task. Once again, the three education groups showed fairly similar effectiveness, though this time with one statistically meaningful difference. Participants with higher education achieved an effectiveness rate of 31.5%, while those with secondary education scored slightly lower at 29.2%. The group with elementary education again performed the best, reaching 38.5% correct responses. Unlike in the first task, here a statistically significant difference emerged, but only **between participants with secondary and elementary education**, as confirmed by the Pearson chi-square test ($\chi^2=5.18$, $p=0.023$) and Fisher's exact test ($p=0.045$). No other group comparisons showed significance. In terms of efficiency, all groups worked at similar speeds. Median response times were:

- *Higher education*: 111 seconds
- *Secondary education*: 89 seconds
- *Elementary education*: 90 seconds

Despite the variation, the Mann–Whitney U test confirmed that none of these differences were statistically significant. Overall, this task shows that while pulsation-based maps may help participants across all education levels, education level did not meaningfully influence speed, and only one pair of groups differed in accuracy.

In the third task for map variant C—**identifying the most frequent earthquake depth**—the overall accuracy increased substantially compared to the earlier tasks. All education groups performed well, though some differences in correctness appeared. Participants with higher education achieved the best results, with an effectiveness rate of 81.5%. Those with secondary education followed closely at 77.1%, while participants with elementary education scored 61.5%. Despite these noticeable differences, neither the Pearson chi-square test nor Fisher's exact test indicated any statistically significant differences between the education groups. Response times showed a similar pattern of consistency. Median response times were:

- *Higher education*: 56 seconds
- *Secondary education*: 58 seconds
- *Elementary education*: 50 seconds

These differences are small, and the Mann–Whitney U test confirmed that none were statistically significant. Overall, this task demonstrated that regardless of education level, participants were able to interpret the color-based depth information efficiently and with high accuracy—suggesting that this particular visual variable (color) was especially intuitive and effective for users across all backgrounds.

The final task in map variant C asked participants to **identify the year with the highest number of earthquake locations**. As with several earlier tasks in this variant, participants across all education levels performed reasonably well, though some differences emerged. Those with higher education achieved the best accuracy, correctly answering 69.9% of the time. Participants with secondary education followed with 62.5%, while the group with elementary education scored 46.1%, making them the least accurate of the three. Despite these differences in correctness, neither the Pearson chi-square test nor Fisher's exact test indicated that the groups differed significantly in effectiveness. When looking at efficiency, the response times varied more noticeably. Median search times were:

- *Higher education*: 101 seconds
- *Secondary education*: 112 seconds
- *Elementary education*: 68 seconds

Interestingly, the elementary education group—although the least accurate—was the fastest. Statistical testing confirmed one significant difference: the Mann–Whitney U test showed that **participants with elementary education were significantly faster than those with secondary education** ($p = 0.046$). No other differences in response time reached statistical significance. Overall, the results suggest that while education levels did not substantially affect accuracy, they may influence search strategies or the time taken to complete the task, particularly for this timeline-based question.

#### 4.3.2.4 Map Variant D

Map variant D, which used static point symbols and served as a reference against the dynamic variants, began with the task of **identifying and counting the earthquake locations with the highest magnitude**. Across education levels, participants performed at similar—though not identical—levels. Participants with higher education achieved an effectiveness of 23.7%, while those with secondary education scored slightly lower at 18.2%. Interestingly, the group with elementary education performed the best, with 28.0% correct responses. Despite these differences, neither the Pearson chi-square test nor Fisher's exact test indicated that the groups differed significantly in accuracy. When looking at efficiency, the three education groups showed increasingly faster median response times:

- *Higher education*: 58 seconds
- *Secondary education*: 37 seconds
- *Elementary education*: 26 seconds

Although this pattern suggests that participants with less formal education completed the task more quickly, the Mann–Whitney U test confirmed that none of these differences were statistically significant. Overall, for this first task in the static-symbol variant, neither accuracy nor speed appeared meaningfully tied to education level.

The second task in map variant D asked participants to **find and count the earthquakes with the highest number of fatalities**. As in the previous task, effectiveness varied somewhat across education groups. Participants with higher education achieved an effectiveness of 23.7%, those with secondary education scored 22.5%, and participants with elementary education reached 20.0%. Although these differences appear modest, statistical testing revealed that **the gap between higher and secondary education groups was significant**, as shown by both the Pearson chi-square test ($\chi^2=5.66$, $p=0.019$) and Fisher's exact test ($p=0.027$). This makes this task one of the very few instances in the study where education level showed a measurable impact on effectiveness. When considering efficiency, however, the pattern was different. Participants with higher education had a median response time of 47 seconds, those with secondary education were slightly slower at 57 seconds, and the fastest responses came from the elementary education group, with a median time of 25 seconds. Despite these differences in speed, the Mann–Whitney U test showed no statistically significant differences. This means that, for this task, education level influenced accuracy, but not response time.

The next task in map variant D focused on **identifying the most frequently occurring earthquake depth**. Here, participants with higher education performed slightly better than the other groups, achieving an effectiveness of 47.5%. Those with secondary education scored almost the same, at 46.8%, while participants with elementary education were notably less accurate, with an effectiveness of 28.0%. Statistical testing showed that this gap mattered: **the difference between the higher education group and the elementary-education group was statistically significant**, as confirmed by the Pearson chi-square test ($\chi^2=7.30$, $p=0.007$) and Fisher's exact test ($p=0.018$). In other words, for this task, education level did influence how accurately participants interpreted the map. When looking at efficiency, however, the differences were small. Participants with higher education had a median response time of 27 seconds, those with secondary education were slightly faster at 23 seconds, and those with elementary education took 36 seconds. According to the Mann–Whitney U test, none of these differences were statistically significant. This indicates that while education level affected accuracy, it did not meaningfully affect speed for this particular task.

The final task in map variant D asked participants to identify the year with the highest number of earthquake locations. Participants with higher education performed best, with 38.8% answering correctly. Those with secondary education followed with 29.5%, while participants with elementary education scored 24.0%. Despite these differences, statistical testing (both the Pearson chi-square test and Fisher's exact test) showed no significant relationship between education level and answer correctness. In other words, participants across all education groups performed similarly on this task. When looking at efficiency, the differences were again small. Participants with higher education had a median response time of 55 seconds, those with secondary education were slightly faster at 49 seconds, and participants with elementary education had a median time of 56 seconds. According to the Mann–Whitney U test, none of these differences were statistically significant. Overall, for this task, education level did not meaningfully influence either accuracy or speed.

When looking at how accurately participants answered the map-based questions, the study compared groups with higher, secondary, and elementary education levels (Figure 4.11). In general, people with higher education tended to perform slightly better, but this pattern was not consistent across all tasks. In some tasks—especially those involving earthquake depth and fatalities—participants with higher education outperformed those with elementary education, and in a few of these cases, the difference was statistically significant. However, for many other tasks, such as identifying earthquakes by magnitude or finding the year with the most events, there were no statistically meaningful differences between any of the education groups. This means that while education level can influence accuracy, its impact depends heavily on the type of task and the symbol design used on the map. In some situations, it matters, and in others, it does not.

When the study looked at response times—that is, how quickly participants solved each task—there were only small differences between the education groups

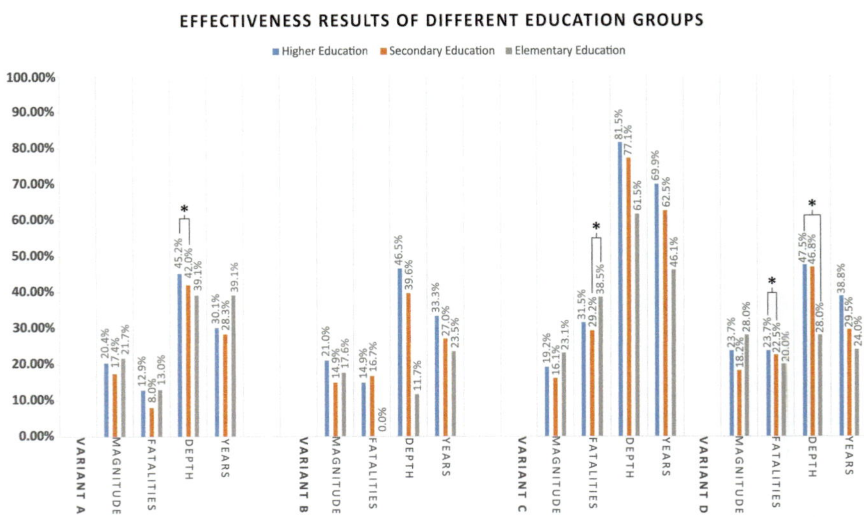


**FIGURE 4.11** Differences in effectiveness among education groups across map-based tasks and map variants. (* denotes statistical significance at $p < 0.05$.)

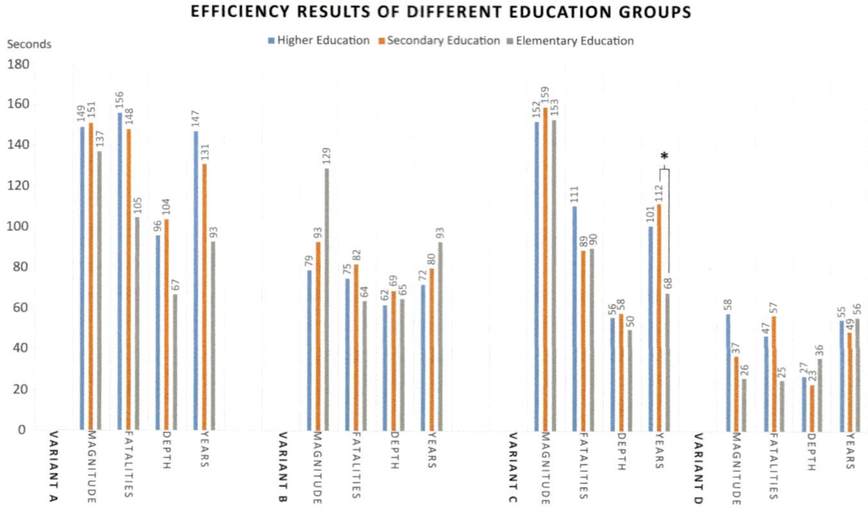


**FIGURE 4.12** Differences in efficiency among education groups across map-based tasks and map variants. (* denotes statistical significance at $p < 0.05$.)

(Figure 4.12). Importantly, most of these differences were not statistically significant. Across static and dynamic map variants:

- People with elementary education sometimes responded slightly faster, especially in simpler tasks.
- Participants with higher and secondary education had very similar response times in most cases.
- The only statistically significant difference appeared in map variant C, in the task that required identifying the year with the most earthquakes. In this specific case, participants with elementary education were significantly faster than those with secondary education ($p = 0.046$).

Overall, the results show that education level had very little influence on efficiency. Participants across all groups completed the tasks in comparable amounts of time, regardless of which map design they were using.

## 4.3.3 How Self-Assessed Map Skills Influenced Performance

### 4.3.3.1 Map Variant A

In map variant A, where earthquake magnitude was shown through vibration, participants' own assessments of their map-reading skills turned out to matter in interesting—and sometimes surprising—ways. For the task of **identifying and counting the strongest earthquakes**, the effectiveness results were not strictly aligned with participants' confidence levels:

- Those who rated their skills as *very good* answered correctly 17.3% of the time.

- Participants who claimed *good* map skills performed similarly, with 16.7% correct responses.
- Surprisingly, the group with *average* skills performed slightly better, reaching 20.0% correctness.

Statistical tests confirmed that this small advantage was meaningful: the Pearson chi-square test ($\chi^2 = 4.77$, $p = 0.03$) and Fisher's exact test ($p = 0.04$) detected a **significant difference between the "good" and "average" groups**. The remaining groups showed mixed results:

- *Low* map-reading skills—21.6% correct
- *Very low* map-reading skills—15.0% correct

Here, the pattern becomes less straightforward, indicating that self-confidence does not necessarily translate to higher accuracy. When looking at response time, the differences between groups became even more noticeable:

- Participants who rated their map skills as *very good* had a median time of 135 seconds,
- while those who rated their skills as *good* were much slower, with a median of 188 seconds.

This difference was statistically significant (Mann–Whitney U test: $p = 0.02$, $z = 2.24$). The remaining groups responded more quickly:

- *Average* skills—159 seconds
- *Low* skills—113 seconds
- *Very low* skills—98 seconds

Here again, statistical tests confirmed meaningful differences. Participants with **good map skills were significantly slower than those with low skills** ($p = 0.002$, $z = 3.07$) **and very low skills** ($p = 0.002$, $z = 3.03$). Similarly, **the average group was significantly slower than both the low** ($p = 0.012$, $z = 2.47$) **and very low groups** ($p = 0.01$, $z = 2.55$).

For the next task in map variant A—**finding and counting all earthquake locations with the highest number of fatalities**—the results showed once again that participants' self-assessed map-reading skills did not neatly predict how well they performed. The correctness rates varied across groups, but not in a pattern that matched participants' confidence:

- *Very good* map skills—10.3% correct
- *Good* map skills—18.6% correct
- *Average* skills—5.8% correct
- *Low* skills—8.1% correct
- *Very low* skills—15.0% correct

Here, participants with good skills performed the best, while those with average skills performed the worst. Interestingly, participants with very low skills performed better than both. Despite these differences, none of them were statistically significant according to the Pearson chi-square test and Fisher's exact test. In other words, the apparent differences in accuracy could easily be due to chance. Response times told a much clearer story:

- *Very good* skills—176 seconds (median)
- *Good* skills—192 seconds
- *Average* skills—156 seconds
- *Low* skills—115 seconds
- *Very low* skills—96 seconds

Participants who rated their map skills as *very good* or *good* generally took more time, while those with *low* or *very low* skills were noticeably faster. Unlike the effectiveness results, several of these differences were statistically significant, according to the Mann–Whitney U test:

- "Very good" vs. "very low"—significant ($p=0.027$, $z=2.20$)
- "Good" vs. "low"—significant ($p=0.028$, $z=2.19$)
- "Good" vs. "very low"—highly significant ($p=0.0005$, $z=3.38$)
- "Average" vs. "very low"—significant ($p=0.003$, $z=2.93$)

Participants who felt confident tended to take longer, while those who rated their skills as low or very low tended to respond more quickly—but not necessarily more accurately. This mirrors a pattern often seen in cognitive tasks: higher confidence participants may spend more time checking or verifying their answers, whereas lower confidence participants respond more quickly but not always carefully.

The next map-based task in variant A asked participants to **identify the most frequently occurring earthquake depth** within the full time span. Across all self-assessed map-reading skill groups, the effectiveness scores were fairly similar:

- *Very good*—34.5% correct
- *Good*—47.5% correct
- *Average*—45.2% correct
- *Low*—41.7% correct
- *Very low*—42.8% correct

Even though the percentages differed slightly, none of these differences were statistically significant based on the Pearson chi-square test or Fisher's exact test. This means that participants' confidence in their map skills did not meaningfully predict how accurate they were on this task. Median response times varied more noticeably:

- *Very good*—95 seconds
- *Good*—129 seconds
- *Average*—108 seconds
- *Low*—87 seconds
- *Very low*—53 seconds

A clear pattern appeared once again—participants with higher self-assessed skills tended to respond more slowly, whereas those who rated their skills as very low responded the fastest. Several of these differences were statistically significant:

- "Good" vs. "very low"—significantly slower ($p=0.007$, $z=2.66$)
- "Average" vs. "very low"—significantly slower ($p=0.003$, $z=2.95$)
- "Low" vs. "very low"—slower ($p=0.036$, $z=2.09$)

Even though accuracy did not differ across groups, participants with lower self-confidence tended to complete the task much faster, while those who rated their skills as "good" or "average" tended to spend more time on the task—possibly reflecting more careful checking or a more deliberate approach.

Another question asked participants to **pinpoint the year with the most earthquakes**—a task that turned out to highlight some clear patterns in how people with different levels of map-reading confidence approached the problem. The task showed a noticeable spread in accuracy across self-assessed map-reading skill groups:

- *Very good*—43.3% correct
- *Good*—45.7% correct
- *Average*—37.4% correct
- *Low*—28.6% correct
- *Very low*—44.2% correct

Although these percentages differ, only one comparison turned out to be statistically significant: **participants who rated their skills as "very good" differed meaningfully from those who rated themselves as "very low."** This was confirmed by the Pearson chi-square test ($\chi^2=4.32$, $p=0.04$) and Fisher's exact test ($p=0.046$). For all other groups, confidence in one's map-reading skills did not reliably predict accuracy on this task. Median response times showed a clearer pattern:

- *Very good*—116 seconds
- *Good*—168 seconds
- *Average*—135 seconds
- *Low*—100 seconds
- *Very low*—92 seconds

Here, participants with higher self-assessed skills took the longest, while those who perceived their skills as low or very low were the fastest. Several of these differences were statistically significant:

- "Good" vs. "low"—"good" group significantly slower ($p=0.003$, $z=2.97$)
- "Good" vs. "very low"—"good" group significantly slower ($p=0.009$, $z=2.57$)
- "Average" vs. "low"—"average" group significantly slower ($p=0.018$, $z=2.36$)
- "Average" vs. "very low"—"average" group significantly slower ($p=0.023$, $z=2.25$)

In summary, while accuracy differences were limited, participants with lower self-assessed map-reading skills consistently completed the task more quickly, whereas those who viewed their skills as "good" or "very good" tended to spend more time checking the map before responding.

#### 4.3.3.2 Map Variant B

After analyzing the results from map variant A (with vibrating point symbols), we now turn to **map variant B**, which uses pictorial motion to represent earthquake activity. In this task, participants were asked to **identify and count the earthquake locations with the highest magnitude**. Accuracy levels across the self-assessed map-reading skill groups were as follows:

- *Very good*—21.3% correct
- *Good*—14.6% correct
- *Average*—19.5% correct
- *Low*—14.7% correct
- *Very low*—15.8% correct

Although the percentages differed slightly, none of these differences were statistically significant. This indicates that participants' confidence in their map-reading abilities did not meaningfully influence how accurately they performed on this task. Median response times showed a somewhat clearer spread:

- *Very good*—78 seconds
- *Good*—94 seconds
- *Average*—99 seconds
- *Low*—77 seconds
- *Very low*—66 seconds

Here, participants who rated themselves as "low" or "very low" tended to respond more quickly, while those who rated themselves as "good" or "average" took longer. However, the Mann–Whitney U test did not reveal any statistically significant differences between the groups. Overall, for this pictorial-motion map, neither accuracy nor response time differed meaningfully across self-assessed skill levels, suggesting that the task placed similar demands on participants regardless of how confident they felt about their map-reading abilities.

The next task in map variant B—the pictorial-motion version—asked participants to **identify and count the earthquake locations with the highest number of fatalities**. In terms of accuracy, the results across self-assessed map-reading skill groups were as follows:

- *Very good*—14.9% correct
- *Good*—16.9% correct
- *Average*—16.1% correct
- *Low*—11.8% correct
- *Very low*—10.0% correct

Overall accuracy was low for all groups, which aligns with the general difficulty observed in variant B tasks—especially compared to variant A, where accuracy levels were typically higher, and differences between groups were clearer. Statistical tests (Pearson $\chi^2 = 6.67$, $p = 0.009$; Fisher's exact $p = 0.049$) indicated only one meaningful difference: **participants who rated their skills as "average" performed significantly better than those who rated themselves as "low."** Response times painted a different picture:

- *Very good*—64 seconds
- *Good*—60 seconds
- *Average*—97 seconds
- *Low*—102 seconds
- *Very low*—188 seconds

Unlike the previous variant B task—where lower skill participants tended to respond faster—this task showed the opposite pattern: the lowest-confidence groups were the slowest, and the highest-confidence groups responded the quickest. This contrasts with variant A, where "very good" and "good" participants often took longer, presumably due to more careful checking. The Mann–Whitney U test confirmed several significant differences:

- "Very good" vs. "low"—"very good" significantly faster ($p = 0.028$, $z = 2.19$)
- "Very good" vs. "very low"—significantly faster ($p = 0.004$, $z = 2.80$)
- "Good" vs. "average"—"good" group faster ($p = 0.013$, $z = 2.46$)
- "Good" vs. "low"—significantly faster ($p = 0.002$, $z = 3.00$)
- "Good" vs. "very low"—significantly faster ($p = 0.0007$, $z = 3.30$)

In summary, while accuracy remained uniformly low across groups, efficiency in variant B showed a clear divide, with higher skill participants responding much more quickly than their lower skill counterparts—an opposite pattern to what was observed in variant A.

In the next task for map variant B, participants were asked to **determine the most common earthquake depth**. Accuracy levels varied across the self-assessed map-reading skill groups:

- *Very good*—40.0% correct
- *Good*—52.4% correct
- *Average*—41.3% correct
- *Low*—36.4% correct
- *Very low*—20.0% correct

Although the percentages show a clear downward trend—from the highest-confidence groups toward the lowest—neither the Pearson chi-square test nor Fisher's exact test indicated statistically significant differences. As in earlier variant B tasks, accuracy appeared relatively insensitive to self-perceived skill level, in contrast to variant A where some tasks produced clearer group differences. Response times, however, again showed a stronger pattern:

- *Very good*—56 seconds
- *Good*—49 seconds
- *Average*—76 seconds
- *Low*—86 seconds
- *Very low*—140 seconds

Here, the highest-confidence participants were the fastest, while the lowest-confidence groups took considerably longer—a pattern consistent with the previous variant B task and opposite to what was often observed in variant A, where confident participants tended to spend more time verifying their answers. The Mann–Whitney U test confirmed several significant differences:

- "Very good" vs. "very low"—"very good" group significantly faster ($p = 0.032$, $z = 2.13$)
- "Good" vs. "average"—"good" group faster ($p = 0.046$, $z = 1.99$)
- "Good" vs. "low"—significantly faster ($p = 0.014$, $z = 2.45$)
- "Good" vs. "very low"—significantly faster ($p = 0.005$, $z = 2.76$)

Overall, this task reinforces a key pattern in variant B: participants with higher self-assessed map-reading skills tended to complete the tasks much more efficiently, even when accuracy did not differ significantly across groups.

The final task in map variant B required participants to **identify the years with the highest number of earthquake locations**. Accuracy levels across the self-assessed map-reading skill groups were as follows:

- *Very good*—34.8% correct
- *Good*—27.2% correct
- *Average*—33.0% correct
- *Low*—26.3% correct
- *Very low*—20.6% correct

Compared with other variant B tasks, the accuracy results here were slightly higher overall, especially for the "very good" and "average" groups. Unlike most variant B tasks—where differences between groups were generally not significant—this task produced a few meaningful contrasts. Statistical tests revealed:

- "Very good" vs. "low"—significant difference ($\chi^2 = 8.73$, $p = 0.008$; Fisher: $p = 0.008$)
- "Good" vs. "average"—significant difference ($\chi^2 = 4.61$, $p = 0.031$; Fisher: $p = 0.039$)

This makes the task one of the few in variant B where self-assessed skill level had a measurable impact on accuracy, bringing it closer to some of the patterns observed earlier in variant A. In terms of efficiency, participants responded at the following median times:

- *Very good*—67 seconds
- *Good*—75 seconds
- *Average*—77 seconds
- *Low*—84 seconds
- *Very low*—148 seconds

These times once again follow the now-consistent variant B pattern: higher confidence groups answered more quickly, while those who rated their skills as low or very low took noticeably longer. However, unlike the prior variant B tasks, none of these differences were statistically significant according to the Mann–Whitney U test. Overall, this final task reinforces two broader observations from variant B: (1) accuracy differences across groups were generally small but occasionally significant, and (2) higher confidence participants consistently completed the tasks faster—even when the differences in speed did not reach statistical significance.

#### 4.3.3.3 Map Variant C

The first task in **map variant C**, which used pulsation, asked participants to **locate and count the earthquake events with the highest magnitude**. Accuracy across the self-assessed map-reading skill groups was as follows:

- *Very good*—15.7% correct
- *Good*—16.7% correct
- *Average*—24.0% correct
- *Low*—6.7% correct
- *Very low*—14.3% correct

Interestingly, the "average" group performed best on this task—a departure from earlier patterns in both variant A (where confidence often aligned with better accuracy) and variant B (where differences were usually minimal). Despite these variations, neither the Pearson chi-square test nor Fisher's exact test indicated significant differences, meaning that none of the group contrasts were strong enough to reach statistical significance. Response times were also fairly consistent across groups:

- *Very good*—153 seconds
- *Good*—159 seconds
- *Average*—157 seconds
- *Low*—180 seconds
- *Very low*—146 seconds

Compared with variants A and B, the times in variant C were noticeably longer overall, reflecting the increased visual and cognitive complexity of the pulsation motion encoding. Still, the Mann–Whitney U test revealed no statistically significant differences between any of the groups. In summary, for this first variant C task, neither accuracy nor response time reliably depended on participants' self-assessed map-reading skill—suggesting that the strong motion signals used in this variant may have leveled the playing field, making the task equally challenging for all participants.

The next task in map variant C, which used pulsating-point symbols, asked participants to **identify and count the locations with the highest number of fatalities**. Accuracy across the self-assessed map-reading skill groups was as follows:

- *Very good*—37.3% correct
- *Good*—29.6% correct
- *Average*—25.4% correct
- *Low*—30.0% correct
- *Very low*—42.9% correct

Unlike the previous variant C task—where the "average" group performed best—accuracy here was more scattered, with the "very low" group unexpectedly achieving the highest result. Despite these contrasts, neither the Pearson chi-square test nor Fisher's exact test indicated statistically significant differences. As a result, no meaningful performance advantage can be attributed to any skill group. Efficiency results showed the following median response times:

- *Very good*—82 seconds
- *Good*—111 seconds
- *Average*—112 seconds
- *Low*—86 seconds
- *Very low*—100 seconds

While the "very good" group responded the fastest, and the "average" and "good" groups were slower, these differences did not reach statistical significance according to the Mann–Whitney U test. Compared with earlier variants, response times in variant C remain relatively moderate—shorter than the first variant C task but still generally longer than most tasks in variant B. Overall, this task reinforces the emerging pattern in variant C: accuracy and response time do not consistently align with self-assessed skill levels, and the strong pulsating-symbol design appears to produce more mixed performance outcomes across all groups.

Another task in map variant C required participants to **determine the most frequently occurring earthquake depth**. This task produced noticeably higher accuracy than the previous variant C tasks. The percentages of correct responses across self-assessed map-reading skill groups were:

- *Very good*—80.4% correct
- *Good*—76.9% correct
- *Average*—83.0% correct
- *Low*—76.7% correct
- *Very low*—57.1% correct

Compared with the earlier variant C tasks—which showed mixed or low accuracy—this task appears to have been much easier for most participants. In fact, the results are more similar to the stronger accuracy patterns seen in variant A, especially in depth-related tasks. Despite the noticeable gap between the highest-performing groups (*very good/average*) and the lowest-performing group (*very low*), neither the Pearson chi-square test nor Fisher's exact test indicated statistically significant differences. Efficiency results showed only modest variation among groups:

- *Very good*—61 seconds
- *Good*—50 seconds
- *Average*—51 seconds
- *Low*—73 seconds
- *Very low*—60 seconds

These times are shorter overall compared with previous variant C tasks, reinforcing the idea that this depth-identification question was relatively straightforward. The Mann–Whitney U test confirmed that none of the timing differences were statistically significant, meaning that all groups completed the task at broadly similar speeds. In summary, this task stands out within variant C: accuracy was high across most groups, response times were faster, and—unlike earlier tasks—performance was relatively consistent. This suggests that the pulsating-point encoding in variant C may have been especially effective for visualizing depth patterns, even if it posed challenges in earlier tasks.

The final task in map variant C asked participants to **determine the year with the highest number of earthquakes**. Accuracy levels across the self-assessed map-reading skill groups were:

- *Very good*—62.7% correct
- *Good*—73.2% correct
- *Average*—57.0% correct
- *Low*—60.0% correct
- *Very low*—85.7% correct

This task produced some of the highest accuracy scores in variant C, especially for the "good" and "very low" groups. The strong performance of the very low-skilled group is unusual compared with earlier tasks in variants A and B, where this group typically scored lower or showed no advantage. Despite these differences, neither the Pearson chi-square test nor Fisher's exact test detected statistically significant differences, meaning that accuracy did not reliably differ between skill groups. Response times, however, revealed clearer contrasts:

- *Very good*—103 seconds
- *Good*—91 seconds
- *Average*—127 seconds (slowest)
- *Low*—97 seconds
- *Very low*—60 seconds (fastest)

Here, the pattern echoes earlier findings from variant B and parts of variant C: participants with lower self-assessed skills often responded faster, while the higher confidence groups tended to take longer. Still, the distinction was not uniform—participants with "average" skills were the slowest overall. The Mann–Whitney U test confirmed several significant differences:

- "Very good" vs. "very low"—"very good" significantly slower ($p=0.033$, $z=2.11$)
- "Good" vs. "average"—"average" group significantly slower ($p=0.017$, $z=2.39$)
- "Average" vs. "very low"—"average" group significantly slower ($p=0.017$, $z=2.35$)

Together with the earlier tasks, this suggests that the powerful pulsating-symbol design of variant C may have supported effective pattern recognition across diverse skill levels—though it did not always translate into consistent response-time behavior.

#### 4.3.3.4 Map Variant D

The final map variant in the study, **variant D**, used static point symbols—a clear contrast to the motion-based designs in variants A, B, and C. The first task in this variant asked participants to **find and count all earthquake locations with the highest magnitude**. Accuracy across the self-assessed map-reading skill groups was as follows:

- *Very good*—18.7% correct
- *Good*—30.2% correct
- *Average*—17.9% correct
- *Low*—17.0% correct
- *Very low*—16.7% correct

Overall, effectiveness in variant D was slightly higher than in some of the more difficult variant C tasks but remained lower than the best-performing tasks in variants A and C. Despite the visible differences—especially the higher score in the "good" group—neither the Pearson chi-square test nor Fisher's exact test identified any statistically significant differences. Response times showed clearer group distinctions:

- *Very good*—60 seconds
- *Good*—37 seconds (fastest)
- *Average*—59 seconds
- *Low*—46 seconds
- *Very low*—44 seconds

Unlike many tasks in variants B and C, where lower self-assessed skill groups often responded faster, variant D presents a different pattern: **the "good" group was the quickest**, while the "average" group was among the slowest. This echoes some of the trends observed in variant A, where higher confidence participants occasionally took more time than necessary. The Mann–Whitney U test identified two significant differences:

- "Good" vs. "average"—"good" group significantly faster ($p=0.046$, $z=1.99$)
- "Average" vs. "low"—"average" group significantly slower ($p=0.03$, $z=2.17$)

In summary, this first task in variant D shows that while accuracy did not differ meaningfully across skill groups, response times did reveal some variation—particularly the slower performance of the "average"-skilled participants compared with both the "good" and "low" groups. The static-symbol design appears to reduce the influence of self-assessed skill on accuracy, placing most groups on a similar footing.

The next task in map variant D, which used static point symbols, asked participants to **identify and count the earthquake locations with the highest number of fatalities**. Accuracy across the self-assessed map-reading skill groups was:

- *Very good*—12.5% correct
- *Good*—22.9% correct
- *Average*—23.3% correct
- *Low*—27.7% correct
- *Very low*—29.2% correct

These values show an interesting pattern: unlike Tasks A–C (and the previous variant D task), the lowest-confidence groups performed the best, though differences were still relatively small. This echoes the occasional reversals observed in variant B and parts of variant C, where "very low" participants sometimes achieved unexpectedly high accuracy. However, neither the Pearson chi-square nor Fisher's exact test indicated that these differences were statistically meaningful. Efficiency results also showed only modest variation:

- *Very good*—51 seconds
- *Good*—58 seconds
- *Average*—62 seconds
- *Low*—59 seconds
- *Very low*—49 seconds

All groups completed the task in roughly similar time, with no consistent performance gradient—very similar to what was observed in some of the earlier variant C tasks. The Mann–Whitney U test confirmed that none of these differences were statistically significant, indicating that self-assessed map-reading skill did not meaningfully influence response time here. In summary, this variant D task shows even more leveling across groups than the previous one: neither accuracy nor efficiency differed significantly, and the static-symbol design appears to place participants of all skill levels on an almost equal footing for this type of question.

The third task in map variant D asked participants to determine **which earthquake depth occurred most frequently across the entire time span**. Accuracy across the self-assessed map-reading skill groups was:

- *Very good*—53.1% correct
- *Good*—45.8% correct
- *Average*—47.0% correct
- *Low*—46.8% correct
- *Very low*—33.3% correct

Compared with the previous variant D tasks, this one produced higher accuracy overall, especially for the "very good" group. The results resemble those in variant C's depth-related task, where depth proved easier to interpret than magnitude or fatality counts. The only statistically significant difference—confirmed by both the Pearson chi-square test ($\chi^2 = 4.13$, $p = 0.042$) and Fisher's exact test ($p = 0.033$)—was **between the "good" and "average" skill groups**. All other differences were not significant. Efficiency results showed similarly small contrasts:

- *Very good*—34 seconds
- *Good*—38 seconds
- *Average*—51 seconds (slowest)
- *Low*—36 seconds
- *Very low*—42 seconds

Here, the "average" group again stood out as the slowest—mirroring a pattern seen in the earlier variant D tasks, where "average" participants often took longer than both higher- and lower confidence groups. Nonetheless, the overall timing differences were small, and the Mann–Whitney U test did not reveal any statistically significant differences. In summary, this depth-identification task in variant D shows that most skill groups performed similarly both in accuracy and speed, with only a minor statistical distinction between the "good" and "average" groups. As in earlier variant D tasks, the static-symbol design seems to equalize performance, producing fewer dramatic differences across self-assessed skill levels than the motion-based variants.

The last task in map variant D asked participants to **identify the year in which the highest number of earthquakes occurred**. Accuracy across the self-assessed map-reading skill groups was as follows:

- *Very good*—40.6% correct
- *Good*—39.4% correct
- *Average*—30.4% correct
- *Low*—23.9% correct
- *Very low*—29.2% correct

These accuracy levels are broadly similar to the previous tasks in variant D—generally moderate and without large contrasts between groups. Unlike several tasks in variants A, B, and C, where at least some group differences reached significance, both the Pearson chi-square test and Fisher's exact test showed no statistically meaningful differences here. Response times showed a similar convergence across groups:

- *Very good*—61 seconds
- *Good*—55 seconds
- *Average*—68 seconds
- *Low*—53 seconds
- *Very low*—57 seconds

As in the earlier variant D tasks, response times differed only slightly, with no clear pattern linking speed to self-assessed map-reading skill. The "average" group was once again the slowest—consistent with earlier D tasks—while both higher and lower confidence groups completed the task more quickly. Still, these differences were too small to be statistically significant according to the Mann–Whitney U test. Compared with the motion-based variants (A, B, and C), variant D produced some of the most uniform performance patterns, suggesting that static symbols may reduce the influence of self-assessed skill level on both task accuracy and efficiency.

Across the different map variants, task accuracy showed how strongly map design can shape performance. Variant A tended to favor participants who felt confident in their map-reading skills, especially in more complex tasks, although some simpler tasks allowed lower skilled users to perform just as well. variant B produced much more uniform results—almost everyone performed similarly—which suggests that its design was easier to interpret regardless of skill level. Variant C generally supported higher skilled participants but also included a few tasks where lower skilled users did surprisingly well, perhaps because certain visual features were more intuitive for them. Variant D, the static map, showed the smallest differences of all, with almost no meaningful variation between groups. Overall, the effectiveness patterns—summarized in Figure 4.13—show that some map designs naturally reduce differences in performance between users.

A similar pattern appeared in task efficiency. Higher skilled participants were usually faster, but not always. In variant A, lower skilled users occasionally completed simpler tasks more quickly, perhaps because they relied on faster, more direct strategies. In variants B and C, differences in timing were generally small, suggesting that these designs reduced navigation demands for everyone. In variant D, higher skilled users tended to be faster overall, though lower skilled participants sometimes finished certain tasks more quickly—likely because the static layout made some tasks especially straightforward. Together, these results—summarized in Figure 4.14—show that map design can either amplify or reduce the influence of user skill, shaping both accuracy and speed in meaningful ways.

### 4.3.4 The Role of Everyday Map-Use Habits

#### 4.3.4.1 Map Variant A

To explore whether everyday map-use habits influenced performance, participants reported how often they typically use maps: every day, several times a week, several times a month, or less than once a month. In **map variant A**, the first task asked

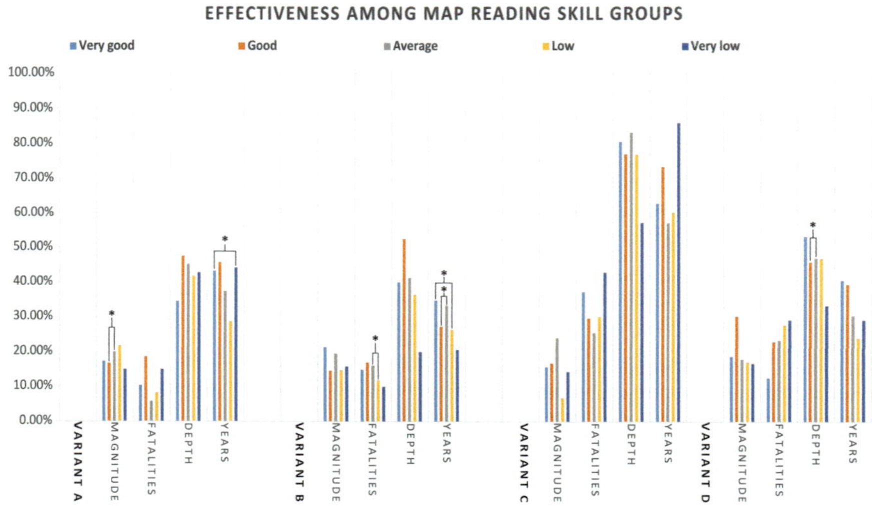


**FIGURE 4.13** Differences in effectiveness among map-reading skill groups across map-based tasks and map variants. (* denotes statistical significance at $p < 0.05$.)

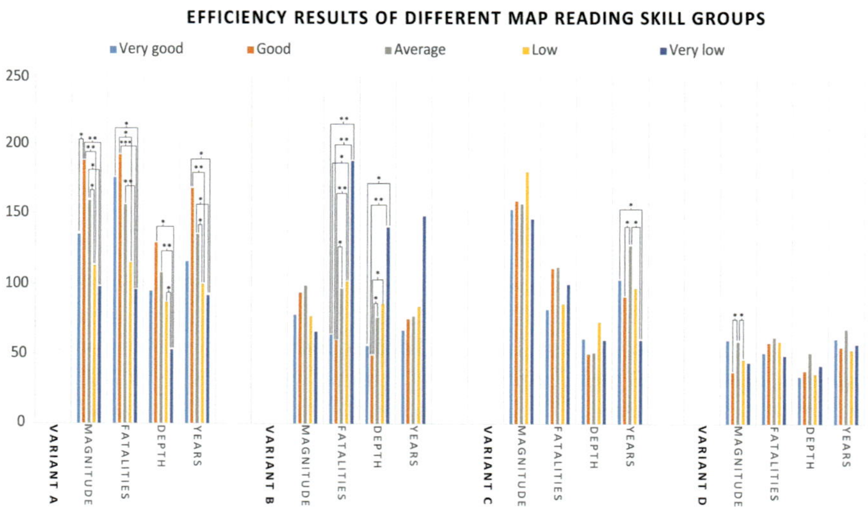


**FIGURE 4.14** Differences in efficiency among map-reading skill groups across map-based tasks and map variants. (* denotes statistical significance at $p < 0.05$; ** denotes statistical significance at $p < 0.005$; *** denotes statistical significance at $p < 0.0005$.)

them to **identify and count the locations with the highest earthquake magnitudes**. Effectiveness varied across the usage groups:

- *Every day*—15.3% correct
- *Weekly*—23.5% correct

- *Monthly*—26.1% correct
- *Less than monthly*—14.6% correct

Although the results differed, only one comparison reached statistical significance: **participants who used maps daily performed significantly differently from those who used maps weekly**, according to both the Pearson chi-square test ($\chi^2=4.67$, $p=0.03$) and Fisher's exact test ($p=0.046$). No other usage group showed meaningful differences in accuracy. Response times showed clearer patterns:

- *Every day*—110 seconds
- *Weekly*—181 seconds
- *Monthly*—184 seconds
- *Less than monthly*—160 seconds

Participants who used maps daily were the fastest group, completing the task much more quickly than all others. The Mann–Whitney U test confirmed three statistically significant differences:

- *Daily* vs. *weekly*—daily users significantly faster ($p=0.01$, $z=2.55$)
- *Daily* vs. *monthly*—strong difference, daily faster ($p=0.0002$, $z=3.60$)
- *Daily* vs. *less than monthly*—strong difference, daily faster ($p=0.0001$, $z=4.35$)

In summary, for this variant A task, daily map users were consistently the fastest, completing the task much more quickly than all other groups. However, daily use did not translate into higher accuracy—in fact, participants who used maps weekly achieved the best effectiveness score, and the only statistically significant difference in accuracy was between the daily and weekly users, with weekly users performing better. This suggests that frequent map use supports efficiency, while moderate, regular use may be more beneficial for this specific task's accuracy.

The next task in map variant A asked participants to **identify and count the locations with the highest number of fatalities**. Effectiveness scores across the different map-use frequency groups were:

- *Every day*—12.5% correct
- *Weekly*—11.8% correct
- *Monthly*—13.3% correct
- *Less than monthly*—6.2% correct

Although the monthly users achieved the highest accuracy and the least frequent users the lowest, neither the Pearson chi-square test nor Fisher's exact test showed statistically significant differences. This means that how often participants used maps in daily life did not meaningfully affect their accuracy for this particular task. Task completion times showed a clearer pattern:

- *Every day*—110 seconds
- *Weekly*—147 seconds

- *Monthly*—183 seconds
- *Less than monthly*—179 seconds

Here, participants who used maps daily were noticeably faster, while the groups with less frequent map use took much longer to finish. The Mann–Whitney U test identified two statistically significant differences:

- *Daily* vs. *monthly*—daily users significantly faster ($p = 0.001$, $z = 3.19$)
- *Daily* vs. *less than monthly*—daily users significantly faster ($p = 0.001$, $z = 3.23$)

In short, while everyday map use did not improve accuracy on this task, it did lead to much faster performance, and the gap in efficiency between daily users and infrequent users was substantial.

The third task in map variant A asked participants to **determine which earthquake depth occurred most frequently**. Accuracy across the different map-use groups was as follows:

- *Every day*—38.4% correct
- *Weekly*—47.1% correct
- *Monthly*—47.7% correct
- *Less than monthly*—44.4% correct

Here, the daily users had the lowest score among all groups, while the weekly and monthly users performed best. The Pearson chi-square test ($\chi^2 = 4.63$, $p = 0.031$) and Fisher's exact test ($p = 0.048$) revealed one statistically significant difference: **participants who used maps every day performed significantly differently from those who used maps less than once a month**. Efficiency results showed more pronounced differences:

- *Every day*—77 seconds
- *Weekly*—101 seconds
- *Monthly*—127 seconds
- *Less than monthly*—115 seconds

Daily map users were the fastest group by a considerable margin, while monthly users took the longest to complete the task. The Mann–Whitney U test identified two significant differences:

- *Daily* vs. *monthly*—daily users significantly faster ($p = 0.019$, $z = 2.33$)
- *Daily* vs. *less than monthly*—daily users significantly faster ($p = 0.012$, $z = 2.51$)

In summary, for this variant A task, everyday map users were much quicker but not necessarily more accurate than those with less frequent map-use habits. Regular, moderate users (weekly and monthly) showed the highest accuracy, while daily users stood out mainly for their speed.

The final task in map variant A asked participants to **identify the year in which the highest number of earthquakes occurred**. Accuracy across the different map-use frequency groups was:

- *Every day*—33.8% correct
- *Weekly*—23.5% correct
- *Monthly*—26.7% correct
- *Less than monthly*—34.1% correct

Interestingly, everyday map users and those who used maps the least both performed similarly well, whereas the weekly and monthly users scored lower. Statistical testing confirmed one significant difference: **participants who used maps every day differed meaningfully in accuracy from those who used maps less than once a month** (Pearson $\chi^2=4.25$, $p=0.039$; Fisher's exact $p=0.048$). Completion times showed a much clearer pattern:

- *Every day*—98 seconds (fastest)
- *Weekly*—160 seconds
- *Monthly*—168 seconds
- *Less than monthly*—141 seconds

Daily users again stood out as the fastest group by a wide margin. The Mann–Whitney U test revealed that **the everyday group was significantly faster than all other groups**:

- *Daily* vs. *weekly*—daily significantly faster ($p=0.008$, $z=2.60$)
- *Daily* vs. *monthly*—strongly significant ($p=0.0007$, $z=3.35$)
- *Daily* vs. *less than monthly*—very strong difference ($p=0.0001$, $z=3.87$)

In summary, for this final variant A task, participants who used maps daily consistently completed the task much more quickly than all other groups and showed one notable accuracy difference compared with the least frequent users. Regular map exposure appears to support faster task execution in variant A, even when accuracy differences remain relatively small.

#### 4.3.4.2 Map Variant B

In map variant B, where earthquake magnitude was shown using a pictorial-motion symbol, participants were asked to **find and count all locations with the strongest magnitude**. Accuracy across the different map-use groups was:

- *Every day*—11.1% correct
- *Weekly*—16.7% correct
- *Monthly*—19.2% correct
- *Less than monthly*—21.2% correct

Here, the less frequent map users actually performed best, while daily users scored the lowest. However, neither the Pearson chi-square test nor Fisher's exact test identified any statistically significant differences, meaning these variations were not strong enough to be considered meaningful. Response times varied somewhat across groups:

- *Every day*—95 seconds
- *Weekly*—120 seconds
- *Monthly*—94 seconds
- *Less than monthly*—68 seconds (fastest)

Participants who used maps less than once a month were the quickest, and monthly users were nearly as fast. Daily and weekly users took longer, though again, the differences between groups did not reach statistical significance. In summary, for this first variant B task, neither accuracy nor speed was strongly linked to how often participants used maps in everyday life. The pictorial-motion design seems to have minimized performance differences between frequent and infrequent map users.

The second task in map variant B, which used pictorial-motion symbols to represent earthquake severity, asked participants to **locate and count the places with the highest number of fatalities**. Accuracy across the map-use groups was:

- *Every day*—10.9% correct
- *Weekly*—13.1% correct
- *Monthly*—12.8% correct
- *Less than monthly*—21.2% correct

Once again, everyday map users achieved the lowest score, while the least frequent users performed best. Surprisingly, this task did reveal some statistically significant differences, despite the overall low accuracy. The Pearson chi-square test ($\chi^2 = 13.45$, $p = 0.0002$) and Fisher's exact test ($p = 0.0039$) showed a significant difference **between *everyday* and *weekly* map users**. A second significant difference appeared **between *weekly* users and *monthly* users** ($\chi^2 = 4.80$, $p = 0.028$). These findings indicate that, for this particular task, map-use habits did influence performance in unexpected ways. Efficiency results were much more consistent:

- *Every day*—82 seconds
- *Weekly*—75 seconds
- *Monthly*—77 seconds
- *Less than monthly*—82 seconds

Response times were tightly clustered, and none of the differences were statistically significant. All groups completed the task at roughly the same speed, suggesting that the pictorial-motion design made the task equally manageable (or equally difficult) for users of all map-use frequencies.

The third task in map variant B asked participants to **identify which earthquake depth appeared most frequently**. Accuracy across the different map-use habits was:

- *Every day*—31.7% correct
- *Weekly*—44.1% correct
- *Monthly*—46.7% correct
- *Less than monthly*—44.2% correct

Here, daily map users again scored the lowest, while the weekly and monthly users achieved the highest accuracy. However, despite these visible differences, neither the Pearson chi-square test nor Fisher's exact test found any of them to be statistically significant. This means that map-use frequency did not clearly influence how accurately participants answered this depth-related task. Response times showed a similarly mixed pattern:

- *Every day*—64 seconds
- *Weekly*—115 seconds
- *Monthly*—62 seconds
- *Less than monthly*—56 seconds

The weekly users took the longest, while the least frequent users were the fastest. Yet, as with accuracy, none of these differences reached statistical significance according to the Mann–Whitney U test. In summary, for this variant B task, neither accuracy nor efficiency meaningfully differed across map-use groups. The pictorial-motion representation appears to have provided a similar level of difficulty—and required similar effort—for participants regardless of how often they use maps in their everyday lives.

The final task in map variant B asked participants to **determine the year in which the highest number of earthquakes occurred**. Accuracy across the map-use groups was:

- *Every day*—80.9% correct
- *Weekly*—71.2% correct
- *Monthly*—73.1% correct
- *Less than monthly*—60.2% correct

Here, every day map users achieved the best results, while those who used maps least often scored the lowest. Most differences were modest, but one comparison reached statistical significance: according to the Pearson chi-square test ($\chi^2=4.02, p=0.045$) and Fisher's exact test ($p=0.051$), **participants who used maps weekly performed significantly differently from those who used them monthly**. Efficiency results showed only small variations:

- *Every day*—92 seconds
- *Weekly*—98 seconds

- *Monthly*—66 seconds (fastest)
- *Less than monthly*—72 seconds

Although monthly users were the quickest and weekly users the slowest, none of these timing differences were statistically significant based on the Mann–Whitney U test. In summary, this final variant B task revealed one noticeable accuracy difference between weekly and monthly users, but otherwise showed similar performance across groups. As in the previous tasks in variant B, the pictorial-motion design resulted in minimal differences in how quickly participants worked, regardless of how often they used maps in their daily lives.

#### 4.3.4.3 Map Variant C

In **map variant C**, which used strong pulsating motion to highlight earthquake activity, the first task asked participants to **find and count all locations with the highest magnitude**. Accuracy across the map-use groups was:

- *Every day*—23.3% correct
- *Weekly*—9.9% correct
- *Monthly*—35.0% correct (highest)
- *Less than monthly*—18.6% correct

Monthly map users performed best on this task, while weekly users scored the lowest. Despite these differences, neither the Pearson chi-square nor Fisher's exact test found any statistically significant contrasts. This means that map-use frequency did not reliably predict accuracy in this task, even though the percentages varied. Response times showed a similar pattern of variation without statistical significance:

- *Every day*—119 seconds
- *Weekly*—146 seconds
- *Monthly*—157 seconds
- *Less than monthly*—148 seconds

Daily users were the fastest, while the other groups took somewhat longer. However, the differences were not strong enough to be statistically meaningful. In summary, for this first variant C task, both accuracy and speed differed across the usage groups, but none of the differences were statistically significant. The visually intense pulsating-symbol design appears to have produced mixed outcomes regardless of how often participants typically use maps.

The second task in map variant C, which used pulsating symbols to represent earthquake characteristics, asked participants to **find and count the locations with the highest number of fatalities**. Accuracy across the map-use frequency groups was:

- *Every day*—33.0% correct
- *Weekly*—30.6% correct
- *Monthly*—40.0% correct (highest)
- *Less than monthly*—23.7% correct

As in the previous task, monthly map users again performed best, while those who rarely used maps scored the lowest. However, none of these differences were statistically significant, meaning that map-use frequency did not clearly influence accuracy in this task. Differences were more noticeable in response time:

- *Every day*—89 seconds
- *Weekly*—84 seconds
- *Monthly*—143 seconds
- *Less than monthly*—127 seconds

Daily and weekly users completed the task quickly, while monthly and less-than-monthly users took considerably longer. These contrasts were strong enough to be statistically meaningful. The Mann–Whitney U test showed two significant differences:

- *Weekly* vs. *monthly*—weekly users significantly faster ($p=0.007$, $z=2.67$)
- *Weekly* vs. *less than monthly*—weekly users significantly faster ($p=0.014$, $z=2.45$)

In summary, while map-use frequency did not affect accuracy in this variant C task, it did influence efficiency. **Participants who used maps more regularly—especially weekly users—completed the task much faster than those who used maps less often**.

The third task in map variant C asked participants to **determine which earthquake depth occurred most frequently**. Accuracy varied across the map-use groups:

- *Every day*—84.5% correct (highest)
- *Weekly*—76.9% correct
- *Monthly*—55.0% correct
- *Less than monthly*—79.7% correct

Daily users performed best, while monthly users scored the lowest. Despite these visible differences, neither the Pearson chi-square test nor Fisher's exact test indicated any statistically significant differences. This means that map-use frequency did not reliably affect accuracy in this pulsating-symbol depth task. Response times were also quite similar:

- *Every day*—64 seconds
- *Weekly*—56 seconds
- *Monthly*—46 seconds (fastest)
- *Less than monthly*—57 seconds

Although the monthly group was the quickest and daily users were somewhat slower, these differences were small and not statistically significant according to the Mann–Whitney U test. In summary, this variant C task showed high accuracy across most map-use groups and very similar response times. The pulsating-symbol design for depth made this task fairly approachable, regardless of how often participants typically use maps in everyday life.

The final task in map variant C asked participants to **determine the year with the highest number of earthquakes**. Accuracy across the map-use frequency groups was:

- *Every day*—64.1% correct
- *Weekly*—66.1% correct
- *Monthly*—55.0% correct (lowest)
- *Less than monthly*—69.5% correct (highest)

Participants who rarely used maps performed surprisingly well, even achieving the highest score. The lowest accuracy came from those who used maps several times a month. Despite this variation, only one comparison reached statistical significance: **participants who used maps weekly performed significantly differently from those who used them monthly** (Pearson $\chi^2=4.85$, $p=0.027$; Fisher's exact $p=0.040$). No other group differences were statistically meaningful. Response times showed modest differences:

- *Every day*—74 seconds
- *Weekly*—80 seconds
- *Monthly*—89 seconds
- *Less than monthly*—83 seconds

Although the monthly users were the slowest and the daily users the fastest, these variations were small. The Mann–Whitney U test confirmed that none of them were statistically significant. In summary, for this final variant C task, accuracy varied across groups, but only one difference reached significance; meanwhile, response times were similar for everyone. The strong pulsating-symbol design of variant C once again appears to have supported fairly consistent performance across different map-use habits.

#### 4.3.4.4 Map Variant D

In **map variant D**, which presented earthquake information using static point symbols, the first task asked participants to **find and count all locations with the highest magnitude**. Accuracy across the different map-use groups was:

- *Every day*—21.8% correct
- *Weekly*—24.5% correct
- *Monthly*—19.7% correct
- *Less than monthly*—20.6% correct

These results were very similar across all groups, and neither the Pearson chi-square test nor Fisher's exact test indicated any statistically significant differences. This suggests that map-use frequency had little influence on accuracy for this static-symbol magnitude task. Median response times followed a similarly uniform pattern:

- *Every day*—47 seconds
- *Weekly*—60 seconds
- *Monthly*—58 seconds
- *Less than monthly*—46 seconds

Daily and infrequent map users were the fastest, while weekly users were slightly slower. However, these differences were small, and the Mann–Whitney U test confirmed that none of them reached statistical significance. In summary, this first variant D task showed minimal differences in both accuracy and speed across all map-use groups. The static-symbol design appears to have made the task equally accessible regardless of how often participants use maps in everyday life.

The second task in map variant D asked participants to **find and count the earthquake locations with the highest number of fatalities**. Accuracy across the map-use frequency groups was:

- *Every day*—22.1% correct
- *Weekly*—25.0% correct
- *Monthly*—24.2% correct
- *Less than monthly*—22.6% correct

Although the percentages are close, two group comparisons reached statistical significance. According to the Pearson chi-square test ($\chi^2 = 3.86$, $p = 0.049$) and Fisher's exact test ($p = 0.054$), there was a meaningful difference **between *daily* map users and those who used maps *less than once a month***. A second significant difference appeared **between *weekly* and *monthly* users** ($\chi^2 = 3.92$, $p = 0.047$; Fisher's $p = 0.056$). These findings suggest that, for this specific static-symbol task, map-use habits played at least some role in effectiveness. Response times were much more uniform:

- *Every day*—54 seconds (fastest)
- *Weekly*—58 seconds
- *Monthly*—59 seconds
- *Less than monthly*—61 seconds

The differences were minimal, and the Mann–Whitney U test confirmed that none of them were statistically significant. All groups completed the task at nearly the same speed. In summary, this second variant D task showed small but notable differences in accuracy between certain usage groups, while efficiency remained essentially the same for everyone.

The third task in map variant D required participants to **identify which earthquake depth occurred most frequently within the dataset**. Accuracy across the different map-use groups was:

- *Every day*—43.6% correct
- *Weekly*—41.5% correct
- *Monthly*—52.2% correct (highest)
- *Less than monthly*—46.7% correct

While the *monthly* users scored slightly higher than the other groups, the differences between all groups were modest. Statistical testing confirmed that none of these differences were significant, meaning that map-use frequency did not meaningfully influence accuracy for this task. Response times also showed only small variations:

- *Every day*—49 seconds
- *Weekly*—32 seconds (fastest)
- *Monthly*—41 seconds
- *Less than monthly*—38 seconds

*Weekly* map users completed the task the quickest, but all groups performed within a similar range, and the Mann–Whitney U test revealed no statistically significant differences. In summary, this third variant D task displayed very similar accuracy and efficiency across all map-use groups, suggesting that the static-symbol design supported consistent performance regardless of participants' everyday map-use habits.

The final task in map variant D asked participants to **identify the year with the highest number of earthquake locations**. Accuracy across the map-use groups was:

- *Every day*—70.0% correct
- *Weekly*—63.5% correct
- *Monthly*—58.2% correct
- *Less than monthly*—70.8% correct

Daily users and the least frequent users performed best, while the monthly users had the lowest accuracy. One statistically significant difference was found: **participants who used maps monthly differed meaningfully from those who used them less than once a month**, according to the Pearson chi-square test ($\chi^2 = 6.31, p = 0.012$) and Fisher's exact test ($p = 0.016$). No other differences reached significance. Response times were generally consistent:

- *Every day*—53 seconds
- *Weekly*—64 seconds
- *Monthly*—61 seconds
- *Less than monthly*—66 seconds

All groups completed the task within a narrow time range, and none of the differences were statistically significant based on the Mann–Whitney U test. In summary, for this final variant D task, accuracy differed for one pair of groups, but efficiency remained similar across all map-use habits. The static-map design once again supported comparable task completion speeds regardless of how often participants used maps in everyday life.

Across all four map variants, the effectiveness results (Figure 4.15) show that participants were generally most successful in tasks that required identifying the most

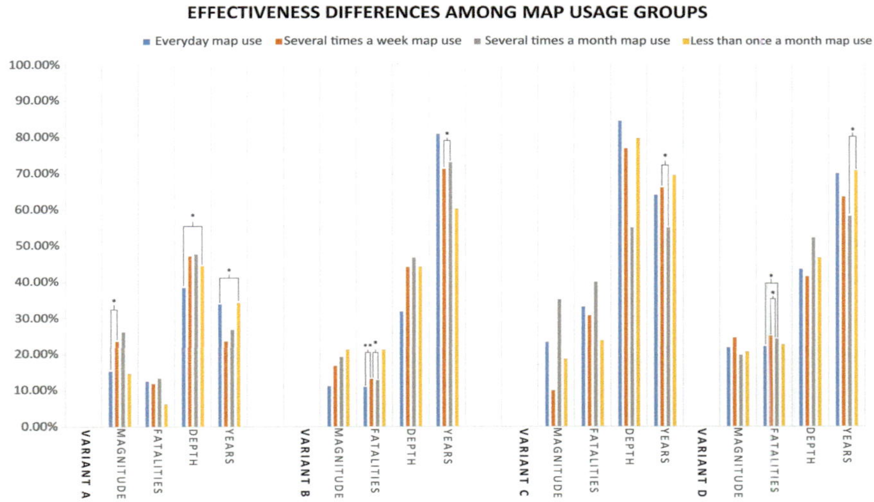


**FIGURE 4.15** Differences in effectiveness among self-reported map usage groups across map-based tasks and map variants. (* denotes statistical significance at $p < 0.05$; ** denotes statistical significance at $p < 0.005$.)

common earthquake depth or the year with the highest number of earthquakes—tasks that were conceptually simpler and visually clearer. However, the assumption that frequent map use leads to better performance was not consistently supported. In fact, several statistically significant results—particularly in variants A (vibrating symbols) and B (pictorial motion)—revealed that participants who *did not* use maps daily sometimes achieved higher accuracy than those who used them more often. This suggests that certain dynamic visualizations may actually reduce the advantage of habitual map users, perhaps because motion-based symbols demand a different type of visual attention. Meanwhile, **variant C**, which used pulsating symbols, **produced some of the highest overall task success rates** and relatively stable performance across usage groups, indicating that this design may strike a better balance between visual emphasis and interpretability.

Efficiency patterns (Figure 4.16) tell a complementary story. The clearest and most frequent differences in response times appeared in variant A, where participants who used maps daily consistently completed tasks the fastest—often significantly faster than all other groups. This trend of quicker performance among daily map users appeared across most tasks, even when effectiveness did not improve at the same rate. When comparing accuracy and speed together, a notable trade-off becomes visible: **frequent map users tended to respond faster but were not always more accurate**, while **less frequent users often worked more slowly but achieved equal or better effectiveness** on some tasks. This combined pattern suggests that map-use habits may influence *how* people approach map-reading tasks—frequent users may rely on quick, habitual strategies, whereas infrequent users may engage in more cautious and deliberate processing.

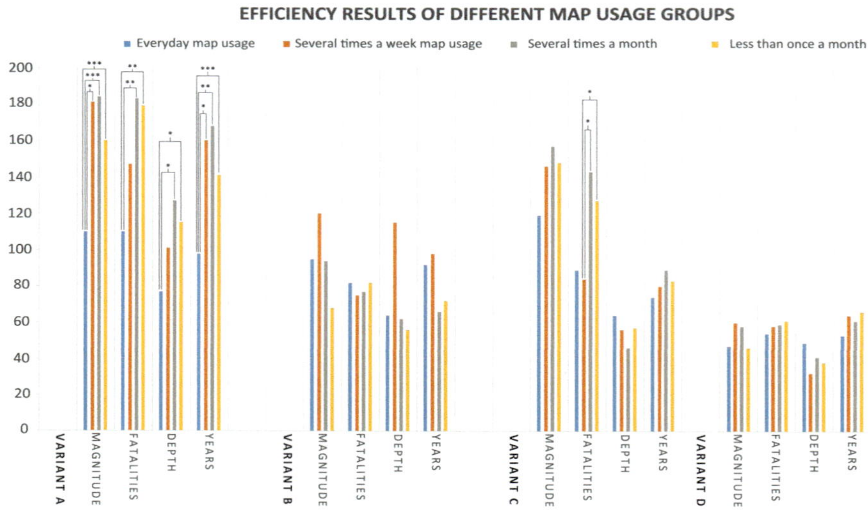


**FIGURE 4.16** Differences in effectiveness among self-reported map usage groups across map-based tasks and map variants. (* denotes statistical significance at $p < 0.05$; ** denotes statistical significance at $p < 0.005$; *** denotes statistical significance at $p < 0.0005$.)

## 4.3.5 How Confident Participants Felt About Their Answers

### 4.3.5.1 Map Variant A

For the first task in **map variant A**, participants were asked to **find and count all earthquake locations with the highest magnitude**. After answering, they rated how confident they felt about their response: *very sure*, *sure*, *neutral*, *unsure*, or *very unsure*. Interestingly, confidence did not reliably predict accuracy:

- *Very sure*: 6.7% of participants chose this category, but only 11.8% of them answered correctly.
- *Sure*: 12.2% of participants; 29.0% correct.
- *Neutral*: 32.3% of participants; 13.4% correct.
- *Unsure*: 21.2% of participants; 11.1% correct.
- *Very unsure*: 27.2% of participants; 27.9% correct.

Despite these differences, neither the Pearson chi-square test nor Fisher's exact test found any statistically significant differences in accuracy. In other words, how confident participants felt did not necessarily match how well they performed—a pattern commonly seen in map reading, where visual complexity can undermine people's self-evaluations. Where confidence did matter was in response time. Strongly confident participants took the longest:

- *Very sure*: 243 seconds
- *Sure*: 250 seconds
- *Neutral*: 130 seconds

- *Unsure*: 188 seconds
- *Very unsure*: 121 seconds

A clear pattern emerges: both the most confident and the confident participants spent the longest time on the task, while the neutral and very unsure groups were considerably faster. Several statistically significant differences appeared:

- *Very sure* vs. *neutral* ($p = 0.008$, $z = 2.61$)
- *Very sure* vs. *very unsure* ($p = 0.022$, $z = 2.27$)
- *Sure* vs. *neutral* ($p = 0.0003$, $z = 3.57$)
- *Sure* vs. *very unsure* ($p = 0.002$, $z = 3.05$)
- *Unsure* vs. *neutral* ($p = 0.001$, $z = 3.15$)
- *Unsure* vs. *very unsure* ($p = 0.006$, $z = 2.72$)

In summary, confidence level strongly influenced how long participants spent on the task, but it had no meaningful connection to accuracy in this variant A task. **Participants who felt very sure or sure tended to spend significantly more time searching**, suggesting different types of hesitation—either overchecking or struggling—while those with neutral confidence worked faster overall.

The second task in map variant A asked participants to **find and count the earthquake locations with the highest number of fatalities**. As before, participants rated how confident they felt about their answers. Accuracy across these confidence levels showed considerable variation, although most groups performed poorly overall:

- *Very sure*: 7.1% of participants; 16.7% correct
- *Sure*: 10.2% of participants; 23.1% correct
- *Neutral*: 31.1% of participants; 6.4% correct
- *Unsure*: 24.8% of participants; 12.7% correct
- *Very unsure*: 26.0% of participants; 6.2% correct

Here, the "sure" group performed best, while both the *neutral* and *very unsure* groups had the lowest accuracy. A statistically significant difference appeared **between the very sure and unsure groups** ($\chi^2 = 11.25$, $p = 0.0008$; Fisher's $p = 0.019$), suggesting a meaningful gap in performance between these two levels of confidence. Efficiency results showed a different story. Response times were:

- *Very sure*: 178 seconds
- *Sure*: 172 seconds
- *Neutral*: 145 seconds
- *Unsure*: 172 seconds
- *Very unsure*: 130 seconds

Although the *very unsure* group responded the fastest and the *very sure* group the slowest, these differences were not statistically significant. In other words, confidence level did not meaningfully affect how quickly participants completed the task. In summary, this second variant A task showed that confidence was weakly related to accuracy (with one significant difference), but it did not influence response time. Participants' feelings about their performance did not reliably match how quickly—or how accurately—they worked with the vibrating-symbol map.

The third task in map variant A required participants to **identify the most common earthquake depth**. Confidence levels varied, and this time the accuracy pattern was quite different from the earlier tasks. Here is how each confidence group performed:

- *Very sure*: 10.2% of participants; 76.9% correct
- *Sure*: 9.0% of participants; 60.9% correct
- *Neutral*: 31.1% of participants; 34.2% correct
- *Unsure*: 23.2% of participants; 52.5% correct
- *Very unsure*: 24.8% of participants; 28.6% correct

This was one of the few tasks where higher confidence actually aligned more closely with higher accuracy. Participants who were very sure or sure performed notably better than those in the *neutral* or *very unsure* groups. Even the *unsure* group performed reasonably well. However, despite these visible distinctions, the statistical tests showed no significant differences in accuracy across the confidence categories. Efficiency results revealed clearer and statistically meaningful differences:

- *Very sure*: 128 seconds
- *Sure*: 97 seconds
- *Neutral*: 126 seconds
- *Unsure*: 89 seconds
- *Very unsure*: 77 seconds (fastest)

Here, confidence had a strong relationship with response time. Participants who felt *very unsure* completed the task the fastest, while the *very sure* and *neutral* groups took the longest. The Mann–Whitney U test confirmed two significant differences:

- *Very unsure* vs. *very sure* ($p=0.015$, $z=2.42$)
- *Very unsure* vs. *neutral* ($p=0.016$, $z=2.40$)

In summary, for this variant A task, accuracy tended to increase with confidence, but only up to a point, and the statistical tests did not confirm clear group differences. In contrast, response times showed a clearer divide: **the most confident and neutral participants took the most time, whereas very unsure participants worked surprisingly quickly**, possibly relying on quicker, less deliberative strategies.

The final task in map variant A asked participants to **identify the year in which the most earthquakes occurred**. As in the previous tasks, participants rated how confident they felt about their answers. The accuracy patterns across confidence levels were as follows:

- *Very sure*: 6.7% of participants; 52.9% correct
- *Sure*: 12.2% of participants; 45.2% correct

- *Neutral*: 31.1% of participants; 25.6% correct
- *Unsure*: 24.8% of participants; 31.7% correct
- *Very unsure*: 24.4% of participants; 21.3% correct

Overall, the "very sure" and "sure" groups performed the best, while the "very unsure" group performed the worst. Despite these visible differences, neither the Pearson chi-square test nor Fisher's exact test found any statistically significant differences. This means that participants' confidence did not reliably predict how accurate they were in identifying the correct year. However, confidence did play a strong role in how long participants spent on the task:

- *Very sure*: 294 seconds (slowest)
- *Sure*: 195 seconds
- *Neutral*: 133 seconds
- *Unsure*: 133 seconds
- *Very unsure*: 117 seconds (fastest)

Once again, the pattern shows that the most confident participants took the longest, while those who were less certain often finished more quickly. The only statistically significant difference emerged **between the sure and neutral groups** ($p=0.032$, $z=2.14$), indicating that these two groups had meaningfully different response times. In summary, confidence levels in this final variant A task did not predict accuracy, but they did influence efficiency. Participants who felt *very sure* tended to spend much more time verifying their answers, while the least confident participants completed the task more quickly—possibly reflecting quicker guessing or reduced rechecking of the vibrating-symbol map.

#### 4.3.5.2 Map Variant B

The first task in **map variant B**, where participants were asked to **find and count earthquake locations with the highest magnitude**, confidence levels varied widely—and so did accuracy. Participants reported how confident they felt about their answers, and the results looked like this:

- *Very sure*: 8.1% of participants; only 4.2% correct
- *Sure*: 15.5% of participants; 26.1% correct
- *Neutral*: 27.0% of participants; 16.4% correct
- *Unsure*: 21.3% of participants; 12.7% correct
- *Very unsure*: 27.7% of participants; 14.8% correct

What stands out is that being *very sure* did not translate into better performance—this group actually had the lowest accuracy. On the other hand, the "sure" group performed the best, though their accuracy was still modest. Statistically, two group comparisons showed meaningful differences:

- *Very sure* vs. *unsure* ($\chi^2=11.5$, $p=0.0007$; Fisher's exact $p=0.043$)
- *Sure* vs. *neutral* ($\chi^2=5.89$, $p=0.015$; Fisher's exact $p=0.033$)

Both were significant according to the Pearson chi-square and Fisher's exact tests. This means that some confidence groups genuinely differed in accuracy, even though overall performance remained low for everyone. When we look at efficiency, the pattern becomes even clearer:

- *Very sure*: 128 seconds (slowest)
- *Sure*: 97 seconds
- *Neutral*: 87 seconds
- *Unsure*: 94 seconds
- *Very unsure*: 78 seconds (fastest)

Here, the most confident participants spent the most time, while the least confident finished fastest—a pattern also seen in variant A. Two statistically significant differences appeared in response times:

- *Very sure* vs. *neutral* ($p=0.027$, $z=2.20$)
- *Very sure* vs. *very unsure* ($p=0.024$, $z=2.24$)

These results suggest that in this particular task, confidence was not a reliable indicator of accuracy—but it was closely tied to how long participants spent on the task. The motion-enhanced symbols in variant B may have increased uncertainty, leading confident participants to overcheck their answers, while less confident participants responded more quickly, possibly relying on quicker impressions or guesses.

The second task in map variant B asked participants to **find and count all earthquake locations with the highest number of fatalities**. Confidence levels varied across the sample:

- *Very sure*: 7.4% of participants; 22.7% correct
- *Sure*: 19.6% of participants; 27.6% correct
- *Neutral*: 30.1% of participants; 10.1% correct
- *Unsure*: 17.9% of participants; 11.3% correct
- *Very unsure*: 25.0% of participants; 9.4% correct

Although the "sure" group performed the best and neutrally confident participants performed the worst, none of these differences reached statistical significance according to the Pearson chi-square and Fisher's exact tests. In other words, confidence did not meaningfully predict accuracy for this task. Efficiency showed a similar pattern, with only modest variation across groups:

- *Very sure*: 95 seconds
- *Sure*: 96 seconds
- *Neutral*: 77 seconds
- *Unsure*: 73 seconds
- *Very unsure*: 65 seconds

Here, participants with lower confidence tended to respond more quickly, but the differences were small. According to the Mann–Whitney U test, no statistically significant differences were found between any of the confidence groups. Overall, for this task in variant B, confidence levels neither predicted accuracy nor influenced the time participants spent completing the pictorial-motion-based fatality count.

The third task in map variant B asked participants to **identify the most common earthquake depth across the full time span**. Confidence levels varied, and accuracy followed a clear downward trend as confidence decreased:

- *Very sure*: 10.8% of participants; 81.2% correct
- *Sure*: 17.2% of participants; 64.7% correct
- *Neutral*: 29.0% of participants; 25.6% correct
- *Unsure*: 18.2% of participants; 40.0% correct
- *Very unsure*: 24.3% of participants; 29.2% correct

Although accuracy differed substantially—particularly between the *very sure/sure* groups and the *neutral/very unsure* groups—neither the Pearson chi-square test nor Fisher's exact test found any statistically significant differences. This means that, statistically speaking, confidence did not reliably predict who would get the depth question correct. Efficiency patterns were also fairly even across groups:

- Very sure: 61 seconds
- Sure: 63 seconds
- Neutral: 53 seconds
- Unsure: 80 seconds
- Very unsure: 71 seconds

Even though *unsure* participants tended to take a bit longer, none of these timing differences were statistically significant according to the Mann–Whitney U test. In short, for this depth-identification task in variant B, confidence levels did not meaningfully influence either accuracy or response time.

The final task in map variant B asked participants to **identify the year in which the most earthquakes occurred**. Confidence levels varied across the sample, and accuracy followed a clear gradient:

- *Very sure*: 7.8% of participants; 47.8% correct
- *Sure*: 17.9% of participants; 47.2% correct
- *Neutral*: 29.7% of participants; 25.6% correct
- *Unsure*: 17.9% of participants; 34.0% correct
- *Very unsure*: 29.4% of participants; 16.3% correct

Two comparisons revealed statistically significant differences in effectiveness:

- *Very sure* vs. *unsure* ($\chi^2 = 5.32$, $p = 0.021$; Fisher's exact $p = 0.036$)
- *Sure* vs. *unsure* ($\chi^2 = 4.16$, $p = 0.041$; Fisher's $p = 0.049$)

This indicates that participants who were "sure" or "very sure" performed meaningfully better than those who were unsure of their answers. Efficiency patterns showed more variation:

- *Very sure*: 93 seconds
- *Sure*: 144 seconds (slowest)
- *Neutral*: 66 seconds (fastest)
- *Unsure*: 71 seconds
- *Very unsure*: 74 seconds

Three statistically significant timing differences appeared:

- *Sure* vs. *neutral*: $p=0.011$, $z=2.54$
- *Sure* vs. *unsure*: $p=0.002$, $z=2.99$
- *Sure* vs. *very unsure*: $p=0.003$, $z=2.89$

Interestingly, **the "sure" group—despite performing well in accuracy—took the longest to respond, significantly longer than all lower confidence groups**. This suggests that their confidence may have driven more deliberate checking, whereas neutral or unsure participants tended to complete the task more quickly, relying on faster judgments of the pictorial-motion symbols.

#### 4.3.5.3 Map Variant C

The first task in **map variant C** asked participants to **find and count all earthquake locations with the highest magnitude**, shown here using pulsating-point symbols. Confidence levels varied, and accuracy followed an uneven pattern:

- *Very sure*: 6.3% of participants; 15.8% correct
- *Sure*: 25.1% of participants; 6.8% correct
- *Neutral*: 25.4% of participants; 19.5% correct
- *Unsure*: 23.4% of participants; 19.7% correct
- *Very unsure*: 19.8% of participants; 28.3% correct

Although accuracy differed across groups, neither the Pearson chi-square test nor Fisher's exact test showed any statistically significant differences. This means confidence levels did not meaningfully predict who performed better in this magnitude-counting task. Efficiency results showed clearer distinctions. Median response times were as follows:

- *Very sure*: 137 seconds
- *Sure*: 155 seconds
- *Neutral*: 96 seconds (fastest)
- *Unsure*: 147 seconds
- *Very unsure*: 119 seconds

Two comparisons showed statistically significant differences in response times:

- *Neutral* vs. *sure*: $p=0.001$, $z=3.22$
- *Neutral* vs. *unsure*: $p=0.008$, $z=2.62$

These findings suggest that **participants who felt *neutrally* confident completed the pulsating-symbol task more efficiently than both the *sure* and *unsure* groups**—possibly because they approached the task with a balanced level of caution and speed.

The second task in map variant C asked participants to **identify and count all earthquake locations with the highest number of fatalities**. Confidence levels varied widely, and accuracy reflected this variation:

- *Very sure*: 6.3% of participants; 68.4% correct
- *Sure*: 16.5% of participants; 48.0% correct
- *Neutral*: 30.0% of participants; 9.9% correct
- *Unsure*: 16.5% of participants; 44.0% correct
- *Very unsure*: 30.7% of participants; 26.9% correct

These differences were noticeable, and one comparison showed a meaningful gap in performance. Accuracy differed significantly between the *sure* vs. *neutral* group. ($\chi^2=4.81$, $p=0.028$; Fisher's exact $p=0.032$). This indicates that **participants who felt sure performed significantly better than those who felt neutrally confident**. Efficiency patterns added another layer to the results:

- *Very sure*: 57 seconds (fastest)
- *Sure*: 84 seconds
- *Neutral*: 89 seconds
- *Unsure*: 86 seconds
- *Very unsure*: 111 seconds (slowest)

Two of these timing differences reached statistical significance:

- *Very unsure* vs. *very sure*: $p=0.038$, $z=2.07$
- *Very unsure* vs. *neutral*: $p=0.023$, $z=2.26$

These results suggest that the **most confident participants completed the pulsating-symbol fatality task far more quickly, while the *very unsure* group not only performed less accurately but also required the most time**—possibly reflecting both difficulty interpreting the visual pulsation and hesitation during decision-making.

The third task in map variant C asked participants to **identify the most frequently occurring earthquake depth**. Confidence levels were high for this task, and accuracy reflected that pattern:

- *Very sure*: 42.6% of participants; 80.6% correct
- *Sure*: 26.1% of participants; 91.1% correct

- *Neutral*: 18.5% of participants; 69.6% correct
- *Unsure*: 3.0% of participants; 100.0% correct
- *Very unsure*: 9.2% of participants; 42.8% correct

Even though accuracy varied—particularly between the confident and *very unsure* groups—neither the Pearson chi-square test nor Fisher's exact test found any statistically significant differences. This means that, statistically, confidence levels did not predict who was more or less accurate in identifying the correct depth. Efficiency scores showed similarly small differences:

- *Very sure*: 54 seconds
- *Sure*: 55 seconds
- *Neutral*: 64 seconds
- *Unsure*: 55 seconds
- *Very unsure*: 66 seconds

Because response times were relatively close across all confidence groups, no statistically significant differences were detected using the Mann–Whitney U test. Overall, for this task, both accuracy and efficiency were consistently high among participants who felt confident, and even the unsure group completed the task quickly and accurately.

The final task in map variant C asked participants to **identify the year in which the most earthquakes occurred within the given time span**. Confidence levels varied, and accuracy showed a clear pattern:

- *Very sure*: 10.9% of participants; 100.0% correct
- *Sure*: 40.9% of participants; 70.2% correct
- *Neutral*: 26.7% of participants; 55.6% correct
- *Unsure*: 6.9% of participants; 71.4% correct
- *Very unsure*: 14.5% of participants; 40.9% correct

Two differences in effectiveness reached statistical significance:

- *Sure* vs. *unsure*: $\chi^2 = 4.20$, $p = 0.04$; Fisher's exact $p = 0.05$
- *Sure* vs. *very unsure*: $\chi^2 = 4.53$, $p = 0.033$; Fisher's exact $p = 0.044$

This means that participants who felt "sure" performed significantly better than those who felt *unsure* or *very unsure*. Efficiency patterns, however, were much more uniform:

- *Very sure*: 87 seconds
- *Sure*: 78 seconds
- *Neutral*: 83 seconds
- *Unsure*: 63 seconds
- *Very unsure*: 70 seconds

Despite these slight differences, no statistically significant differences in response time were found using the Mann–Whitney U test. In other words, participants across all confidence levels completed the pulsating-symbol task at similar speeds, even though their accuracy varied.

#### 4.3.5.4 Map Variant D

The first task in **map variant D** asked participants to **find and count all earthquake locations with the highest recorded magnitude**, using a static-symbol display. Confidence levels varied across the sample, but accuracy remained fairly similar:

- *Very sure*: 2.1% of participants; 14.3% correct
- *Sure*: 14.5% of participants; 20.4% correct
- *Neutral*: 20.8% of participants; 21.4% correct
- *Unsure*: 27.0% of participants; 23.1% correct
- *Very unsure*: 35.6% of participants; 23.1% correct

Although accuracy differed slightly between groups, neither the Pearson chi-square test nor Fisher's exact test found any statistically significant differences. This means confidence levels did not meaningfully predict who performed better on this static-map magnitude task. Efficiency results followed a similarly consistent pattern:

- *Very sure*: 27 seconds (fastest)
- *Sure*: 47 seconds
- *Neutral*: 59 seconds
- *Unsure*: 59 seconds
- *Very unsure*: 57 seconds

Because the differences in response times were minimal, the Mann–Whitney U test revealed no statistically significant differences. Overall, for this first static-map task in variant D, neither accuracy nor efficiency varied meaningfully across confidence levels.

The second task in map variant D asked participants to **find and count all earthquake locations with the highest number of fatalities**. Confidence levels varied widely, and accuracy showed noticeable—but uneven—differences:

- *Very sure*: 8.0% of participants; 25.9% correct
- *Sure*: 14.2% of participants; 27.1% correct
- *Neutral*: 26.1% of participants; 13.6% correct
- *Unsure*: 13.0% of participants; 29.5% correct
- *Very unsure*: 34.7% of participants; 20.5% correct

Interestingly, **the *unsure* group performed better than the *very unsure* group**, and this difference was statistically significant (Pearson $\chi^2=7.91$, $p=0.005$; Fisher's $p=0.011$). No other confidence comparisons reached significance, meaning that accuracy differences were limited to this specific pair. Efficiency followed a slightly different pattern:

- *Very sure*: 58 seconds
- *Sure*: 63 seconds

- *Neutral*: 64 seconds
- *Unsure*: 50 seconds
- *Very unsure*: 50 seconds

Participants who were *unsure* or *very unsure* completed the task the fastest, while those with higher confidence took longer. Two significant differences emerged: *neutral* vs. *unsure* ($p=0.048$, $z=1.97$) and *neutral* vs. *very unsure* ($p=0.035$, $z=2.11$). Overall, this task showed that **lower confidence groups were both faster and, in one comparison, more accurate than their very unsure counterparts**.

The third task in map variant D required participants to **identify the most frequently occurring earthquake depth**. Confidence levels varied considerably, and accuracy patterns were uneven across groups:

- *Very sure*: 13.5% of participants; 64.4% correct
- *Sure*: 17.4% of participants; 80.0% correct
- *Neutral*: 21.2% of participants; 22.5% correct
- *Unsure*: 20.7% of participants; 47.8% correct
- *Very unsure*: 27.2% of participants; 33.0% correct

Although the *sure* group achieved the highest accuracy and the *neutral* group the lowest, none of these differences reached statistical significance according to the Pearson chi-square and Fisher's exact tests. Efficiency results showed a similarly mixed pattern:

- *Very sure*: 43 seconds
- *Sure*: 44 seconds
- *Neutral*: 32 seconds (fastest)
- *Unsure*: 40 seconds
- *Very unsure*: 38 seconds

Completion times differed modestly, with *neutrally* confident participants responding the fastest, but statistical testing revealed no significant differences among any of the confidence groups. Overall, both accuracy and response time distributions suggest that for this static-symbol task, confidence levels did not meaningfully predict performance.

The final task in map variant D required participants to **identify the year with the highest number of earthquakes**. Confidence levels varied widely, and accuracy declined steadily with lower confidence:

- *Very sure*: 4.8% of participants; 68.7% correct
- *Sure*: 15.2% of participants; 64.0% correct
- *Neutral*: 19.7% of participants; 40.0% correct
- *Unsure*: 27.9% of participants; 17.4% correct
- *Very unsure*: 32.4% of participants; 20.6% correct

The most notable contrast appeared **between the *neutral* and *unsure* groups**, a difference that was statistically significant according to both the Pearson chi-square test ($\chi^2 = 8.82$, $p = 0.003$) and Fisher's exact test ($p = 0.002$). No other confidence comparisons reached significance. Efficiency results, represented by median response time, were highly consistent across confidence levels:

- *Very sure*: 58 seconds
- *Sure*: 67 seconds
- *Neutral*: 57 seconds
- *Unsure*: 61 seconds
- *Very unsure*: 59 seconds

Because these time differences were small, no statistically significant differences in response time were detected using the Mann–Whitney U test. Overall, this task showed that higher confidence generally aligned with higher accuracy, but confidence had little connection to the speed of completing the task.

Across all confidence groups and map variants, a consistent pattern emerges: confidence is generally aligned with higher accuracy, especially in tasks requiring identification (e.g., the most frequent earthquake depth or the year with the most earthquakes). Among all map designs, map variant C, which used pulsating-point symbols, proved to be the most effective overall (Figure 4.17). In nearly every task, participants who felt "very sure" or "sure" delivered the highest accuracy, particularly in tasks that required recognizing the most frequent values or interpreting temporal patterns.

In contrast, effectiveness in counting tasks (e.g., counting locations with the highest magnitude or fatalities) tended to be lower across all variants, and the gap between confidence groups was noticeably smaller. Both B and C showed clearer distinctions

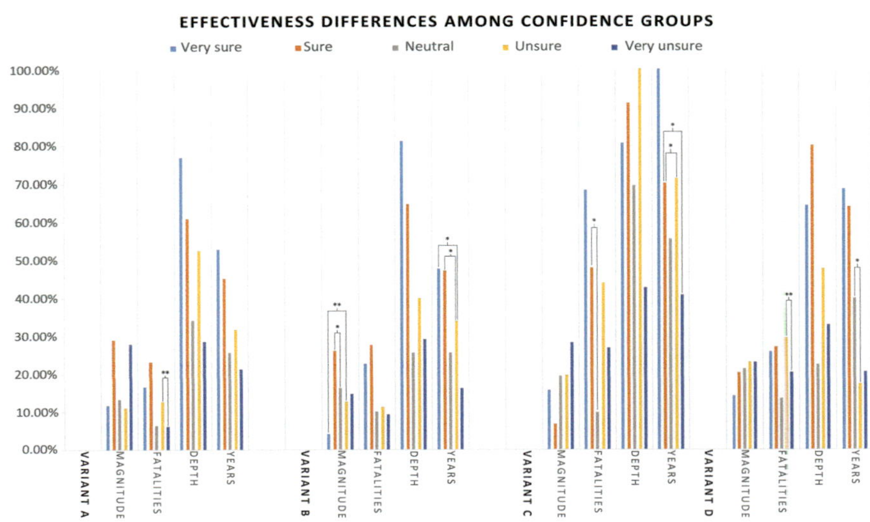


**FIGURE 4.17** Differences in effectiveness among self-reported confidence groups across map-based tasks and map variants. (* denotes statistical significance at $p < 0.05$; ** denotes statistical significance at $p < 0.005$.)

where "sure" participants outperformed "neutral" ones—suggesting that dynamic symbols support confident decision-making—but these differences remained modest. Perhaps the most interesting deviation occurred in variant A, where some of the "unsure" or "very unsure" groups unexpectedly achieved accuracy comparable to more confident participants. This suggests that the vibrating symbols used in variant A may have introduced uncertainty that dominated skill or confidence, making judgments less predictable. Variant D (static symbols) showed moderate, stable accuracy across confidence groups but did not outperform variant C. Overall, **the effectiveness results strongly support variant C as the design that best translates participant confidence into correct performance**.

The efficiency analysis paints a different picture, showing that higher confidence often coincided with slower response times, particularly in the more demanding dynamic variants (Figure 4.18). The most striking example is variant A, where the "very sure" and "sure" participants took the longest to respond—despite not achieving the highest accuracy. This suggests that vibrating symbols may have caused participants to double-check or hesitate, leading to "overthinking" rather than improved performance. In contrast, participants with lower confidence often responded faster, sometimes achieving relatively good accuracy (e.g., unsure groups in B and C for identification tasks). This indicates that rapid, impression-based judgments occasionally worked well, especially when the map design supported immediate visual recognition.

The most efficient variant overall was variant D (static symbols). Across all tasks and all confidence groups, response times in variant D were consistently shorter and showed minimal differences between groups. However, this efficiency did not come with higher accuracy: variant D was not more effective than variant B and clearly less effective than variant C, highlighting that speed alone does not guarantee better

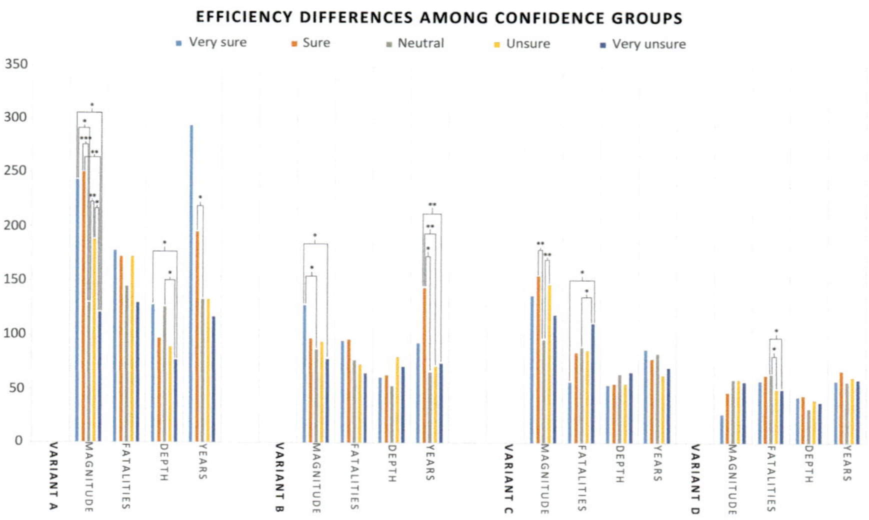


**FIGURE 4.18** Differences in efficiency among self-reported confidence groups across map-based tasks and map variants. (* denotes statistical significance at $p<0.05$; ** denotes statistical significance at $p<0.005$; *** denotes statistical significance at $p<0.0005$.)

task performance. A recurring trend across variants is that counting tasks, especially those involving motion-based cues (e.g., velocity in variant B), were the most challenging. These tasks produced the highest variability in both accuracy and completion time, particularly among less confident participants who required more time but did not achieve higher correctness. Taken together, the results show a trade-off:

- Variant D—fastest but not the most accurate
- Variant A—slowest and often least predictable
- Variant B—moderate performance but difficulty with motion-based counting
- Variant C—best balance of accuracy and reasonable completion time, and most consistent with participants' confidence

Thus, **variant C stands out as the recommended design for practical applications** where both accuracy and user trust (confidence) are important, without imposing unnecessary time costs.

### 4.3.6 What Participants Thought About the Map Designs

Participants were also asked how much they liked the style of the maps they worked with. For **map variant A**, 5.1% of participants said they liked the style very much, and 24.0% said they liked it. Almost half of the group (47.2%) had no opinion, while 15.4% disliked the style and 7.9% strongly disliked it. Only 0.4% did not answer.

For **map variant B**, opinions were similar. A total of 5.2% liked the map style very much, and 30.0% liked it. Another 37.9% had no opinion. On the negative side, 16.6% disliked the style and 10.3% strongly disliked it.

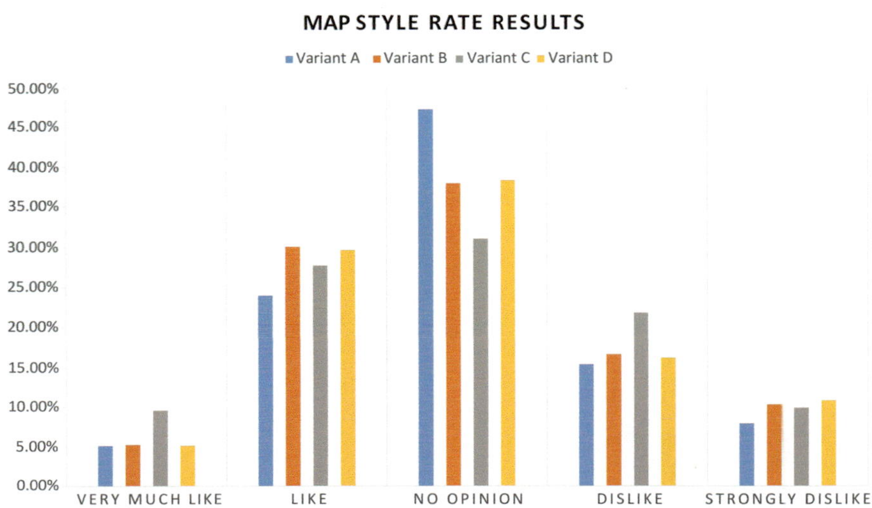


**FIGURE 4.19** Results of participants' ratings of the map styles presented to them and used for task completion.

**Map variant C received the most positive feedback overall**. Here, 9.6% of participants liked the map style very much—the highest percentage among all variants—and 27.7% liked it. About one-third (31.0%) had no opinion. However, a notable share of participants also disliked (21.8%) or strongly disliked (9.9%) the style.

For the static-**map variant D**, 5.1% of participants liked the style very much, and 29.6% liked it. A total of 38.3% expressed no opinion, while 16.2% disliked and 10.8% strongly disliked the map's appearance.

Overall (Figure 4.19), the most common response across all map variants was having no opinion about the map style. Still, in every variant, more participants liked the map style than disliked it. This suggests that, in general, dynamic-symbol maps were perceived just as positively as the static version, and the aesthetic reception of all four map designs remained broadly consistent.

# 5 Performance and Visual Attention on Animated Maps

## *Eye-Tracking Results*

This chapter presents the findings from the laboratory eye-tracking experiment, focusing on how participants performed map-reading tasks using dynamic point symbols and how their visual attention was distributed during those tasks. Although eye-movement measures play an important role in the analyses that follow, the chapter begins by reporting task effectiveness and efficiency—mirroring the structure of the online questionnaire results. This enables a direct comparison between the two study settings, helping to clarify how animated map symbols affect both user performance and visual behavior.

## 5.1 HOW ACCURATE PARTICIPANTS PERFORMED

In the eye-tracking experiment, participants completed a set of tasks that required them to identify different earthquake attributes—such as magnitude, number of fatalities, depth, and the year with the most events. The results clearly show that the way information is encoded with dynamic point symbols has a noticeable impact on how accurately people can search for and interpret earthquake data. Across the tasks, animated symbols often led to better performance than static symbols, and in several cases, the results were at least comparable. Importantly, many of these differences were statistically significant, meaning they were not just random fluctuations but reliable effects of the map design.

A key takeaway is that well-designed dynamic symbols can be successfully used in earthquake mapping without disrupting the perception of other visual variables like color or size. This suggests that animated point symbols have strong potential beyond earthquakes as well—for example, in visualizing volcanic activity, air-quality changes, or patterns in access to services or medical facilities.

### 5.1.1 Finding the Strongest Earthquakes (Motion-Based Search)

In this task, participants had to **find and count all earthquake locations with the highest magnitude**. Their accuracy is summarized by how far their answers deviated from the correct number (0 = correct, ±1 = close, ±2 = moderately off, ±3 or more = far off). Across all map variants, no one made an error larger than ±2, showing that the task was manageable regardless of symbol design.

DOI: 10.1201/9781003728535-5

Variant A—Vibrating symbols:

- 29.6% (8people) gave the correct answer.
- 48.1% (13 people) were very close (±1).
- 25.9% (7 people) were off by ±2.

Variant B—Pictorial motion:

- 45.8% (11 people) answered correctly—the highest accuracy of all variants.
- 29.2% (7 people) were within ±1.
- 25.0% (6 people) were within ±2.

Variant C—Pulsing symbols:

- Only 25.0% (5 people) found the exact number.
- Another 25.0% (5 people) were close (±1).
- 50.0% (10 people) missed by ±2—indicating this encoding was harder for counting tasks.

Variant D—Static reference map:

- 33.3% (10 people) were correct.
- 33.3% (10 people) were close (±1).
- 33.3% (10 people) were off by ±2.

Overall, map variant B—the pictorial-motion design—led to the highest accuracy, suggesting that motion-based encoding was effective in helping participants spot the strongest earthquakes. In contrast, the pulsing symbols in variant C seemed to make the counting task more difficult, with half the participants missing the correct answer by two steps.

The statistical analysis showed a clear difference in how well participants performed depending on the type of animation used. According to the Pearson chi-square test ($\chi^2 = 5.41$, $p = 0.020$) and Fisher's exact test (two-tailed $p = 0.038$), participants using the pictorial-motion map (variant B) performed significantly better than those using the pulsating-symbol map (variant C). This difference is meaningful: in variant C, half of the participants missed the correct answer by ±2 classes, while variant B produced the largest share of fully correct responses across all variants. In other words, when it comes to identifying the strongest earthquakes in a controlled eye-tracking setting, the pictorial-motion style clearly helped users interpret the data more accurately. These results are summarized in Figure 5.1.

However, the picture shifts slightly when the evaluation is broadened to include answers that were close to correct (±1 deviation). Under this more forgiving scoring, variant A becomes the top performer with 74.1% effectiveness, closely followed by variant B, which reaches 73.9%. The pulsation-based map (variant C), which had performed best in the online survey, now shows the lowest effectiveness in

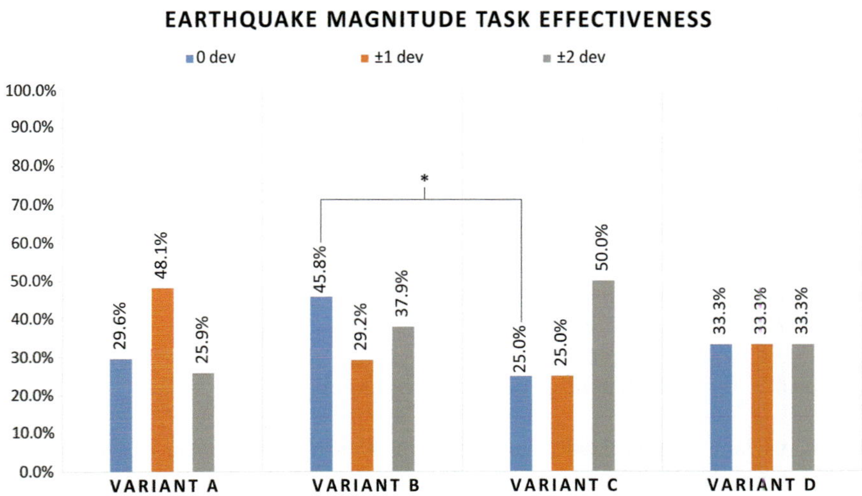


**FIGURE 5.1** Earthquake magnitude search-and-counting task results in the context of different map variants and response accuracy. (*denotes statistical significance at $p<0.05$.)

the laboratory experiment at 50.0%. The static map (variant D) falls in the middle with 65.5% effectiveness. Despite these numerical differences, neither the Pearson chi-square test nor Fisher's exact test found statistically significant differences among the map variants for this inclusive scoring method. This suggests that while symbol design can strongly influence perfect accuracy, many participants across all variants were still able to produce near-correct estimates.

### 5.1.2 Finding the Deadliest Earthquakes (Size-Based Search)

This task asked participants to **find and count all earthquake locations with the highest number of fatalities**. Accuracy was again measured by how far each answer deviated from the correct number of locations (0 = correct, ±1 = close, ±2 = moderately off, ±3 or more = far off).

Variant A—Vibrating symbols:

- 37.0% (10 people) gave the correct number.
- Another 37.0% were close (±1).
- 14.8% (4 people) were off by ±2.
- 11.1% (3 people) missed by ±3 or more.

Variant B—Pictorial motion—This was the strongest-performing variant:

- 58.3% (14 people) answered correctly.
- 25.0% (6 people) were within ±1.
- 12.5% (3 people) were within ±2.
- Only 1 person (4.2%) had an error of ±3 or more.

Variant C—Pulsing symbols:

- 40.0% (8 people) answered correctly.
- 45.0% (9 people) were very close (±1).
- 5.0% (1 person) missed by ±2.
- 10.0% (2 people) missed by ±3 or more.

Variant D—Static symbols (reference map):

- 40.0% (12 people) provided a correct count.
- 26.7% (8 people) were within ±1.
- 13.3% (4 people) were within ±2.
- 20.0% (6 people) had large errors (±3 or more).

Overall, variant B again stood out, producing the highest number of fully correct responses and the fewest large errors. Variants C and D performed similarly, while variant A showed a wider spread of inaccuracies, including more errors in the ±3 range compared to other dynamic designs.

Overall, the results show that map variant B, which used moving pictorial icons, was the most effective design for helping participants find and count the earthquake locations with the highest number of fatalities (Figure 5.2). This dynamic motion cue made the relevant earthquakes easier to spot, leading to the highest share of fully correct answers. Interestingly, this outcome differs slightly from the online survey, where variant B ranked second behind the pulsating-symbol map (variant C). The remaining map designs, including the static map, performed at similar levels, and none of the differences between variants were statistically significant.

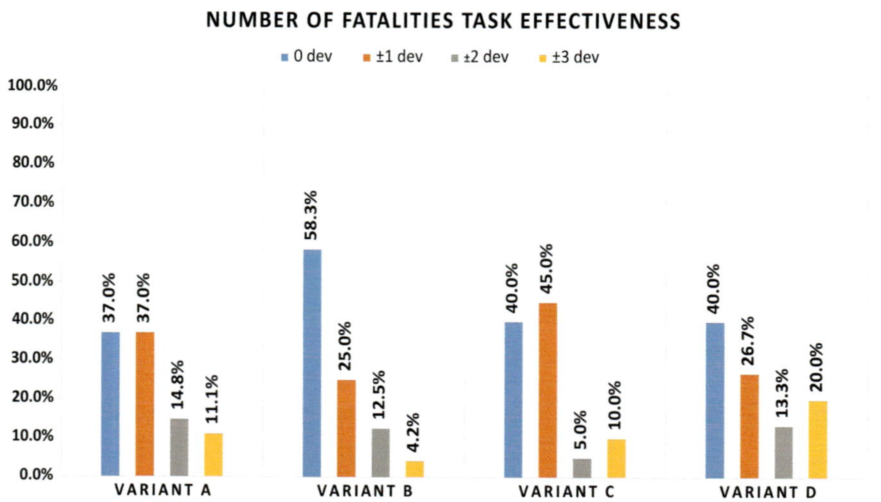


**FIGURE 5.2** The results of the search-and-counting task for the highest number of fatalities in the context of different map variants and response accuracy.

When the analysis was broadened to include answers that were "close" (±1 from the correct value), variants B and C again emerged as the most effective, with effectiveness levels of 83.3% and 84.2%, respectively. Variant A also performed well at 74.1%, while the static map showed the lowest performance at 66.7%. As in the strict evaluation, these differences were not statistically significant. Together, the findings suggest that dynamic symbolization—especially motion-based designs—can support more accurate identification of fatal earthquake clusters, although clear performance gaps between the variants were not strong enough to be confirmed statistically.

### 5.1.3 Identifying the Most Common Earthquake Depth (Color-Based Search)

This task required participants to **identify the most frequently occurring earthquake depth**. Accuracy was measured by how far each answer deviated from the correct depth (0 = correct, ±1 = close, ±2 = moderately off, ±3 or more = far off).

Variant A—Vibrating symbols:

- 92.6% (25 people) gave the correct answer (0).
- 7.4% (2 people) were close (±1).
- No participants provided answers deviating by ±2 or ±3.

Variant B—Pictorial motion:

- 83.3% (20 people) answered correctly.
- 12.5% (3 people) were within ±1.
- 4.2% (1 person) was moderately off (±2).
- No answers deviated by ±3 or more.

Variant C—Pulsing symbols:

- 90.0% (18 people) identified the correct depth.
- 10.0% (2 people) were slightly off (±1).
- No participants provided larger deviations.

Variant D—Static symbols (reference map):

- 80.0% (24 people) answered correctly — the lowest accuracy among the variants.
- 10.0% (3 people) were within ±1.
- 3.3% (1 person) deviated by ±2.
- 3.3% (1 person) was far off (±3 or more).
- One participant did not provide an answer.

Overall, this task produced very high accuracy across all designs, with variants A and C performing best. The static map (variant D) showed slightly weaker performance

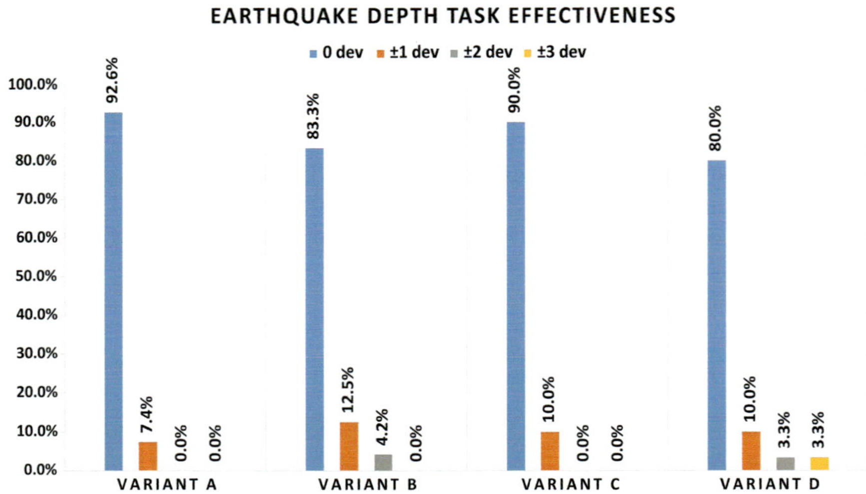


**FIGURE 5.3** The results of the searching and indicating the most frequent earthquake depth in the context of different map variants and response accuracy.

and included the only large-error responses, but differences between variants were small given the overall high success rate.

Overall, the results show that map variant A was the most effective, although variants C and B performed almost equally well, with only small differences in accuracy. Even the static map (variant D) produced strong results, and neither Pearson's chi-square test nor Fisher's exact test detected statistically significant differences between any of the mapping solutions. This indicates that all four variants supported participants very well in identifying the most common earthquake depth. When applying a more inclusive scoring approach—treating answers within a ±1 deviation as correct—effectiveness became nearly perfect across all variants, as shown in Figure 5.3. Compared with the online survey questionnaire, the laboratory eye-tracking experiment produced substantially higher accuracy, suggesting that controlled experimental conditions may improve users' ability to interpret color-encoded depth information.

### 5.1.4 Identifying the Year with the Most Earthquakes (Temporal-Based Search)

This task asked participants to **identify the year in which the highest number of earthquakes occurred**. Accuracy was evaluated, as previously, based on how far their answers deviated from the correct value (0 = correct, ±1 = close, ±2 = moderately off, ±3 or more = far off).

Variant A—Vibrating symbols:

- 92.6% (25 people) answered correctly (0 deviation).
- 0% were within ±1 or ±2.
- 3.7% (1 person) were incorrect by ±3 or more.
- One participant did not answer.

Variant B—Pictorial motion:

- 74.1% (20 people) gave the correct answer.
- 0% were within ±1.
- 4.2% (1 person) missed by ±2.
- 8.3% (2 people) were off by ±3 or more.
- One participant did not answer.

Variant C—Pulsing symbols:

- 70.0% (14 people) answered correctly.
- 5.0% (1 person) were within ±1.
- 5.0% (1 person) were within ±2.
- 20.0% (4 people) had large errors (±3 or more).

Variant D—Static symbols:

- 90.0% (27 people) provided the correct answer.
- 0% were within ±1 or ±2.
- 10.0% (3 people) were off by ±3 or more.

In summary, map variants A and D emerged as the most effective for identifying the year with the most earthquakes, as illustrated in Figure 5.4. These two variants produced the highest proportion of fully correct responses and the fewest large errors. In contrast, variant C—highly successful in the online survey—performed noticeably worse in the controlled laboratory experiment, showing the highest share

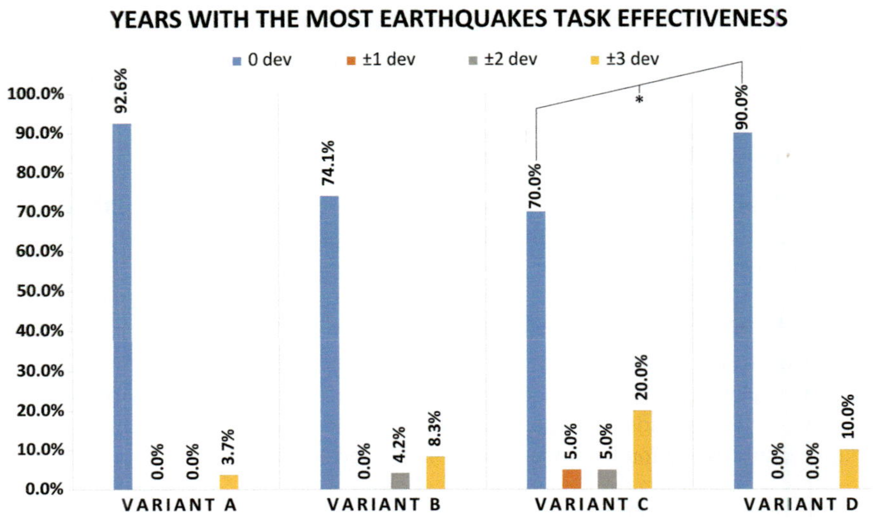


**FIGURE 5.4** The results of searching for and identifying the year with the most earthquakes in the context of different map variants and response accuracy. (*denotes statistical significance at $p < 0.05$.)

of large deviations (±3 or more). This reversal suggests that the pulsation symbol, while intuitive in online settings, may be more visually demanding or distracting when viewed under stricter, eye-tracked conditions. A statistically significant **difference in accuracy was confirmed between variant C and the static reference map** (variant D), supported by the Pearson chi-square test ($\chi^2 = 5.18$; $p = 0.022$) and Fisher's exact test ($p = 0.048$). Considering answers deviated by one class does not alter the interpretation, as such cases were rare. Overall, these findings indicate that temporal information may be easier to interpret when the visual encoding is stable (static) or minimally dynamic, and that certain animated designs may introduce cognitive load that affects temporal judgments more than other types of tasks.

## 5.2 HOW FAST PARTICIPANTS COMPLETED TASKS

After examining how accurately participants identified earthquake characteristics, the next step is to assess how efficiently they performed these tasks, especially when interacting with different forms of animated and static point symbols. In this part of the analysis, efficiency is evaluated using the same two measures applied in the online survey: average response time and median response time. However, since the Shapiro–Wilk test showed that response times were not normally distributed for several map variants, the median is treated as the more reliable indicator of task efficiency.

The following subsections, therefore, focus on median response times for each map variant, highlighting which designs allowed participants to complete tasks the fastest—and, equally important, whether those differences were meaningful from a statistical perspective. Where relevant, $z$ values and $p$ values from the Mann–Whitney U test are reported, but only when the $p$ value approached the significance threshold ($p \approx 0.05$). In cases where results clearly showed no statistical differences, the narrative simply states that no significant effects were observed. Together, these efficiency results complement the earlier findings on accuracy, allowing for a more complete understanding of how dynamic and static map symbols support (or hinder) quick visual search and interpretation during real-time interaction with animated maps.

### 5.2.1 Time to Find the Strongest Earthquakes

Across all map variants, participants showed a consistent pattern: **those who answered correctly tended to take longer**, but these differences were not statistically significant.

#### 5.2.1.1 Variant A—Vibrating Symbols

Participants took 108 seconds on both the average and a median to complete the task. Those who answered correctly needed even more time—119 seconds on average and a median of 135 seconds. Despite this difference, the Mann–Whitney U test showed no significant distinction between correct and incorrect responders.

#### 5.2.1.2 Variant B—Pictorial-Motion Symbols

Response times were slightly longer here, with an average of 124 seconds and a median of 137 seconds. Correct responders again took more time (average 148 seconds, median 149 seconds), while incorrect responders were faster (median 99 seconds). The difference approached significance but did not meet the threshold ($p=0.066$, $z=1.83$).

#### 5.2.1.3 Variant C—Pulsing Symbols

This was the fastest dynamic solution, with an average of 85 seconds and a median of 78 seconds. Still, correct responders needed more time (average 117 seconds, median 127 seconds) than those who were incorrect (median 75 seconds). The efficiency difference was close to significance but did not reach statistical significance ($p=0.087$, $z=1.71$).

#### 5.2.1.4 Variant D—Static Symbols

The static map produced the overall shortest response times, an average of 73 seconds and a median of 61 seconds. Unlike the dynamic variants, correct and incorrect responders performed nearly identically. The Mann–Whitney U test again confirmed no significant difference.

In summary (Figure 5.5), **map variant C turned out to be the most efficient of all dynamic symbol designs** in the laboratory experiment. This contrasts with the online questionnaire, where variant B had been the fastest dynamic option. The overall most efficient map, in both the lab and the online test, was the static variant D—most likely because static symbols are more familiar and easier to process at a glance. Several statistically significant differences in response time were found between map variants:

- A vs. D ($p=0.0009$, $z=3.23$)
- B vs. C ($p=0.01$, $z=2.52$)
- B vs. D ($p=0.0004$, $z=3.44$)

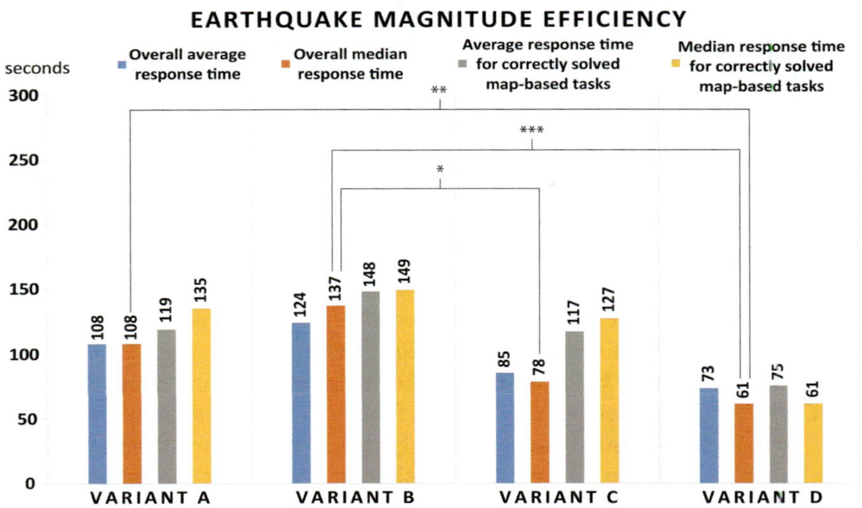


FIGURE 5.5 Participants' response times (efficiency) for map variations in tasks involving finding and counting the highest magnitude earthquakes. (*denotes statistical significance at $p<0.05$; **denotes statistical significance at a level of $p<0.005$; ***denotes statistical significance at a level of $p<0.0005$.)

These results show that some symbol types slowed participants down more than others, especially vibrating symbols (A) and pictorial motion (B). However, no significant difference was found between C and D, and the actual time gap between them was very small. This suggests that **pulsating symbols (C) are nearly as efficient as the static solution**, making variant C a strong dynamic alternative—especially when balancing efficiency and user performance.

### 5.2.2 Time to Find the Deadliest Earthquakes

#### 5.2.2.1 Variant A—Vibrating Symbols

In map variant A, where participants counted earthquake locations with the highest fatalities, the average response time was 68 seconds, with a median of 65 seconds. Participants who answered correctly took slightly longer on average (77 seconds, median 67 seconds). However, this difference was small, and the Mann–Whitney U test showed no statistically significant time difference between correct and incorrect responses.

#### 5.2.2.2 Variant B—Pictorial-Motion Symbols

In map variant B, the efficiency pattern was similar. Participants had an average response time of 67 seconds and a median of 62 seconds. Those who answered correctly were only slightly slower (average 72 seconds, median 63 seconds). Incorrect answers showed nearly identical timing, and again, no statistically significant difference was found.

#### 5.2.2.3 Variant C—Pulsing Symbols

Map variant C, which used pulsating point symbols, was once more the most efficient dynamic design. Both the average and median response times were 51 seconds, making it the fastest among all dynamic variants. Participants who answered correctly required only slightly more time (average 58, median 53 seconds), but as before, the difference between correct and incorrect responses was small and not statistically significant.

#### 5.2.2.4 Variant D—Static Symbols

In map variant D, participants completed the task with an average time of 56 seconds and a median of 47 seconds, making it the second-fastest overall. Correct respondents took slightly longer (average 59, median 52 seconds), while incorrect respondents were somewhat quicker (median 46 seconds). The Mann–Whitney U test again indicated no significant difference between the groups.

Overall, **map variant C (pulsating symbols) was once again the most efficient design**, with participants completing the task faster than in any other dynamic variant (Figure 5.6). Variants A and B were noticeably slower, and this was confirmed statistically: **response times in variant B were significantly longer compared to both variant C** ($p=0.035$, $z=2.09$) **and the static variant D** ($p=0.017$, $z=2.38$). Because no significant differences were found between variants C and D, and their median response times were very close, it can be concluded that the **pulsating design is just as efficient as the static map**—even though it uses animation. This reinforces the strong performance of variant C across both effectiveness and efficiency measures.

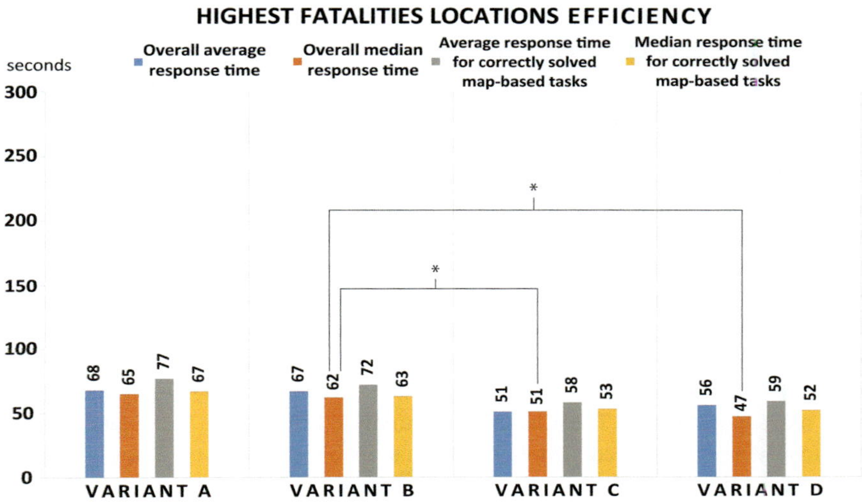


**FIGURE 5.6** Participants' response times (efficiency) for map variations in tasks involving finding and counting the highest number of fatalities. (*denotes statistical significance at $p < 0.05$.)

### 5.2.3 Time to Identify Earthquake Depth

In the task where participants had to identify the most frequently occurring earthquake depth, all four map variants produced fast response times, and efficiency differences between correct and incorrect answers were minimal.

#### 5.2.3.1 Variant A—Vibrating Symbols

Map variant A showed an average response time of 49 seconds and a median of 41 seconds. Participants who answered correctly performed almost identically, with a median of 41 seconds. No statistically significant differences were observed between correct and incorrect responses.

#### 5.2.3.2 Variant B—Pictorial-Motion Symbols

For map variant B, the average response time was 55 seconds, with a median of 50 seconds. Correct responders showed nearly the same timing—56 seconds on average and a 50-second median. As in variant A, the Mann–Whitney U test did not reveal any significant efficiency differences.

#### 5.2.3.3 Variant C—Pulsing Symbols

Map variant C again emerged as the most efficient option—both within the dynamic group and static counterpart. The average response time was 49 seconds, and the median was 41 seconds. Interestingly, those who answered correctly were even faster, with an average of 40 seconds and a median of just 32 seconds. Despite this trend, the efficiency difference between correct and incorrect responses was not statistically significant.

#### 5.2.3.4 Variant D—Static Symbols

Finally, map variant D produced very similar results, with both the average and median response times at 46 seconds. Correct responders showed nearly identical efficiency, with an average of 45 seconds and a median of 46 seconds. As with the other variants, no statistically significant differences were found.

Overall, **map variant C once again proved to be the most efficient, standing out clearly from the other dynamic designs**. The difference between variants C and B was statistically significant (Mann–Whitney U: $p = 0.017$, $z = 2.37$), showing that the pulsating symbols supported noticeably faster responses than the pictorial-motion symbols. Another important finding is that the **static map variant (D) was just as efficient as the dynamic variants**. Since no statistically significant differences were found between variant D and any of the dynamic designs, this indicates that animation did not slow users down—even though static symbols are generally more familiar to most participants. These results, summarized in Figure 5.7, suggest that for tasks involving color-based search, dynamic symbols can perform equally well—or even better—than traditional static representations.

### 5.2.4 Time to Identify the Most Active Year

In the task requiring participants to identify the year with the highest number of earthquakes, response times varied noticeably across map variants, but the general pattern was similar to previous tasks: dynamic symbols did not slow participants down, and the pulsating-symbol variant remained the most efficient.

#### 5.2.4.1 Variant A—Vibrating Symbols

Participants needed 105 seconds on both average and median. Those who answered correctly were only slightly slower (average 106 seconds, median 107 seconds).

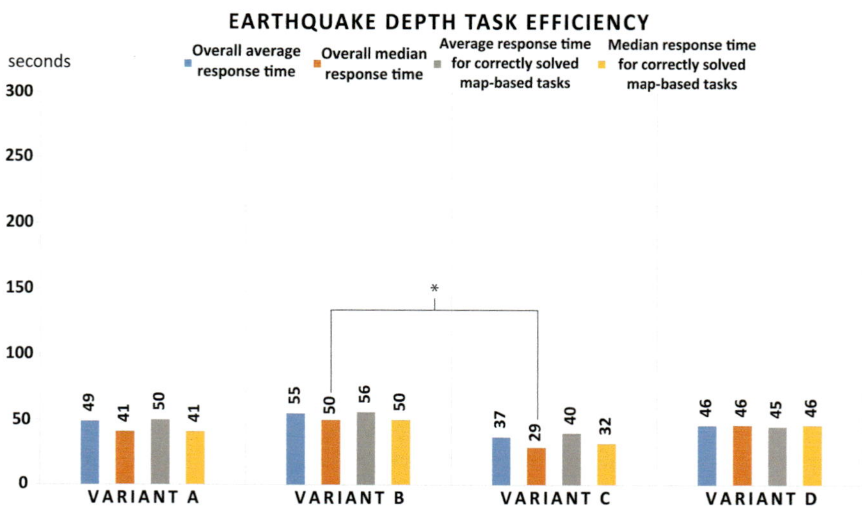


**FIGURE 5.7** Participants' response times (efficiency) for map variations in tasks involving finding the most frequently occurring earthquake depth. (*denotes statistical significance at $p < 0.05$.)

The Mann–Whitney U test showed no significant difference between correct and incorrect responses.

### 5.2.4.2 Variant B—Pictorial-Motion Symbols

The average response time was 105 seconds, but the median was shorter at 90 seconds. Participants who answered correctly needed a bit more time (average 114 seconds, median 99 seconds). **Those who answered incorrectly were much faster** (median 55 seconds), and because this group was small, the statistical comparison relied on the asymptotic result, showing a significant difference ($p=0.025$, $z=2.24$).

### 5.2.4.3 Variant C—Pulsing Symbols

This variant was again the most efficient. Participants needed an average of 79 seconds, with a median of 59 seconds. Correct responses took slightly longer (average 92 seconds, median 79 seconds), while incorrect responses were faster (median 49 seconds). However, the difference did not reach statistical significance ($p=0.076$, $z=1.77$).

### 5.2.4.4 Variant D—Static Symbols

Participants required an average of 93 seconds and a median of 80 seconds. Correct responses were somewhat slower (average 99 seconds, median 90 seconds). Incorrect responses, on the other hand, were very fast (median 43 seconds). Because this group was small, the asymptotic *p* value was used, revealing a statistically significant difference between correct and incorrect responders ($p=0.038$, $z=2.07$).

As shown in Figure 5.8, **map variant C (pulsating symbols) once again proved to be the most efficient** for identifying the year with the most earthquakes. This result contrasts with the online survey, where the static map (variant D) was the fastest for

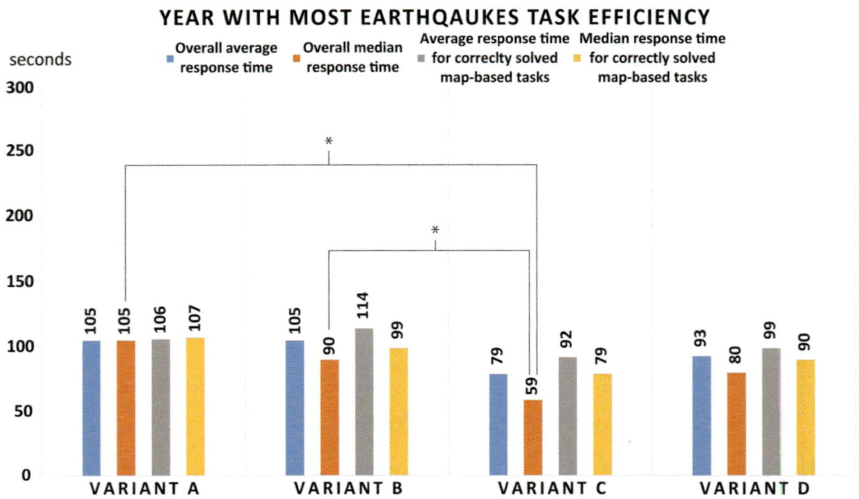


FIGURE 5.8 Participants' response times (efficiency) for map variations in tasks involving finding the year in which the most earthquakes occurred. (*denotes statistical significance at $p<0.05$.)

this type of temporal search task. The laboratory experiment, therefore, suggests that when users view animated symbols under controlled conditions, pulsation can help them more quickly detect temporal patterns. Several efficiency differences reached statistical significance—most notably between variants A and C ($p=0.032$, $z=2.14$) and between variants B and C ($p=0.039$, $z=2.06$). These findings reinforce that pulsating symbols offer a clear speed advantage over both vibrating and pictorial-motion designs. However, **because no significant difference emerged between variant C and the static variant D, the two can be considered similarly efficient in practice**. Taken together, the results show that pulsation is the strongest-performing dynamic solution, achieving the speed benefits of static symbols while still offering the additional expressiveness of animation.

## 5.3 HOW PARTICIPANT CHARACTERISTICS SHAPED THE RESULTS IN EYE-TRACKING EXPERIMENT

Building on the performance and efficiency findings presented earlier, the next step in the analysis was to examine whether individual participant characteristics influenced how users interacted with the maps. However, unlike the large online survey, the eye-tracking experiment involved a relatively small and fairly uniform group of participants. Nearly all were university students with similar levels of experience in map reading, and most evaluated their own map-reading skills as good. Other categories, such as "low," contained only one or two individuals, making meaningful statistical comparisons impossible.

Because of this homogeneity, the eye-tracking experiment could not examine the full range of user characteristics analyzed in the online study. Instead, the focus shifted toward factors that were both well-represented and potentially meaningful for interpreting visual behavior. For this reason, three participant-related aspects were selected for further analysis:

- Declared gender, to determine whether men and women differed in performance or response speed.
- Self-reported confidence, to see whether confidence aligns with effectiveness and efficiency under laboratory conditions.
- Evaluation of map style, to understand whether participant preferences differed from those observed in the online study.

Together, these characteristics allow for a deeper understanding of how users approached the tasks—complementing the eye-movement metrics explored later in this chapter.

### 5.3.1 Differences Between Men and Women

#### 5.3.1.1 Map Variant A

In the first task of map variant A, men and women showed noticeable differences in how accurately they **identified the earthquake locations with the highest magnitude**. Male participants performed very poorly under the strict scoring criterion, with only 8.3% providing the exact correct number. In contrast, female participants performed much better, with 40.0% achieving a fully correct answer. When the more inclusive scoring approach was applied—counting answers that were off by one class (±1) as effectively correct—the gap between men and women became much smaller. Under this relaxed criterion, women still held a slight advantage (75.0% correct) compared to men (73.3% correct), but the difference was minimal. Importantly, these differences did not reach statistical significance. Both the Pearson chi-square test and Fisher's exact test showed that the performance gap could not be reliably interpreted as a gender effect.

Efficiency results were similarly aligned across genders. The median response time for men was 103 seconds, while for women it was 111 seconds—a small difference that was not statistically significant according to the Mann–Whitney U test. Together, these findings suggest that although women appeared to perform better in this specific task under variant A, the sample size and statistical tests indicate that men and women behaved similarly in terms of both accuracy and task completion time.

In the same map variant (A), the second task asked participants to **find and count all earthquake locations with the highest number of fatalities**. Here, the pattern between men and women shifted slightly compared to the previous task. Under the strict scoring criterion, men performed somewhat better, with 41.7% providing the correct answer, compared to 33.3% of women. However, once the more inclusive criterion was applied—where answers within one class of the correct value (±1) were treated as correct—the difference between men and women nearly disappeared. In this broader evaluation, men achieved 73.3% accuracy, while women scored 75.0%, making their performance essentially equivalent. As in the previous task, none of these differences reached statistical significance. Both the Pearson chi-square and Fisher's exact tests confirmed that the observed variations could not be attributed to gender-related differences in performance.

Efficiency results showed a small gap, with men completing the task slightly faster (median 57 seconds) compared to women (median 68 seconds). Yet again, the Mann–Whitney U test indicated that this difference was not statistically significant. Overall, for the fatalities task in variant A, both men and women performed similarly in terms of accuracy and response time, reinforcing the pattern observed across most tasks in the eye-tracking experiment.

For the third task in map variant A, participants were asked to **identify the most frequently occurring earthquake depth**. This task turned out to be one of the easiest, and both men and women performed very well. Male participants reached an accuracy rate of 83.3%, while all female participants answered correctly (100%). When applying the more inclusive scoring criterion—counting answers within one class deviation as correct—both groups achieved perfect accuracy, confirming that gender did not meaningfully affect effectiveness in this task.

A clearer difference appeared in efficiency. **Men completed the task significantly faster than women**, with median response times of 23 seconds and 50 seconds, respectively. This difference was statistically significant ($p=0.021$, $z=2.26$). Interestingly, this reflects a broader pattern observed across the experiment: **participants who completed tasks very quickly often showed lower accuracy**, whereas those who took more time tended to be more accurate—especially in the most challenging tasks.

In the final task of map variant A, participants were asked to **identify the year with the highest number of earthquakes**. Both men and women performed very well. Male participants achieved an accuracy of 91.7%, while all female participants answered correctly (100%). When applying the more inclusive scoring criterion—counting responses within one class deviation as correct—both groups reached 100% accuracy, confirming that this temporal task was well understood regardless of gender.

In terms of efficiency, men completed the task with a median response time of 91 seconds, while women required 109 seconds. However, this difference was not statistically significant according to the Mann–Whitney U test. This result fits into the broader pattern observed throughout variant A: **women tended to be slightly more accurate, while men often completed tasks faster**, though differences rarely reached statistical significance. This reinforces the earlier observation that speed and accuracy do not always align.

#### 5.3.1.2 Map Variant B

In map variant B, which used pictorial-motion symbols to represent earthquake magnitude, the first task required participants to **find and count all locations with the strongest earthquakes**. The performance differences between men and women were substantial. Only 9.1% of male participants identified the correct number, while female participants reached 75.0% accuracy. However, because each group was small, neither the Pearson chi-square test nor Fisher's exact test detected statistically significant differences. When using the more inclusive scoring approach—accepting answers within one class of deviation—men achieved 54.5% accuracy, whereas women maintained a high level of performance at 91.7%. Still, the statistical tests did not indicate significant group differences.

Efficiency results followed the familiar pattern observed in earlier tasks: men responded faster, with a median response time of 106 seconds, compared to 145 seconds for women. Yet this time difference was not statistically significant. Overall, variant B again reflected a common trend in the experiment: **women tended to be more accurate, while men tended to respond more quickly**, reinforcing the recurring tension between speed and accuracy seen across multiple tasks.

The second map-based task in variant B required participants to **find and count all earthquake locations with the highest number of fatalities**. Unlike the first task—where women clearly outperformed men—performance here was more balanced. Male participants achieved 66.7% accuracy, while female participants reached 50.0%. When using the more inclusive scoring method (accepting answers within one class deviation), both groups performed equally well at 83.3%, and statistical tests again confirmed no significant differences.

Efficiency was also similar across genders. Men completed the task with a median response time of 56 seconds, while women required 64 seconds. This small difference was not statistically significant. As in earlier tasks, **men tended to be slightly faster, but this did not translate into meaningfully higher or lower accuracy**.

Continuing with map variant B, the pattern of comparable accuracy but differing response speed between genders became even more apparent in the third task.

This task required participants to **identify the most frequently occurring earthquake depth** within the displayed spatiotemporal range. Both men and women performed equally well, each achieving 83.3% accuracy, and when the more inclusive scoring method was applied, performance rose to 91.7% for men and 100.0% for women. As in the previous tasks, these differences were not statistically significant.

Where the groups differed was in efficiency. Men completed the task far faster, with a median response time of 35 seconds, while women required 73 seconds. This difference was statistically significant (Mann–Whitney U test: $p=0.020$, $z=2.28$). Once again, the emerging pattern mirrors earlier results: **men tended to respond more quickly, but their speed did not consistently translate into higher accuracy**, whereas women responded more slowly but generally maintained stable performance.

This pattern of balanced effectiveness but differing response speed continued into the final task of map variant B, which required participants to **identify the year with the highest number of earthquakes**. Both men and women again achieved high effectiveness, scoring 81.8% and 91.7%, respectively. When the more inclusive scoring method was applied, the results remained unchanged, and neither the Pearson chi-square test nor Fisher's exact test revealed statistically significant differences between the groups.

What did differ—this time reversing the earlier trend—was efficiency. **Women completed the task substantially faster**, with a median response time of 77 seconds, while men required 121 seconds. This difference was statistically significant (Mann–Whitney U test: $p=0.032$, $z=2.12$). Notably, this is the only task in variant B where **women outperformed men in terms of speed**, showing that the speed–accuracy pattern between genders is not fixed but varies depending on the task and visual encoding method.

#### 5.3.1.3 Map Variant C

Continuing into **map variant C**—where pulsation served as the dynamic visual cue—the first task required participants to **find and count all earthquake locations with the highest magnitude**. Once again, gender differences emerged in effectiveness, but without reaching statistical significance. Women outperformed men, with 42.8% correctly identifying the number of strongest earthquakes compared to 16.7% of men. When the more inclusive scoring criterion was applied, this advantage persisted, as 71.4% of women provided accurate responses compared to 41.7% of men. Yet, despite these numerical differences, neither Pearson's chi-square test nor Fisher's exact test indicated statistical significance.

In terms of efficiency, however, the groups performed almost identically. The median response time was 80 seconds for men and 76 seconds for women, and the Mann–Whitney U test confirmed that this difference was not statistically significant. This task, therefore, continues the broader pattern seen across earlier map variants: **women tend to be more accurate** in magnitude-based searches.

The second task in variant C—this time focused on **counting earthquake locations with the highest number of fatalities**—showed a similar pattern. **Women again outperformed men in overall accuracy**, with 42.8% answering correctly compared to 33.3% of men. When applying the more inclusive criterion that accepts responses within one class deviation, both groups improved substantially, reaching 83.3% accuracy for men and 85.7% for women. Despite these differences, neither Pearson's chi-square nor Fisher's exact test indicated statistical significance, suggesting that the observed gaps may reflect individual variation rather than systematic group differences.

Efficiency results further reinforced this interpretation. Both groups completed the task at nearly the same speed, with median response times of 50 seconds for men and 51 seconds for women. The Mann–Whitney U test confirmed the absence of statistical significance. Taken together, these findings mirror the broader trend observed across earlier tasks: **women often achieve higher accuracy in these search-and-count tasks**, yet efficiency remains virtually identical across genders, leading to no meaningful statistical distinctions.

The third task in variant C—**identifying the most frequently occurring earthquake depth**—produced high accuracy levels across both groups, reflecting the relative simplicity of this task. Men achieved an accuracy rate of 91.7%, while women scored 87.5%, and once the more inclusive approach was applied, both groups reached 100% correctness. This mirrors results from earlier variants, where depth identification consistently emerged as the easiest task. Efficiency differences were modest. Men completed the task slightly faster, with a median response time of 26 seconds, compared to 36 seconds for women. Still, the Mann–Whitney U test did not confirm this as a statistically significant difference. As seen in previous tasks, **higher speed did not systematically correspond to higher accuracy**, and once again, the gender groups performed at a similar overall level despite small variations in timing.

The final task in variant C—**identifying the year with the most earthquakes**—again reflected the wider pattern seen throughout this map variant: **women consistently outperformed men in effectiveness**, even when the symbolic encoding required temporal interpretation. Women achieved an accuracy rate of 87.5%, while men reached 58.3%. When the more inclusive criterion was applied, women's effectiveness remained unchanged, and men's performance rose modestly to 66.7%. Still, these differences did not reach statistical significance according to the Pearson chi-square and Fisher's exact tests.

In terms of efficiency, men completed the task more quickly, with a median response time of 55 seconds, compared to 86 seconds for women. Yet, as in many of the previous tasks, this speed difference did not translate into a statistically significant result. This continues the broader trend across the experiment: **gender-related differences in performance appeared numerically but rarely reached statistical significance**, and faster responses did not consistently align with higher accuracy.

#### 5.3.1.4 Map Variant D

As the analysis moved from dynamic symbol designs to the static reference solution, the overall pattern observed across previous variants remained noticeable. **Women continued to outperform men in effectiveness, while both groups showed comparable efficiency**. In map variant D—the traditional, nonanimated baseline—participants completed the same task of **finding and counting all earthquake locations with the highest magnitude**. Here, 25.0% of men and 38.5% of women identified the correct number of earthquakes. When the more inclusive criterion was applied, allowing answers within one class deviation, women again maintained higher effectiveness (76.9%) compared to men (56.2%). As in earlier results, however, these differences did not reach statistical significance according to the Pearson chi-square test and Fisher's exact test.

In terms of efficiency, **both groups performed at a similar pace**, with median response times of 68 seconds for men and 61 seconds for women. The Mann–Whitney U test confirmed that this small difference was not statistically significant, reinforcing the broader observation that while women were often more accurate, speed differences between groups were minimal and not statistically meaningful.

A similar pattern of small but nonsignificant gender differences appeared in the next task. In the second task for map variant D, participants were asked to **find and count all earthquake locations with the highest number of fatalities**. Unlike previous tasks—where women typically showed higher accuracy—men outperformed women in this case, with effectiveness scores of 50.0% compared to 28.6%. When the more inclusive ±1 deviation criterion was applied, the gap narrowed substantially: men reached 68.8%, and women 64.3%, again with no statistically significant differences according to the Pearson chi-square and Fisher's exact tests.

Efficiency followed the same trend observed earlier. Both groups responded at nearly identical speeds, with men requiring 48 seconds and women 47 seconds to complete the task. The Mann–Whitney U test confirmed that this minimal difference was not statistically significant, reflecting once more that while **accuracy sometimes varied between genders, their task completion times remained consistently similar** across the static and dynamic mapping solutions.

The third task in variant D continued this pattern of generally strong performance from both groups. In this task, participants were asked to **identify the most frequently occurring earthquake depth**—a task that had already proven to be one of the easiest across all map variants. **Both men and women performed very well**, with effectiveness scores of 80.0% and 85.7%, respectively. When the more inclusive ±1 deviation criterion was applied, effectiveness rose to 93.3% for men and 92.8% for women. As in the earlier tasks, these small differences did not reach statistical significance based on either the Pearson chi-square test or Fisher's exact test.

Efficiency results mirrored this closeness in performance. Median response times were highly comparable: 45 seconds for men and 47 seconds for women. The Mann–Whitney U test showed no statistical significance, confirming that **both genders completed this task at nearly the same pace**.

The final task in variant D asked participants to **identify the year with the highest number of earthquakes**. As in earlier temporal tasks, both groups performed very well, with 87.5% of men and 92.8% of women correctly selecting the target year. When the more inclusive ±1 deviation criterion was applied, the effectiveness levels remained unchanged, reinforcing that both groups handled this temporal search task with similar accuracy. Neither the Pearson chi-square test nor Fisher's exact test indicated any statistically significant difference between men and women.

Efficiency results again reflected this general similarity. Although the median response time was somewhat shorter for men (73 seconds) than for women (94 seconds), the Mann–Whitney U test showed that this difference was not statistically significant. Together, the results from all four tasks in variant D suggest that **static point symbols support comparable performance across genders**.

Taken together, the effectiveness results reveal a consistent—though not statistically confirmed—pattern: **women generally performed better than men across the majority of tasks** (Figure 5.9). This trend was especially visible in tasks involving the identification of the strongest earthquakes, where the gap between groups was the largest. Compared to the online survey, where differences between genders were minimal, the laboratory setting brought these contrasts into sharper focus. At the same time, it is important to emphasize that **overall effectiveness was much higher in both groups during the eye-tracking experiment**, likely due to better focus, controlled conditions, and the absence of external distractions. Nonetheless, no task showed statistically significant gender differences, meaning that observed patterns should be interpreted as tendencies rather than definitive group distinctions.

When examining efficiency, the picture becomes more balanced. Response times for men and women were largely comparable, with only a few exceptions reaching statistical significance (Figure 5.10). The most notable case occurred in variant B's depth-identification task, where men responded significantly faster. However, this speed advantage consistently aligned with a recurring pattern observed throughout both studies: **participants who responded faster were not necessarily more accurate**, especially in more demanding tasks. Interestingly, variant C—based on pulsating symbols—proved to be the most efficient dynamic method in the laboratory experiment, contrasting with the online questionnaire, where variant B led in efficiency. Overall, **the controlled laboratory environment contributed to shorter response times across all variants**, particularly in variant A, where efficiency improved dramatically compared to the online setting.

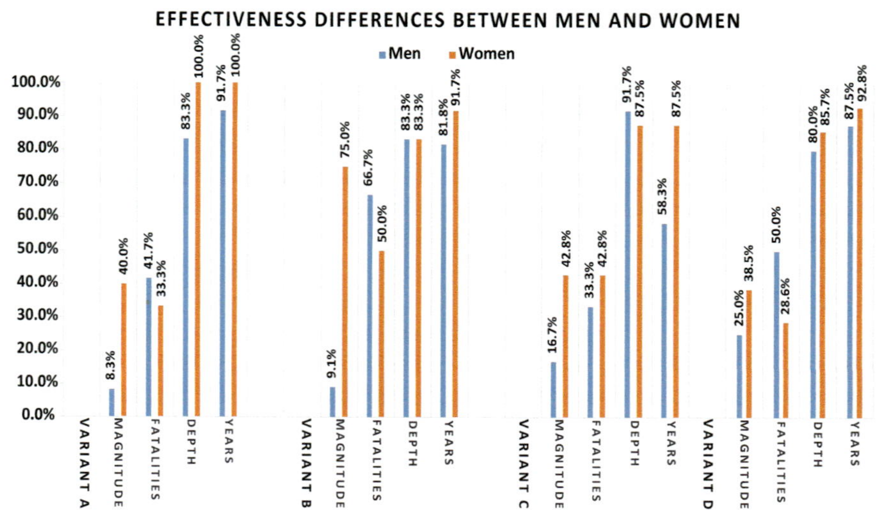


**FIGURE 5.9** Bar chart illustrating the differences in effectiveness between men and women across various map variants in specific map-based tasks under the eye-tracking laboratory conditions.

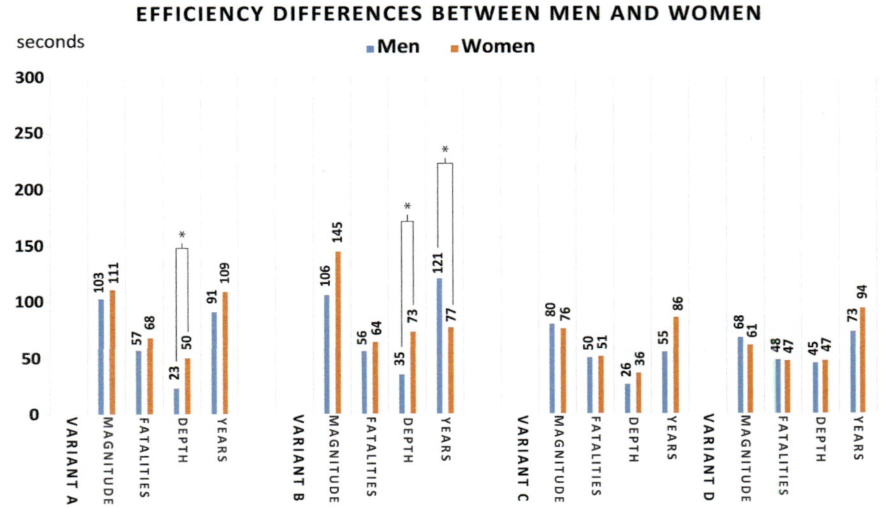


**FIGURE 5.10** Bar chart illustrating the differences in efficiency between men and women across various map variants in specific map-based tasks under the eye-tracking laboratory conditions. (*denotes statistical significance at $p < 0.05$.)

These findings reinforce that while gender differences may appear in specific tasks or symbol types, **they do not substantially affect overall performance**, and both men and women demonstrated high levels of accuracy and comparable speed in interpreting dynamic and static map symbols in the eye-tracking experiment.

### 5.3.2 How Confident Participants Felt About Their Answers

Because the laboratory experiment involved a much smaller group of participants than the online survey, the analysis of response confidence needed to be simplified. To ensure meaningful comparisons, confidence levels were grouped into broader categories:

- Confident (combining "very sure" and "sure")
- Neutral
- Not confident (combining "unsure" and "very unsure")

For this part of the study, we need to focus primarily on how these two broader confidence groups aligned with effectiveness (how accurately participants performed), rather than calculating statistical significance for every comparison. This approach provides a clear, easy-to-read picture of how participants' feelings about their answers related to their actual performance under the laboratory conditions.

#### 5.3.2.1 Map Variant A

In the first task of **map variant A**, where participants had to **locate and count all earthquakes with the strongest magnitude**, the relationship between confidence and performance turned out to be far from straightforward. About one-third of participants (33.3%) rated themselves as *very sure* or *sure*, and within this confident group, 44.4% gave the correct answer. A smaller group (18.5%) felt *neutral* about their response, and only 20.0% of them were correct. Nearly half of the participants (48.1%) fell into the *unsure* or *very unsure* category, and their accuracy was the lowest, with only 15.4% identifying the correct number of locations. However, when we apply a more forgiving interpretation of accuracy—where answers within one class deviation are also treated as correct—the picture shifts. Under this broader criterion, two-thirds (66.7%) of the confident group answered correctly, 60.0% of the neutral group did so, and even the unsure group performed surprisingly well, with 84.6% classified as correct. This again reflects a recurring pattern observed earlier: **being less confident does not necessarily mean performing worse**, and in some cases, less confident participants even matched or surpassed the more confident ones.

When it comes to efficiency, the confident group completed the task with a median response time of 81 seconds, while neutrally confident participants took considerably longer at 149 seconds. Those who felt *unsure* or *very unsure* fell between these two groups, with a median of 111 seconds. Despite these differences in timing, the Mann–Whitney U test confirmed that none of the differences in response time were statistically significant, meaning that confidence levels did not reliably predict how quickly participants finished the task.

In the second task for map variant A, where participants had to **identify and count all earthquake locations with the highest number of fatalities**, confidence again played an interesting but not entirely consistent role. Nearly half of the participants (48.1%) described themselves as *very sure* or *sure*, yet only 38.5% of this confident group provided the correct answer. When the more inclusive scoring method was applied, however, their success rate rose sharply to 84.6%, suggesting that confident participants were often close to the correct solution even if not exact.

A large portion of the sample (40.7%) reported *neutral* confidence. Their accuracy was similar to that of the highly confident group: 36.4% answered correctly, and 72.7% were considered correct when including responses within one class deviation. Once again, this indicates that neutral confidence did not translate into clearly better or worse performance. The smallest group—participants who were *unsure* or *very unsure*—made up just 11.2% of the sample. Only 33.3% of them provided the correct answer, and interestingly, this percentage did not improve at all when the inclusive approach was applied. This suggests that uncertain participants tended to make larger, less precise errors on this particular task.

When examining response times, the differences between confidence groups were modest. The highly confident participants completed the task with a median of 66 seconds, those with neutral confidence were slightly faster at 51 seconds, and the unsure group required 71 seconds. Despite these variations, the Mann–Whitney U test confirmed that none of the differences in efficiency reached statistical significance, meaning that **confidence levels again did not reliably predict how quickly or effectively participants performed the task**.

The third task in map variant A, which required **identifying the most frequently occurring earthquake depth**, showed a noticeably clearer separation between confidence groups than the previous tasks. A large majority of participants (88.9%) described themselves as *very sure* or *sure* of their answer, and this confidence closely aligned with performance: 95.8% of these participants identified the correct depth. Under the inclusive approach, all of them (100%) were considered correct. In contrast, the much smaller group of *neutral* participants (11.1%) achieved a considerably lower accuracy of 66.7%—a meaningful difference from the confident group, even though the small sample size prevented statistical significance. Still, when the inclusive approach was applied, even this group reached 100% correctness.

Efficiency results added another layer to the interpretation. The confident participants completed the task in 44 seconds, while neutrally confident participants finished more than twice as fast, needing only 20 seconds. This difference was statistically significant (Mann–Whitney U test: $p=0.034$, $z=2.12$). This means that in this particular task, **greater confidence and greater accuracy were accompanied by longer response times**, highlighting that the relationship between confidence, accuracy, and speed can vary depending on the type of map-based task.

The final map-based task in variant A asked participants to **identify the year in which the most earthquakes occurred**. The largest group—those who felt *very sure* or *sure* of their response—represented 63.0% of all participants, and nearly all of them (94.1%) correctly identified the target year. Using the more inclusive approach did not alter this outcome. Participants who reported feeling *neutral* about their answer made up 25.9% of the sample, and every one of them correctly solved the task. Interestingly, the smallest group—those who described themselves as *unsure* or *very unsure* (11.1%)—also achieved a 100% accuracy rate, despite their low confidence.

In terms of efficiency, the *very sure/sure* group required the most time (117 seconds), followed by the *unsure/very unsure* group (90 seconds). The *neutral* participants were the fastest, completing the task in just 70 seconds. As in the previous tasks, these time differences did not reach statistical significance. This pattern again highlights a key observation from variant A: **high accuracy was achieved across all confidence levels**, and longer response times did not necessarily lead to more accurate answers.

### 5.3.2.2 Map Variant B

In **map variant B**, the first task—**identifying and counting all locations of the strongest earthquake**—showed a confidence–performance pattern that was broadly similar to variant A, yet with some notable differences. Participants who felt *very sure* or *sure* constituted the largest group (45.8%), and 45.4% of them provided the correct answer. When the more inclusive approach was applied, their effectiveness rose to 81.8%, closely mirroring the improvement seen in variant A's high-confidence group. Participants who felt *neutral* represented 25.0% of the sample, and only one-third (33.3%) identified the correct number of earthquakes; this increased to 66.7% when $\pm 1$ deviations were accepted. Interestingly, the *unsure* and *very unsure* group—also 25.0% of the sample—performed slightly better than the neutral group, with 50.0% answering correctly and 66.7% meeting the inclusive criterion. This contrasts with variant A, where low-confidence participants generally scored lower than those who felt neutral or sure.

Response times also revealed a clearer divergence than in variant A. The *very sure/sure* group completed the task the fastest, with a median of 117 seconds.

The *neutral* group required slightly more time (123 seconds), while the *unsure/very unsure* participants were the slowest, taking 154 seconds on average. Unlike in variant A—where confidence did not reliably predict speed—variant B showed a statistically significant **difference in efficiency between high- and low-confidence groups** ($p=0.027$, $z=2.16$). Overall, while the confidence–accuracy relationship remained somewhat inconsistent—as seen throughout both variants—variant B exhibited a clearer link between confidence and response speed, **with lower-confidence participants working more slowly without achieving higher accuracy**.

The second task in map variant B—**finding and counting all earthquake locations with the highest number of fatalities**—continued the overall pattern observed in the first task, but with a more balanced relationship between confidence and accuracy. Participants who reported being *very sure* or *sure* formed the largest group (62.5%), and two-thirds of them (66.7%) identified the correct number of locations. When the more inclusive criterion was applied, their success rate rose to 80.0%, closely mirroring the improvement seen in the previous task. The *neutral-confidence* group was much smaller (16.7%), yet half of these participants answered correctly—and notably, all of them were considered correct when the inclusive approach was used. This contrasts with the first task in variant B, where neutral confidence was associated with lower performance. Meanwhile, participants who felt *unsure* or *very unsure* accounted for 20.8% of the sample, with 40.0% answering correctly. Under the inclusive scoring, however, even this low-confidence group reached 80.0% effectiveness, matching the performance of the high-confidence participants.

In terms of efficiency, the results were more uniform than in the previous task. Both the high-confidence and low-confidence groups completed the task in essentially the same median time (61 seconds), suggesting that confidence did not influence speed as strongly here as it did in the first task. Neutral participants were slower, taking a median of 80 seconds, but the difference was not statistically significant. Overall, for this task, **confidence levels did not translate into clear differences in either accuracy or efficiency**, reinforcing the recurring observation that confidence is only loosely—if at all—connected to actual performance in these map-based search tasks.

The third task in map variant B—**identifying the most common earthquake depth**—showed a clearer relationship between confidence and effectiveness than the previous two tasks. A large majority of participants (70.8%) felt *very sure* or *sure* about their answer, and nearly all of them (94.1%) correctly identified the most frequent depth. Unlike other tasks, the inclusive scoring did not improve this group's performance, as almost everyone in it had already provided the correct value. A smaller portion of participants (20.8%) reported *neutral* confidence, and their performance was also strong: 80.0% correctly indicated the most common depth. Under the inclusive approach, this group achieved a perfect result, suggesting that even when participants were not entirely confident, they were still capable of identifying the correct visual cue. The *unsure* and *very unsure* participants represented only 8.3% of the sample and were the least effective group, with none of them selecting the correct answer outright. However, the inclusive analysis showed that all of them were off by just one class, meaning their responses were still close to the correct value. This underscores an important pattern observed throughout the study: low confidence did not necessarily lead to completely inaccurate results—participants often came close even when uncertain.

In terms of efficiency, the high-confidence group once again provided the fastest responses, with a median time of 46 seconds. Surprisingly, the *unsure* and *very unsure* group was the second fastest at 50 seconds, while *neutrally* confident participants required the most time (74 seconds). Despite these differences, none reached statistical significance. This task reinforces a recurring theme in the eye-tracking experiment: **confidence and accuracy often align, but response time does not always follow the same pattern**—quick answers can be correct, and slower responses do not necessarily yield better outcomes.

The final task in map variant B asked participants to **identify the year in which the highest number of earthquakes occurred**. As in the earlier tasks, confidence played a notable role in how participants approached the problem. The largest portion of respondents—62.5%—reported feeling *very sure* or *sure* about their answer, and their performance reflected this confidence: an impressive 93.3% correctly identified the target year. The inclusive scoring did not further improve this result, as nearly all answers were already fully correct. In contrast, only 16.7% of participants reported *neutral* confidence, and this group struggled with the task: just 33.3% provided the correct answer, and even under the inclusive approach, their effectiveness did not improve. This pattern stands out because, in previous tasks, *neutral*-confidence participants often achieved middling or reasonably strong results. Here, however, they appeared less able to interpret the temporal pattern shown on the map. Interestingly, the *unsure* and *very unsure* participants—who made up 20.8% of the sample—performed surprisingly well, with all of them correctly identifying the target year. This unexpected outcome is consistent with a broader trend in the study: **low confidence does not always predict poor performance**. In several tasks, unsure participants either matched or exceeded the accuracy of those who felt more certain, suggesting that participants sometimes underestimated their own success.

The efficiency results further underline this complexity. Participants who felt *very sure* or *sure* took the longest to complete the task, with a median response time of 123 seconds. The quickest responses came from the *neutrally* confident group, who averaged just 71 seconds, while the *unsure* and *very unsure* group took 88 seconds. Despite these differences, none were statistically significant, reinforcing that response speed did not reliably track with confidence or accuracy. Once again, this task illustrates the overarching theme of the eye-tracking study: **confidence, effectiveness, and efficiency interact in nuanced ways**. High confidence was often—but not always—aligned with accuracy, while faster responses did not guarantee correctness, and slow responses were not necessarily a sign of careful or successful reasoning.

#### 5.3.2.3 Map Variant C

In **map variant C**, where earthquakes were represented using pulsating point symbols, the first task asked participants to **locate and count all earthquakes with the highest magnitude**. Compared to variants A and B, confidence and performance were more weakly aligned here—a pattern that became characteristic of variant C.

More than half of the participants (55.0%) described themselves as *very sure* or *sure* about their answers. Yet, despite this confidence, fewer than one-third (27.3%)

correctly identified the number of earthquakes. When the inclusive scoring approach was applied, their accuracy improved to 54.5%, but the result still lagged behind the performance seen in other map variants for the same task. This mismatch between confidence and accuracy reflects a broader issue observed in variant C: pulsation appeared to make magnitude changes harder to judge precisely. Participants who felt *neutral* about their answers made up 15.0% of the sample. Their effectiveness was slightly higher, with 33.3% correctly identifying the number of earthquakes. Interestingly, when using the inclusive approach, every participant in this group was considered correct—meaning all errors were small, deviating by just one class. This suggests that although neutrally confident participants underestimated their performance, they were closer to the correct answer than they realized. The *unsure* and *very unsure* group comprised a relatively large portion of respondents (30.0%), and they performed the least accurately: only 16.7% answered correctly. Even under the inclusive approach, their accuracy rose only to 33.3%. This reinforces the idea that, for this particular task, pulsating symbols may have hindered precise search and counting, especially for participants who already felt uncertain.

Efficiency differences among the three groups were small. *Very sure* and *sure* participants completed the task in a median of 83 seconds, the *neutral* group in 94 seconds, and the *unsure* and *very unsure* participants in 75 seconds. None of these differences were statistically significant, and unlike in variant B—where response time increased clearly with lower confidence—variant C showed no clear relationship between confidence and speed. Overall, this task revealed one of the clearest limitations of the pulsation-based design. **Confidence and performance were loosely connected, accuracy was relatively low across all confidence levels, and response times did not help explain the difficulties participants experienced**. This stands in contrast to variants A and B, where confidence groups showed more predictable behavior.

The second task in variant C—**finding and counting earthquake locations with the highest number of fatalities**—revealed a similar pattern of mismatched confidence and accuracy, though with a surprising twist compared to the magnitude task.

A large majority of participants (70.0%) reported feeling *very sure* or *sure* of their answers. Yet, despite this confidence, only about one-third of them (35.7%) identified the correct number of fatal earthquakes. When applying the inclusive scoring approach, however, the picture improved significantly: most of their errors were small, and the effectiveness of this group rose sharply to 85.7%. This suggests that participants had a general sense of where fatal earthquakes were located but struggled with exact counting when using pulsating symbols. In contrast, none of the participants declared *neutral* confidence. Instead, the remaining 30.0% fell into the *unsure* or *very unsure* group—and their performance turned out to be unexpectedly strong. A striking 83.3% of them correctly identified the number of locations with the most fatalities, and this result remained unchanged even after applying the inclusive approach. In other words, the least confident participants actually performed the best, further reinforcing the idea that pulsation may have distorted participants' ability to assess their own accuracy.

Efficiency patterns did not help explain these results. The median response time for the *very sure* and *sure* group was 50 seconds, while the *unsure* and *very unsure* group completed the task in 53 seconds—almost identical. This lack of difference, combined with the strong performance of the unsure group, supports a recurring observation in variant C: **pulsating symbols made confidence a poor predictor of performance**.

The third task in variant C—**identifying the most common earthquake depth**—stood out as one of the easiest tasks across the entire experiment, and this was reflected in both confidence levels and accuracy.

As in previous tasks, most participants (75.0%) reported being *very sure* or *sure* of their answers, and this confidence aligned closely with their performance: nearly all of them (93.3%) correctly identified the most frequent earthquake depth. When applying the inclusive approach, which counts near-misses as correct, this group reached a perfect 100% accuracy. A smaller group (10.0%) reported *neutral* confidence. Only half of them answered correctly at first, but closer inspection showed that all errors were minor—within one class—so their accuracy also rose to 100% under the inclusive approach. Interestingly, the least confident participants—those who were *unsure* or *very unsure* (15.0%)—performed perfectly from the start, with 100% correct responses even before applying the inclusive criterion.

Efficiency results once again demonstrated that confidence did not predict response time. The *very sure* and *sure* group completed the task in 36 seconds on average, while the *neutral* group took noticeably longer at 59 seconds. The *unsure* and *very unsure* group, despite lower confidence, were the fastest overall, completing the task in just 24 seconds. However, the differences among these groups were not statistically significant. Taken together, this task continued a pattern already seen in variant C: pulsating symbols made the task easy to solve, but **participants' confidence did not reliably reflect either their accuracy or the time they spent completing the task**.

The final task in map variant C required participants to **identify the year in which the most earthquakes occurred**. As in earlier tasks with pulsating symbols, confidence levels varied, yet performance remained generally strong. Most participants (60.0%) reported being *very sure* or *sure* of their answer. Two-thirds of this group (66.7%) correctly identified the target year, and the inclusive approach—counting answers within one class deviation as correct—raised their effectiveness modestly to 75.0%. The second-largest group consisted of *neutrally* confident participants (30.0%), who performed even better: 83.3% of them answered correctly, and none improved under the inclusive scoring because their few errors were larger than ±1. The smallest group—participants who felt *unsure* or *very unsure*—made up 10.0% of all respondents. Only half of them (50.0%) identified the correct year, and those who missed it did so by two or more classes, meaning the inclusive approach offered no additional improvement.

Efficiency results again reflected a recurring pattern from variant C: **lower confidence did not necessarily lead to slower responses**. The *very sure* and *sure* group completed the task in 73 seconds on average, *neutrally* confident participants did so more quickly at 57 seconds, and the *unsure* or *very unsure* group was the fastest overall, finishing in just 44 seconds. However, none of these differences were statistically significant.

#### 5.3.2.4 Map Variant D

In **map variant D**—the static reference map—the first task required participants to **find and count all earthquake locations with the highest magnitude**. Confidence levels were quite distributed across the sample, and effectiveness varied accordingly.

A smaller portion of participants (23.3%) reported being *very sure* or *sure* of their answer, yet less than one-third of them (28.6%) counted the locations correctly. Still, most of their errors were minor, and once the inclusive scoring approach was applied, their effectiveness increased substantially to 71.4%. The *neutrally* confident group accounted for 26.7% of all respondents. Only a quarter of them (25.0%) answered correctly, but—similar to the high-confidence group—most of their mistakes were close to the correct value, raising their accuracy to 62.5% under the inclusive criterion. Interestingly, the largest portion of participants (50.0%) reported being *unsure* or *very unsure*. Despite their low confidence, 40.0% of them successfully identified the correct number of locations, and the inclusive approach raised this to 66.7%. As observed in previous map variants, **confidence was not a reliable predictor of accuracy**, and low-confidence participants often performed comparably to, or even slightly better than, more confident groups once small deviations were taken into account.

Efficiency results followed a similar pattern. The *very sure* and *sure* group completed the task in 58 seconds on average, while the *neutral* group required noticeably more time—94 seconds. The *unsure* and *very unsure* group fell between these extremes at 61 seconds. However, despite these differences, the Mann–Whitney U test showed no statistically significant variation in response times. Overall, this task reinforces the recurring insight from earlier map variants: **static symbols did not necessarily make confident participants more accurate, nor did low confidence translate into poor performance**.

The second task in map variant D asked participants to **find and count all earthquake locations associated with the highest number of fatalities**. Confidence levels again varied widely, and—as in earlier tasks—the relationship between confidence and performance was far from straightforward.

A majority of participants (60.0%) reported being *very sure* or *sure* of their answers. Within this group, 44.4% provided the correct count, and most remaining errors were small. Once the inclusive scoring method was applied, their effectiveness increased notably to 77.8%. A much smaller portion of participants (16.7%) reported *neutral* confidence, yet this group performed surprisingly well, with 60.0% answering correctly. Because their incorrect responses deviated only slightly from the correct value, they showed no improvement under the inclusive approach. The *unsure* and *very unsure* participants accounted for 23.3% of the sample, and—as expected—their baseline effectiveness was the lowest, at 14.3%. Under the inclusive scoring approach, their accuracy increased to 28.6%, but still remained well below the levels observed in the more confident groups. This pattern reflects the general trend observed throughout the study: very low confidence is occasionally associated with weaker performance, although not consistently across all map variants.

Efficiency results add another layer to this pattern. Despite their low effectiveness, the *unsure* and *very unsure* group was the fastest, completing the task in just 36 seconds on average. The *very sure* and *sure* group needed more time—53 seconds—while the *neutral* group took 41 seconds. Yet, as in the previous task, these time differences were not statistically significant, indicating that **response speed did not reliably reflect confidence or accuracy**. Overall, this task again highlights that confidence alone is not a dependable predictor of performance, and that quicker responses, particularly from uncertain participants, do not necessarily correspond to better accuracy.

In the third map-based task of variant D—**identifying the most common earthquake depth**—participants again demonstrated generally high accuracy. A large majority (80.0%) described themselves as *very sure* or *sure*, and most of them (87.5%) correctly selected the most frequent depth category. Under the inclusive scoring approach, which treats answers off by one class as correct, their effectiveness rose to nearly perfect (91.7%). A small proportion of participants (10.0%) reported *neutral* confidence, and just one-third of them (33.3%) answered correctly. However, the inclusive approach showed that all their errors were minor, raising their effectiveness to 100%. The remaining 10.0% of participants—those who were *unsure* or *very unsure*—performed unexpectedly well: every participant in this group (100%) correctly identified the most common depth.

When considering response times, all confidence groups showed similar levels of efficiency. *Very sure* and *sure* participants completed the task in an average of 47 seconds, the *neutral* group in 38 seconds, and the *unsure* or *very unsure* participants in 52 seconds. As in earlier tasks, these small differences were not statistically significant, reinforcing a pattern observed repeatedly across the experiment: **confidence is not a reliable indicator of speed or accuracy**, and participants with low confidence can still deliver correct—and timely—responses.

The final task in map variant D asked participants to **identify the year in which the most earthquakes occurred**. Half of all participants (50.0%) reported feeling *very sure* or *sure* of their answer, and most of them (86.7%) selected the correct year. The more inclusive scoring approach did not change this result, indicating that their errors—when present—were not minor. A similarly large portion of the sample (43.3%) reported *neutral* confidence, and this group performed even slightly better: 92.3% correctly identified the target year, with no improvement under the inclusive approach. The smallest group, those who were *unsure* or *very unsure* (6.7%), also performed flawlessly, with all participants correctly indicating the year with the most earthquakes.

Response times revealed a more distinct pattern. *Neutral* participants were the fastest group, completing the task in 65 seconds on average, followed by the *unsure* and *very unsure* participants at 76 seconds. The *very sure* and *sure* group took considerably longer, averaging 105 seconds. This mirrors trends observed in earlier tasks, where higher confidence did not necessarily correspond to quicker responses. In this case, the difference between the sure and unsure groups was statistically significant, as confirmed by the Mann–Whitney U test ($p = 0.0034$, $z = 3.52$), reinforcing the broader finding that **greater confidence often coincided with longer search times—not greater efficiency**.

Overall, the effectiveness results (Figure 5.11) show that confidence levels varied slightly across the map variants, with variants A and B performing particularly

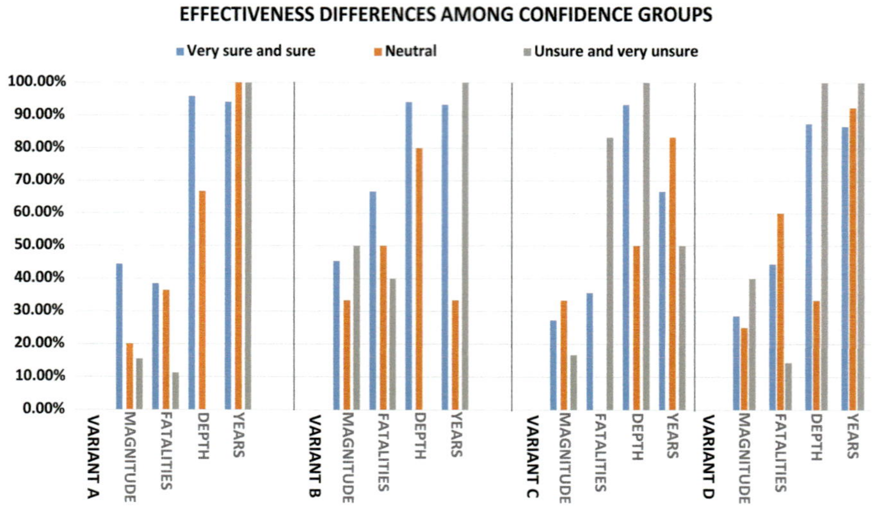


**FIGURE 5.11** Differences in effectiveness among self-reported confidence groups across map-based tasks and map variants in the laboratory eye-tracking procedure.

well among participants who felt very sure or sure of their answers, especially in the tasks that required counting earthquakes by magnitude or fatalities. However, the results also show that confidence was not always a reliable indicator of performance: in several tasks, participants who felt unsure performed just as well as those who felt confident. Importantly, dynamic point symbols (used in variants A, B, and C) did not reduce accuracy or confidence when compared with traditional static symbols (variant D). This suggests that **animated mapping techniques can be used in earthquake visualizations without negatively affecting how well users understand the data**.

Efficiency results (Figure 5.12) paint a slightly different picture. Variants A and B tended to be more effective, but they also required more time, especially for tasks involving motion interpretation, such as identifying the strongest earthquakes. In contrast, map variant C, which used pulsating symbols, proved nearly as fast as the static map in variant D. This is an important finding: it shows that symbol pulsation, a more subtle animation technique, can deliver efficiency levels comparable to traditional, nonanimated maps. In other words, **pulsating symbols offer the benefits of animation without slowing users down**.

### 5.3.3 How Participants Rated the Map Styles

Building on the relationships between confidence, effectiveness, and efficiency, it is also important to consider how participants felt about the maps they worked with. User preference is especially relevant in the context of animated symbols, since favorable aesthetics can increase engagement and support broader map usability.

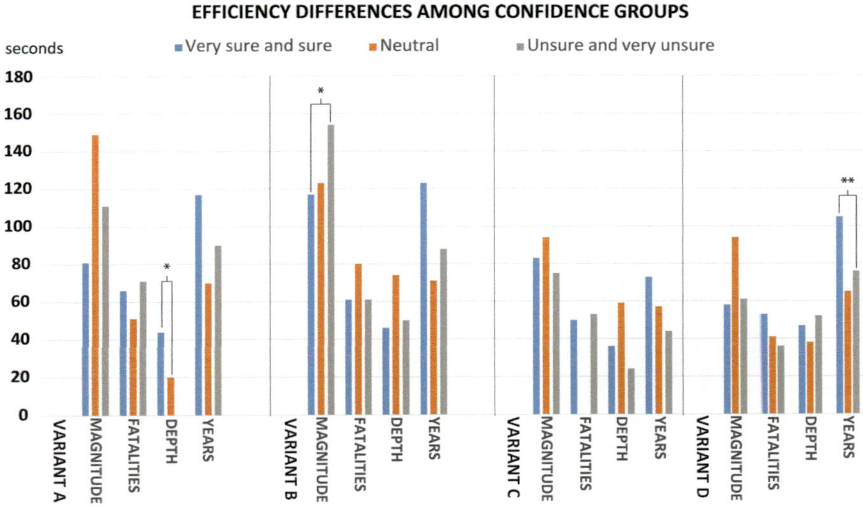


**FIGURE 5.12** Differences in efficiency among self-reported confidence groups across map-based tasks and map variants in the laboratory eye-tracking procedure. (*denotes statistical significance at $p < 0.05$; ** denotes statistical significance at $p < 0.005$.)

Therefore, the final part of this section focuses on how participants rated the visual style of each map variant.

In map variant A, which used vibrating symbols, two-thirds of participants (66.7%) stated that they *liked* or *very much liked* the map style. Only 18.5% *disliked* it, and none *strongly disliked* the design; 14.8% had *no opinion*. A similar pattern was observed in map variant B with pictorial motion: 65.2% rated the style *positively*, while 17.4% *disliked* it, and another 17.4% expressed *no opinion*. Map variant C, featuring pulsating symbols, received the most positive feedback, with 70.0% of participants rating it favorably and only 20.0% expressing negative views. The remaining 10.0% had *no opinion*.

In contrast, the static map variant D was the least appreciated. Only about half of the participants (51.7%) rated the style positively, while nearly one-third (31.0%) *disliked* it, and 17.2% expressed *no opinion*. Compared with the dynamic variants, this reflects a much more polarized perception of the classic static map design.

Overall, participants clearly preferred dynamic map styles over the static ones. Pulsating symbols (variant C) were the most well-received, followed by vibrating (A) and pictorial motion (B). The static variant (D) received the weakest evaluation, with the highest proportion of negative feedback. As shown in Figure 5.13, these results differ from the online questionnaire findings, where opinions were more evenly distributed and dominated by neutral responses. In the controlled laboratory setting, participants appeared more confident in expressing stronger preferences, particularly in favor of dynamic symbol designs.

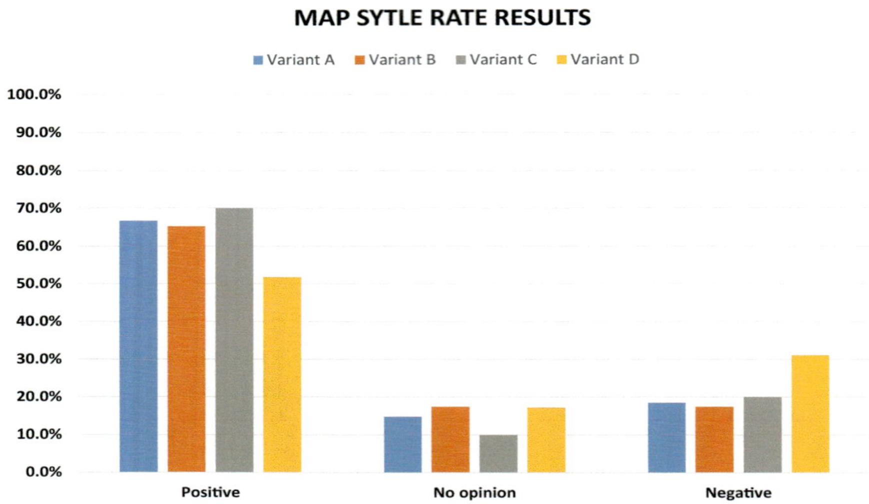


**FIGURE 5.13** Results of participants' ratings of the map styles presented to them and used for task completion in the eye-tracking laboratory experiment.

## 5.4 WHAT EYE MOVEMENTS REVEAL

Having explored how participants performed on the tasks—as well as how they felt about their answers and the map designs, the analysis now shifts to what is often the most revealing layer of a map-reading study: eye movements. While accuracy and response time show what participants achieved, eye-tracking helps explain how they arrived there.

Eye-tracking metrics provide a window into users' visual strategies, showing where attention is drawn, how long it lingers, and how efficiently it moves between elements on the map. In this study, most eye-movement indicators were analyzed using the median, since their distributions were typically non-normal—an expected characteristic in behavioral data. One exception is the total number of fixations, which the eye-tracking software exports as a fixed count per participant; therefore, both the mean and median were considered for this metric.

To understand how users interacted with the animated and static symbols, several key measures of visual behavior were analyzed:

- **Total number of fixations**—how many distinct gaze points were registered
- **Median fixation duration**—how long a participant typically focused on one spot
- **Saccadic amplitude (fixation distance)** —how far the eyes moved between fixations
- **Median saccadic velocity**—how quickly these eye movements occurred
- **Viewing time on the legend and temporal legend**—how often and how long participants needed to reference the explanations

Together, these metrics help uncover the deeper mechanics of how people process earthquake maps—whether dynamic symbols invite more exploration, whether static ones encourage quicker scanning, and how often participants rely on the legend when interpreting visual variables such as motion, size, color, or time. In the following subsections, these eye-movement patterns are examined in detail, revealing how different mapping techniques shape users' visual attention—and ultimately, their performance.

### 5.4.1 How Many Fixations Participants Made

Table 5.1 summarizes the fixation count results using averages, medians, and standard deviations. These values provide insight into how intensively participants visually searched each map, depending on both the task type and the symbol variant.

Across all variants, the task that required **identifying the year with the most earthquakes** consistently produced the highest number of fixations. This suggests that interpreting temporal information was more demanding than identifying spatial or attribute-based patterns. The most extensive visual search was observed in map variants A and B, both showing an average of 310.7 fixations, indicating that these symbol designs required considerable visual effort in temporal tasks.

In contrast, tasks involving **earthquake depth** produced the lowest fixation counts. Variant C showed the fewest fixations across the entire experiment (112.8 on average), suggesting that depth information encoded with pulsation was especially

**TABLE 5.1**
**Crosstabulation of Fixation Count Statistics According to Each Map Variant and Map-Based Task**

| Map Variant | Task | Average Fixations | Standard Deviation | Median Fixations |
|---|---|---|---|---|
| A | Magnitude | 289.52 | 105.84 | 300 |
| | Fatalities | 203.48 | 114.10 | 183 |
| | Depth | 149.07 | 99.44 | 150 |
| | Years | 310.65 | 134.02 | 330.5 |
| B | Magnitude | 327.26 | 121.65 | 327 |
| | Fatalities | 188.25 | 55.56 | 185.5 |
| | Depth | 157.13 | 82.93 | 134 |
| | Years | 310.65 | 154.78 | 280 |
| C | Magnitude | 231.32 | 81.72 | 228 |
| | Fatalities | 156.00 | 75.31 | 161 |
| | Depth | 112.75 | 66.66 | 90.5 |
| | Years | 241.80 | 137.45 | 205 |
| D | Magnitude | 231.83 | 111.23 | 197 |
| | Fatalities | 177.63 | 83.64 | 145 |
| | Depth | 146.79 | 74.66 | 142 |
| | Years | 300.93 | 173.88 | 259 |

easy to interpret. Variants A and D also produced relatively low fixation counts (149.1 and 146.8), confirming that color-based depth encoding generally required less visual exploration.

For the two counting tasks—**highest magnitude and most fatalities**—fixation counts fell between these extremes. Variant B produced the highest number of fixations for the magnitude task (327.3), while variant A recorded the highest for the fatalities task (203.5). These elevated values suggest that these specific symbol–task combinations required more detailed or effortful searching.

A clear trend emerged when comparing variants:

- **Variant C** consistently required fewer fixations, especially for the fatalities and depth tasks, indicating that its pulsating symbols may have supported more intuitive perception.
- **Variant B** often showed the highest fixation counts, particularly for magnitude and year tasks, suggesting that the pictorial-motion symbols demanded more cognitive processing.
- **Variant D** yielded moderate fixation counts overall, but showed the highest variability in the temporal task (SD = 173.9), meaning that some participants struggled much more than others.

Together, these results show that the number of fixations—and, therefore, visual effort—varies notably across dynamic symbol designs, with pulsation generally guiding viewers more efficiently and pictorial motion requiring more extensive search.

The statistical comparison of fixation counts conducted using the Mann–Whitney U test and summarized in Figure 5.14, provides additional insight into how symbol design influenced visual search behavior. Significant differences appeared across many task–variant combinations, showing that participants did not visually process all map designs in the same way. The most frequent significant contrasts emerged in the depth task and the year task, suggesting that **both color-based depth encoding and temporal interpretation elicited greater variability in visual effort than the magnitude or fatalities tasks**.

A particularly consistent pattern involved variant C, which showed numerous significant differences relative to the other variants. This indicates that **pulsating symbols had a distinct, measurable effect on where and how participants directed their gaze**. Interestingly, differences did not arise solely across variants—some also occurred within the same variant. For example, in variant C, the depth task differed significantly from the magnitude, fatalities, and year tasks, meaning that even when the map style stayed constant, the task type itself still shaped visual search behavior.

Taken together, these findings show that **dynamic symbol design does influence how participants explore a map**, but not uniformly across tasks. The tasks involving color (depth) and temporal reasoning (year) were especially sensitive to differences in symbol style, while tasks tied to motion-based or size-based attributes showed more stable visual behavior. In short, the visual-attention data reveal that dynamic point symbols—particularly pulsation—change how users search, scan, and process earthquake information, often in subtle but statistically meaningful ways.

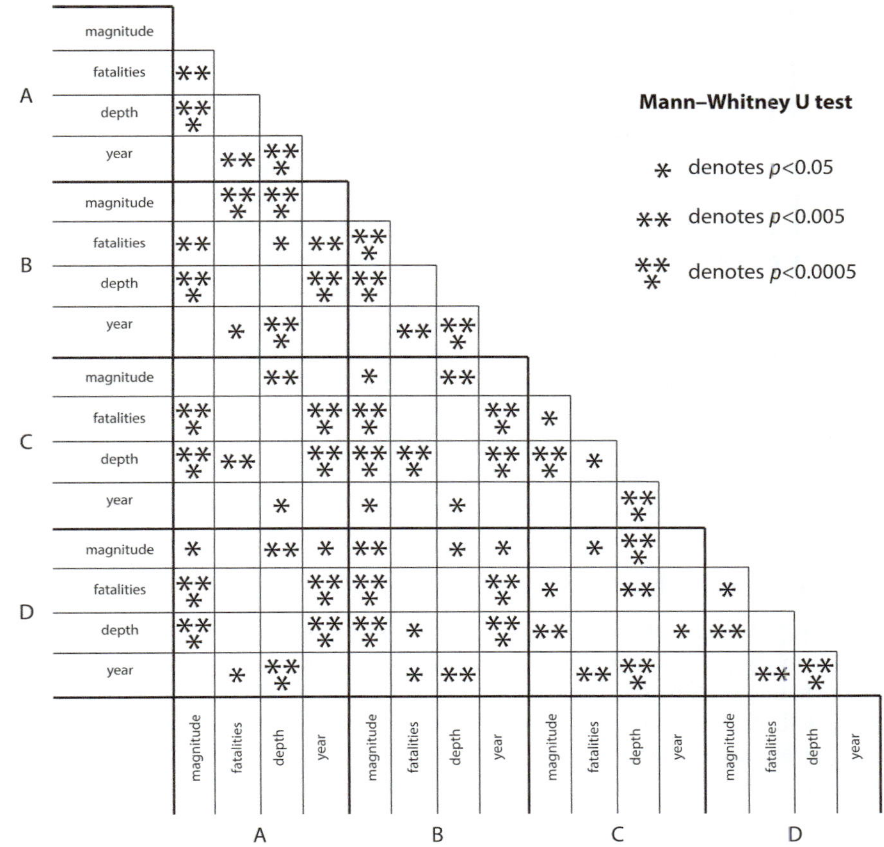


**FIGURE 5.14** Statistical comparison of fixation counts across four map variants (A: vibrating symbols, B: pictorial motion, C: pulsation, D: static symbols) and four search tasks (magnitude, fatalities, depth, year), analyzed using the Mann–Whitney U test. Each cell compares two combinations of map variant and task.

When these fixation patterns are compared with task effectiveness, several important relationships emerge. In **the magnitude task**, the high fixation count observed in variant C did not translate into strong performance. This suggests that the **pulsating symbols may have produced unnecessary visual effort**, possibly even mild cognitive overload—without improving users' ability to judge earthquake strength. In contrast, variant B's **pictorial motion appears to have guided attention more efficiently**, supporting quicker and more accurate comparisons.

A similar pattern appears in **the fatalities task**. Although fixation counts varied widely across variants, the differences in effectiveness were relatively small. This implies that **the task itself required substantial visual scanning regardless of symbol design**, although variant B again offered a slight performance advantage, likely because its motion-based encoding helped highlight extreme values.

In **the depth task**, fixation count varied, but effectiveness remained consistently high. This indicates that **color-based depth encoding was visually straightforward and easy to interpret**. Here, symbol animation mattered far less—the visual variable (color) carried the cognitive load effectively on its own. Additional fixations in some variants were likely redundant rather than helpful.

The most striking contrast appears in **the year-identification task**. Variant C not only produced the highest number of fixations but also the lowest task performance. This alignment strongly suggests that the **pulsating symbols disrupted temporal reasoning** by diverting attention away from the timeline or temporal cues. Meanwhile, both static symbols (variant D) and pictorial motion (variant B) appear to have supported more stable, structured visual exploration, helping participants focus on the timeline rather than the distracting dynamics of the points.

Taken together, these observations reinforce a broader pattern: **more fixations do not necessarily mean better performance**, and in some cases, too much visual activity on the map may direct attention away from crucial elements. Dynamic symbols can guide or misguide users depending on the task, highlighting the need to align symbol animation with the cognitive demands of the map-based task.

### 5.4.2 How Long Fixations Lasted

While fixation count tells us *how much* participants looked around the map, fixation duration reveals *how demanding* each moment of visual processing was. Across the three dynamic symbol variants (A–C), the median fixation duration stayed relatively stable, with no statistically significant differences between tasks or map variants. This consistency suggests that the presence of motion, whether vibration, pictorial animation, or pulsation, did not necessarily make each fixation cognitively heavier. In other words, **dynamic behavior changed *where* people looked, but not *how long* they needed to process what they saw**.

However, a different pattern emerged in **the static map variant** (D). Here, fixation durations were significantly shorter in several tasks, including the magnitude, fatalities, and depth tasks. Statistical testing confirmed these differences, indicating that processing static symbols required less cognitive effort per fixation than interpreting dynamic ones. This aligns well with earlier findings: **static maps are deeply familiar to users**, and without motion-driven distractions or novelty effects, participants can process information more quickly and efficiently at each glance. Overall, Table 5.2 and Figure 5.15 highlight a clear trend that dynamic symbols influence visual search strategy (as seen in fixation count), but static symbols reduce cognitive load during each fixation. This distinction helps explain why some dynamic variants led to higher fixation counts without improving task accuracy—participants looked more, but not necessarily more effectively.

The combination of short fixation durations, low fixation counts, and consistently high accuracy, seen most clearly in the static map variant (D), points to what can be considered optimal cognitive processing. Participants needed fewer fixations, processed each one quickly, and still performed well. In contrast, some dynamic variants produced longer fixations and more extensive scanning, patterns typically associated with increased cognitive load or less efficient visual processing. Yet, this

**TABLE 5.2**
**Crosstabulation of Median Fixation Duration Across Each Map Variant and Map-Based Task**

| Map Variant | Task | Median Fixation Duration [ms] |
|---|---|---|
| A | Magnitude | 227 |
| | Fatalities | 222.5 |
| | Depth | 227 |
| | Years | 247.5 |
| B | Magnitude | 237.5 |
| | Fatalities | 236 |
| | Depth | 236.25 |
| | Years | 237 |
| C | Magnitude | 227 |
| | Fatalities | 226 |
| | Depth | 218.75 |
| | Years | 242.25 |
| D | Magnitude | 202 |
| | Fatalities | 201 |
| | Depth | 202 |
| | Years | 221.75 |

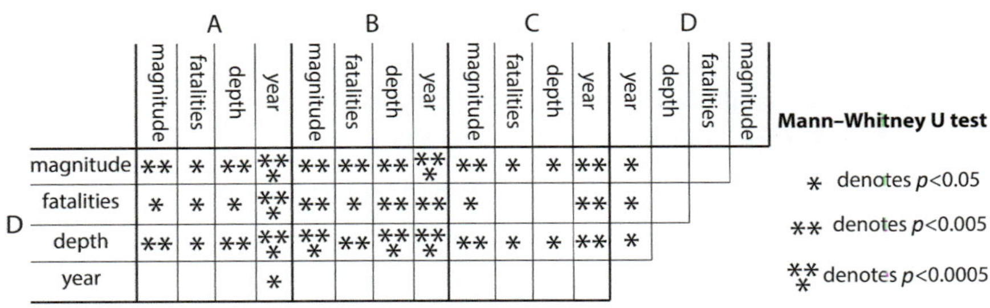


**FIGURE 5.15** Statistical comparison of median fixation duration across four map variants (A: vibrating symbols, B: pictorial motion, C: pulsation, D: static symbols) and four search tasks (magnitude, fatalities, depth, year), analyzed using the Mann–Whitney U test. Each cell compares two combinations of map variant and task.

efficiency did not always translate directly into better task performance. Variants B and C, despite higher fixation counts or longer processing times, still achieved higher overall effectiveness in several tasks. The static variant only surpassed the others in the temporal task (identifying the year with the most earthquakes). This shows that **while static symbols may support efficient visual processing, dynamic symbols—when well-designed—can still guide attention effectively** and support successful task completion.

### 5.4.3 How Far the Gaze Moved (Saccade Distance)

Median fixation distance—also called *saccadic amplitude* or *saccade distance*—captures how far the eyes travel between fixations. Shorter distances suggest that participants examined the map locally and in detail, while longer distances indicate that the eyes jumped across larger portions of the map, often signaling global search strategies or uncertainty. Because this measure is expressed in degrees of visual angle, it directly reflects how widely users scanned the map interface.

As shown in Table 5.3, the patterns differed noticeably across map variants and task types. **Variant A**, which used vibrating symbols, consistently produced longer fixation distances, especially in tasks such as identifying fatalities and depth categories. This suggests that the vibration effect may have introduced visual noise, pulling users' gaze across larger areas of the map and potentially reducing visual stability.

In contrast, both **variant B** (pictorial motion) and **variant D** (static symbols) encouraged a more focused and localized visual search. Participants made shorter saccadic movements, and statistical comparisons confirmed that these variants differed significantly from **variant A** in several tasks. This implies that these symbol designs helped guide users' attention more efficiently, particularly in tasks requiring magnitude interpretation, where the visual cues were clearer and less visually intrusive.

**Variant C** (pulsing symbols) produced the widest and most scattered scanning patterns, particularly for the fatalities and depth tasks. This matches earlier findings that pulsing symbols—despite increasing visual effort—led to the highest accuracy in these tasks. In other words, the **broader gaze movements in variant C seem to reflect deeper**

**TABLE 5.3**
**Crosstabulation of Median Fixation Distance Across Each Map Variant and Map-Based Task**

| Map Variant | Task | Median Fixation Distance (°) |
|---|---|---|
| A | Magnitude | 2.9 |
| | Fatalities | 3.27 |
| | Depth | 2.98 |
| | Years | 2.635 |
| B | Magnitude | 2.3 |
| | Fatalities | 2.98 |
| | Depth | 2.985 |
| | Years | 2.62 |
| C | Magnitude | 2.97 |
| | Fatalities | 3.94 |
| | Depth | 3.645 |
| | Years | 2.855 |
| D | Magnitude | 2.33 |
| | Fatalities | 3.202 |
| | Depth | 3.115 |
| | Years | 2.475 |

**or more thorough visual engagement rather than confusion.** Pulsation demanded more visual exploration but ultimately supported successful task performance.

As shown in Figure 5.16, several of these patterns are confirmed statistically. The figure highlights where differences in median fixation distance between map variants reached significance according to the Mann–Whitney U test, helping to pinpoint which symbol designs prompted broader scanning and which supported more focused visual attention.

### 5.4.4 How Fast the Gaze Moved (Saccade Velocity)

While fixation distance helps explain *where* and *how far* users moved their eyes across the map, another important piece of the visual-attention puzzle is *how quickly* these movements occurred. This is captured by **saccadic velocity**. Saccadic velocity

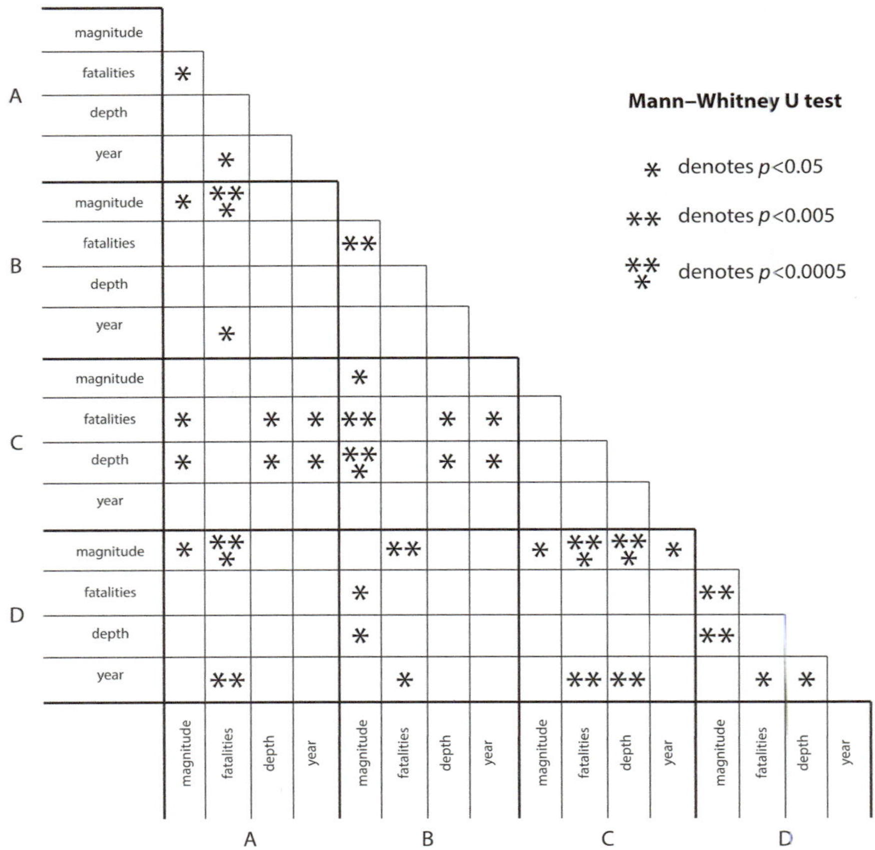


**FIGURE 5.16** Statistical comparison of median fixation distance across four map variants (A: vibrating symbols, B: pictorial motion, C: pulsation, D: static symbols) and four search tasks (magnitude, fatalities, depth, year), analyzed using the Mann–Whitney U test. Each cell compares two combinations of map variant and task.

reflects the speed of eye movements during saccades—the rapid jumps the eyes make between fixations. Faster saccades often signal quick, sometimes impulsive scanning triggered by strong visual stimuli or visual clutter. Slower saccades, in turn, usually indicate more deliberate, carefully guided visual exploration. Because saccades vary considerably between individuals and moments, the median value is the most reliable for interpreting overall gaze behavior. This metric provides another important perspective on how dynamic and static map symbols guide user attention—and how comfortably participants navigated different visual designs.

Across map variants, saccadic velocity revealed clear differences in how dynamically the eyes moved during visual search. **Variant C** (pulsation) consistently produced the highest saccadic velocities, especially in the fatalities and depth tasks (Table 5.4; Figure 5.17). This pattern suggests more rapid—or even erratic—eye movements, likely caused by the strong visual salience and rhythmic motion of pulsing symbols. Interestingly, **despite this more turbulent gaze behavior, variant C also produced high effectiveness**, indicating that fast and wide-ranging eye movements did not prevent participants from successfully completing the tasks.

At the opposite end of the spectrum, **variant D** (static symbols) showed the lowest saccadic velocities, reflecting calmer and more controlled gaze transitions. This likely stems from participants' familiarity with static point symbols, which reduce visual tension and encourage more deliberate scanning—particularly in spatially focused tasks like counting earthquake events. However, these **smoother gaze dynamics did not result in consistently higher correctness**, highlighting that visual comfort does not always translate into better performance. Variant A (vibration) produced moderately high velocities, while variant B (pictorial motion) showed slightly lower and more stable values.

Statistical comparisons reinforce these trends. Significant differences ($p < 0.05$) cluster around **variant C**, mainly in the fatalities and depth tasks when compared with A, B, and D. This confirms that pulsing motion accelerates saccadic movements, likely through visual overstimulation. Significant differences also appear between variants A and D in the magnitude and year tasks, reflecting the more dynamic nature of vibrating symbols. Across multiple tasks—including magnitude, fatalities, and depth—**variants C and D** differ sharply, underscoring the contrast between highly dynamic pulsating symbols and familiar static representations.

**TABLE 5.4**

**Crosstabulation of Median Saccadic Velocity (°/s) Across Each Map Variant and Map-Based Task**

| Variant/Task | Magnitude | Fatalities | Depth | Year |
|---|---|---|---|---|
| **A** | 102.2 | 108.9 | 103.3 | 96.7 |
| **B** | 96.1 | 106.6 | 104.7 | 95.3 |
| **C** | 101.0 | 116.5 | 113.7 | 102.3 |
| **D** | 92.6 | 108.1 | 105.2 | 94.2 |

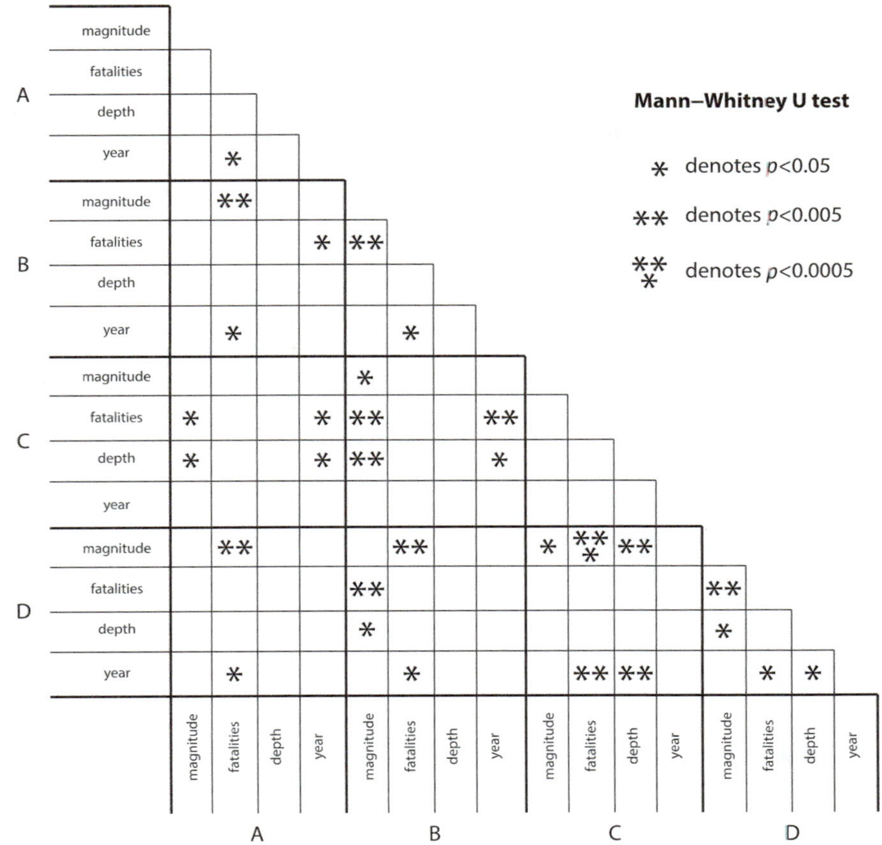


**FIGURE 5.17** Statistical comparison of median saccadic velocity across four map variants (A: vibrating symbols, B: pictorial motion, C: pulsation, D: static symbols) and four search tasks (magnitude, fatalities, depth, year), analyzed using the Mann–Whitney U test. Each cell compares two combinations of map variant and task.

### 5.4.5 How Much Time Participants Spent Looking at the Legend and Temporal Scale

The amount of time participants spent viewing the map legend and the temporal scale provides another important perspective on how they navigated the visual information in each task. Dwell time—the total duration of gaze directed at these reference elements—varied notably across map variants and task types, reflecting differences in how much support participants needed from the legend to interpret dynamic or static symbols during visual search. In some cases, longer dwell times suggested uncertainty or a greater need to decode unfamiliar visual variables. In others, short dwell times indicated that participants quickly understood how the symbols worked, allowing them to focus more on the map itself. By examining these patterns, we gain insight into which symbol designs were intuitive, which demanded extra cognitive effort, and how participants balanced their attention between the data display and its explanatory elements.

Across the four map variants, participants spent very different amounts of time looking at the legend and temporal scale.

**Variants A and B** (vibrating symbols and pictorial motion) had the longest legend dwell times in every task (Table 5.5). For example, participants spent about **20–21 seconds** looking at the legend when identifying earthquake magnitude. This suggests that these animated symbol types required more frequent checking—either because the symbol behavior was unfamiliar or because users wanted to be precise. Interestingly, both variants also produced relatively high task accuracy, especially variant B. This suggests that **spending more time consulting the legend helped participants correctly interpret the moving symbols**. Variants A and B also showed almost the same dwell-time values, meaning both designs encouraged a similar level of legend checking.

**Variant C** (pulsation) was very different. Participants spent much less time on the legend—sometimes as little as **7–8 seconds**, even for complex tasks like identifying the most active year or the most common depth. Statistical tests (Figure 5.18) show that variant C differs strongly from almost all other variants, with many very small $p$ values (often $p < 0.0005$). This confirms that pulsing symbols drew participants' attention away from the legend. Despite this reduced legend use, variant C still produced good task results, but earlier eye-tracking patterns (e.g., faster saccades, more scattered gaze) suggest that the pulsation increased cognitive load. **Users may have relied more on quick visual impressions rather than deliberate checking of the legend**.

**Variant D** (static symbols) produced the shortest legend dwell times overall. Participants barely looked at the legend, likely because they were already familiar with traditional static map symbols. While this may seem efficient, other eye-tracking measures—like low fixation counts and slow gaze movements—show that participants may have skimmed the display quickly without deeply processing all information. Notably, variant D's effectiveness was often lower than that of variant B. This suggests that **"not needing the legend" does not automatically translate into better performance**.

**TABLE 5.5**

**Crosstabulation of Summary Time (Dwell Time) of Viewing Legend and Temporal Legend (s) Across Each Map Variant and Map-Based Task**

| Variant/Task | Magnitude [s] | Fatalities [s] | Depth [s] | Year [s] |
|---|---|---|---|---|
| **A** | 20.4 | 16.2 | 13.5 | 15.6 |
| **B** | 21.0 | 17.6 | 13.6 | 13.3 |
| **C** | 14.6 | 10.3 | 8.0 | 7.0 |
| **D** | 12.4 | 12.4 | 8.9 | 9.5 |

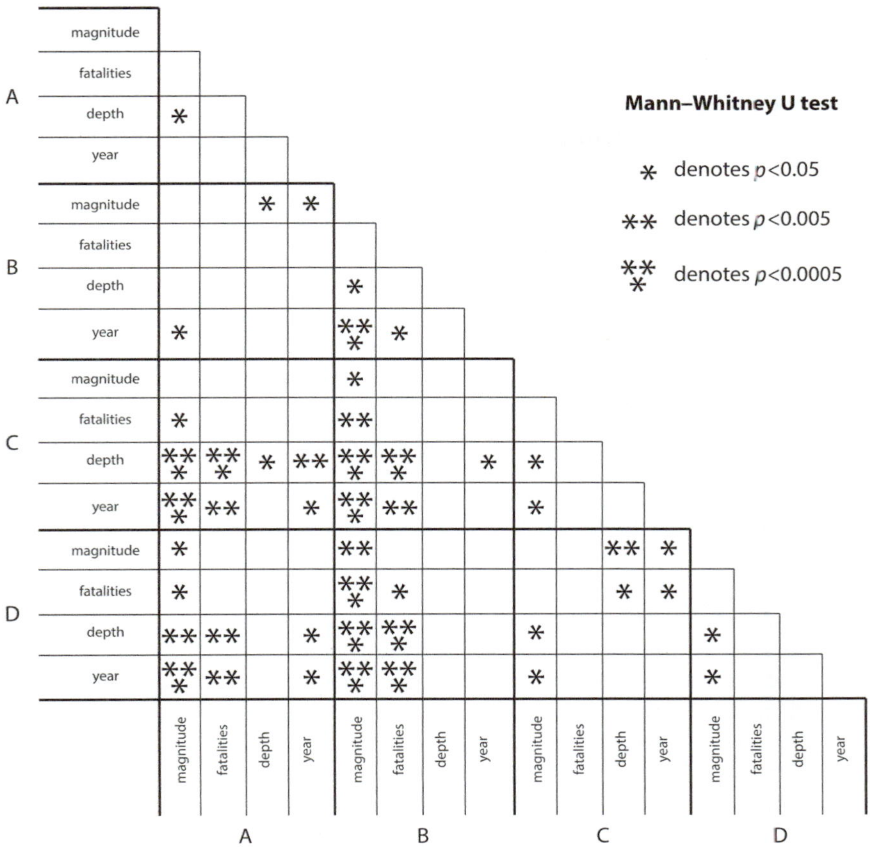


**FIGURE 5.18** Statistical comparison of summary time of watching legend and temporal legend across four map variants (A: vibrating symbols, B: pictorial motion, C: pulsation, D: static symbols) and four search tasks (magnitude, fatalities, depth, year), analyzed using the Mann–Whitney U test. Each cell compares two combinations of map variant and task.

Overall, these patterns show that legend dwell time reflects how users adapt their strategies to different visual designs:

- More attention to the legend (B) —careful checking, higher accuracy, more systematic interpretation.
- Little attention to the legend (C) —more reliance on symbol salience, potentially higher cognitive load.
- Minimal legend use (D) —fast, familiar processing, but not always effective.

In other words, legend dwell time captures not only cognitive effort but also the style of visual engagement—whether users proceed cautiously, rely on intuition, or skim through familiar visuals.

# 6 What We Learned about Animated Symbols

## *Bringing the Two Studies Together*

### 6.1 WHAT THE TWO METHODS TAUGHT US: ONLINE SURVEY VS. EYE-TRACKING

This study used a mixed-methods approach that combined a large-scale online survey with a detailed, in-person eye-tracking experiment. Together, these two methods provided both wide-ranging insights and fine-grained evidence about how people read dynamic point symbols in earthquake maps.

Online surveys are well-known for their ability to reach large and diverse groups of respondents. This makes them especially valuable in cartographic research where public perception, visual preferences, and usability must be assessed across multiple users (Dillman & Bowker, 2001; Mann & Steward, 2000). In this project, the online study allowed us to test multiple symbol types with strong statistical support, revealing broad patterns in how effective and efficient each visual encoding was. Similar approaches have proven useful in earlier cartographic studies exploring user performance, symbol clarity, and visual complexity (Brychtová & Çöltekin, 2017; Schiewe, 2019; Cybulski & Medyńska-Gulij, 2018).

However, online studies also come with well-known limitations. Researchers have limited control over who participates, how carefully respondents follow instructions, or even the technical conditions under which they view the maps. Screen resolution, display size, ambient lighting, and distractions all vary widely and can influence how animated symbols—especially those with subtle motion like blinking or pulsation—are perceived (Wright, 2017; Faliszewska, 2012). Even when demographic data are collected, they may be incomplete or imprecise, further limiting the depth of interpretation.

To address these constraints, the research was complemented with a laboratory-based eye-tracking experiment, which included its own traditional in-person survey (Thompson et al., 2003). While this phase involved far fewer participants—a natural limitation of lab studies—it offered full control over the testing environment, from screen size and viewing distance to lighting and calibration. More importantly, eye-tracking provided access to detailed behavioral indicators such as fixations, saccades, and legend-viewing time. By observing how participants visually processed the symbols, not just what answers they gave, the experiment made it possible to explain certain patterns from the online study in greater depth (Krassanakis & Cybulski, 2021; Fairbairn & Hepburn, 2023).


DOI: 10.1201/9781003728535-6

The eye-tracking methodology added an entirely different layer of understanding to the study. Measures such as fixation duration, fixation distance, and the number of fixations made within areas of interest (AOIs) offered objective, fine-grained indicators of attention, cognitive load, and how visually noticeable different symbol types were (Bylinskii et al., 2017; Duchowski, 2017; Keskin et al., 2018). These metrics allowed us to see how participants interacted with the maps—whether they scanned widely, hesitated over certain symbols, or relied heavily on the legend. Such insights simply cannot be captured through online questionnaires alone.

Importantly, the eye-tracking results also helped explain several patterns that emerged in the online study. When response times were long, eye-movement data made it possible to determine whether participants were struggling to locate the symbol, revisiting the legend, or being distracted by the animation. In this way, the laboratory study served as a diagnostic tool, clarifying why some visual encodings performed better than others.

Of course, the use of in-person eye-tracking came with its own limitations. The biggest constraint was the much smaller sample size, which reduced the generalizability of the findings compared to the large online survey. This trade-off between depth and scale is well recognized in empirical cartography (Ooms et al., 2014; Keskin et al., 2020): eye-tracking provides rich behavioral insight but at the cost of broad population coverage. Even so, the lab data proved essential, offering context that validated and strengthened the online results rather than contradicting them.

Bringing these two methods together follows the direction recommended in modern usability and cartographic research (Kiefer et al., 2017; Slocum et al., 2001). The large-scale online survey exposed broad patterns in symbol recognition, response accuracy, and user preferences across a diverse audience. Meanwhile, the eye-tracking experiment uncovered the perceptual and cognitive mechanisms that shaped those patterns—showing how participants processed motion, color, size, and temporal cues on the map. Viewed as a whole, this mixed-method approach provided a complete picture: not only which dynamic symbols worked best, but why they worked the way they did. This combined insight sets the stage for understanding performance differences more deeply in the next section, where the effectiveness results from both studies are compared.

## 6.2 HOW WELL PARTICIPANTS PERFORMED: COMPARING ACCURACY ACROSS BOTH STUDIES

A core goal of this research was to evaluate how effectively dynamic point symbols communicate earthquake information compared to traditional static symbols. To do this, four types of visual-search tasks were examined: motion-based tasks (identifying earthquake magnitude), size-based tasks (number of fatalities), color-based tasks (earthquake depth), and time-based task (finding the year with the most events). Earthquakes provide a suitable test case because they naturally involve all three major visual variables—motion, size, and color—while also requiring temporal reasoning. Importantly, the findings are not limited to earthquakes: any geographical phenomenon that can be visualized using dynamic point symbols based on these variables (such as volcanic eruptions, traffic accidents, or disease outbreaks) is likely to produce comparable results with high probability, making the conclusions broadly transferable across cartographic applications.

Across both studies, task performance was assessed using two complementary measures: effectiveness, meaning how accurately participants answered each task, and efficiency, meaning how quickly they were able to do so (Garlandini & Fabrikant, 2009; Cybulski, 2020). By comparing results from the broad online sample with those from the detailed eye-tracking sessions, we gain a clearer understanding not only of which symbol types worked best, but also of how consistently they supported accurate interpretation in different testing environments.

### 6.2.1 How Well People Found the Strongest Earthquakes

In the task that required identifying the strongest earthquakes, several motion-based symbol types—pulsation, vibration, and pictorial movement—were compared with a traditional static representation. Results from the online survey showed that participants performed similarly across all variants. In other words, adding motion did not automatically make the task easier or lead to better accuracy than static symbols. This mirrors earlier observations by Harrower (2007), who noted that although animation can successfully draw the eye, it does not always improve the user's ability to extract information. The present findings reinforce that motion is a viable alternative to static encodings like size, but not a universally superior one. Instead, its effectiveness depends on the context of the task and the user's ability to interpret motion as a meaningful visual cue.

These observations are consistent with empirical studies by Cybulski and Wielebski (2019, 2023), which emphasize that the success of motion as a cartographic variable depends not only on its ability to draw attention but also on how intuitively it reflects the underlying data. When the type of motion is not clearly aligned with the phenomenon it represents, its interpretive value decreases—especially in tasks that require careful spatial comparison or counting. In this sense, motion can support the search for the strongest earthquakes, but only when users can easily connect the visual behavior of the symbol with the meaning it encodes.

Building on this understanding, the controlled eye-tracking experiment provided clearer distinctions between different motion types and their effectiveness. While pulsation-based dynamic symbols did not substantially outperform static ones in the online setting, the laboratory data revealed that pictorial motion symbols led to markedly higher accuracy. This finding reinforces the concept of motion congruency (Griffin et al., 2006; Slocum et al., 2000, 2001, 2004), which argues that motion becomes most effective when its form—such as direction, tempo, or expressiveness—metaphorically aligns with the meaning of the data. As noted by Hohnsbein and Mateeff (1998), motion perception is strongest when the visual system can anchor moving elements to patterns that feel physically plausible or familiar. This likely explains why pictorial symbols, which resemble real-world movement, supported stronger performance than rhythmic pulsation, a finding also echoed in recent observations by Ge et al. (2025a, 2025b).

Taken together, these results point to a broader pattern visible across contemporary cartographic research: the effectiveness of motion is highly context-dependent. As highlighted by Cybulski (2024), Fish (2021), Fish et al. (2023), Krassanakis et al. (2016), and Obuhov and Panidi (2024), motion does not operate as a universally beneficial visual variable. Instead, its success depends on how well the perceptual cues embedded in the motion align with users' cognitive strategies, the structure of the task, and the overall visual complexity of the map. The present findings reinforce this perspective, showing that motion-based symbology performs best when its behavior is tailored to the goals of the visualization—providing guidance when helpful, while avoiding unnecessary or distracting dynamics that may hinder interpretation.

This task-specific sensitivity to motion also appeared when examining how different user groups interacted with the symbol types—most notably in the gender-based patterns observed in the eye-tracking experiment. Although no statistically significant differences emerged, a clear directional trend was visible: female participants consistently performed slightly better than male participants when working with dynamic symbols, particularly pictorial motion. This aligns well with earlier findings by Castro-Alonso et al. (2019) and Keskin et al. (2018, 2019), who reported systematic gender-related differences in visual strategy and spatial reasoning (see also Montello et al., 1999). More recent research by Fang et al. (2025) and Jiao (2023) further suggests that women often adopt more stable and systematic scanning patterns on animated maps—an approach that may be especially advantageous for tasks requiring memory integration, close symbol comparison, or ongoing tracking of movement. While this study's gender-related trends did not reach statistical significance—likely due to the smaller laboratory sample—the consistency of the pattern across tasks suggests that perceptual strategies can meaningfully shape how users respond to motion-based symbols. These insights echo broader calls within geovisualization research for more user-sensitive and cognitively informed design approaches (Griffin et al., 2017; Roth et al., 2017), especially when integrating dynamic variables such as motion into thematic maps.

These user-specific patterns become even clearer when paired with participants' confidence ratings. Confidence, much like gender, did not reliably predict accuracy in either the online survey or the eye-tracking experiment. In fact, across multiple dynamic symbol conditions—especially those with strong motion cues—participants who felt "very sure" were often no more accurate than those who reported only moderate or even low confidence. This mirrors long-standing observations by Hegarty et al. (2002, 2006), who demonstrated that perceived clarity in geospatial tasks frequently diverges from actual task performance. One likely explanation, supported by Stankov and Crawford (1997), is that animated symbols can create a false sense of fluency: movement makes patterns appear easier to interpret than they truly are. In this study, this effect was particularly visible in the pulsation and vibration variants, where visual salience heightened participants' sense of certainty without producing a corresponding improvement in accuracy. This disconnect reinforces the importance of evaluating not only how users feel about dynamic symbols, but also how effectively those symbols support correct spatial reasoning under real task demands.

Interestingly, one condition deviated from this broader pattern. In the pulsating symbol variant, participants who reported neutral confidence consistently achieved surprisingly high accuracy. Rather than being disadvantaged by uncertainty, these participants appeared to engage with the task more cautiously and deliberately—avoiding the overconfidence sometimes triggered by salient motion cues, yet not exhibiting the hesitation typically seen among low-confidence respondents. This balanced cognitive stance aligns with observations by Tversky et al. (2002), who argued that moderate confidence often encourages an optimal blend of analytical reasoning and exploratory visual behavior. In the context of this study, neutral-confidence participants may have scrutinized the pulsating symbols more carefully, cross-checking with the legend instead of relying on intuition or first impressions.

Taken together, these confidence patterns highlight that motion-based symbol reading is not governed solely by visual salience—users' cognitive stance and strategy play a central role. This naturally extends into the eye-tracking evidence, which shows that dynamic symbols do more than influence outcomes—they actively shape how people search. From an eye-tracking perspective, motion-based symbols elicited distinct visual behavior profiles. Vibration and pictorial motion (variants A and B) resulted in high fixation counts and long fixation durations, suggesting greater cognitive processing demands. These variants also recorded larger saccadic amplitudes, indicating broader exploratory scanning across the map. Such patterns align with findings by Opach et al. (2014) and De Cock et al. (2022), who note that longer and more spatially dispersed fixations often signal elevated task difficulty or increased need for information integration.

In contrast, pulsation (variant C) led to fewer but longer fixations and notably wider saccades. This pattern suggests that participants' attention was strongly drawn to the highly salient pulsating features, but this came at the cost of reduced awareness of the surrounding context (Liu et al., 2025). As Keil et al. (2024) and Liao et al. (2017) point out, excessively salient cues can dominate attention, causing users to lock onto them prematurely and overlook important spatial detail elsewhere on the map. This mechanism helps explain why pulsation-based designs, despite demanding less overall visual effort, produced mixed effectiveness results—excellent in some tasks (especially color- and size-based) but weaker in tasks requiring identifying magnitude or determining the most active year.

Static representations (variant D), by contrast, generated the shortest fixation durations and the smallest saccadic amplitudes, reflecting compact, localized scanning. Yet these minimal eye-movement metrics were paired with weaker task effectiveness, echoing findings by Ruginski et al. (2022): efficient eye movement alone is not enough. When visual salience is low—as in static symbols—users may need broader search patterns and longer, more deliberate fixations to perform well. Without these compensatory strategies, quick and localized scanning can lead to missed information and reduced accuracy.

Taken together, these findings show that motion can support users in thematic cartography, but only when the type of motion meaningfully matches the data it represents. In the magnitude task—where participants had to locate the strongest earthquakes—**pictorial motion (variant B) proved to be the most effective** in the eye-tracking study. This form of motion behaves in a way that is perceptually "natural," allowing users to intuitively link symbol movement with increasing intensity. In contrast, vibration sometimes scattered attention, and pulsation, although highly salient, often drew users' focus too narrowly onto individual symbols, making it harder to judge magnitude in a broader spatial context. Thus, for magnitude-based visual search, motion works best when its rhythm and form reinforce the meaning of the attribute rather than competing with it.

More generally, these results highlight that **pictorial motion-enhanced symbols can improve engagement** and expand visual search when designed with semantic congruence in mind. But motion is not universally advantageous: depending on its form, it can misdirect attention, create visual clutter, or increase cognitive load. As a result, motion in map design must be applied thoughtfully and validated through user-centered testing.

Earthquakes serve as an exemplary case for studying these relationships because they combine multiple types of information—intensity, depth, fatalities, and time—in a way that mirrors many real-world point-based phenomena. **The lessons learned here extend well beyond seismology**. The same design principles can guide dynamic visualizations of volcanic eruptions, air quality peaks, extreme weather events, or even transportation and environmental observations. In all of these cases, the careful use of pictorial motion, color, and size can help communicate patterns of change more clearly—provided that the chosen visual dynamics support, rather than obscure, the story the map is meant to tell.

### 6.2.2 How Well People Found the Deadliest Earthquakes

While motion was not always the strongest performer in the magnitude task, its value became much more apparent in the fatalities task—especially when the visual encoding matched the meaning of the data. The fatalities task, which required identifying earthquake locations with the highest number of casualties, revealed clear advantages for animated symbols, particularly those using pulsating size. Among all tested designs, the pulsation-based encoding (variant C) consistently achieved high accuracy in the online study and performed competitively in the controlled eye-tracking experiment. These outcomes support broader theories in cartographic visualization, which argue that dynamic representations become especially effective when their perceptual qualities are congruent with the attribute they represent (Cybulski & Medyńska-Gulij, 2018; Pourgolmohammadgolshani, 2022; Ruddro & Mohna, 2023).

Here, pulsation appears to work with the user's cognitive expectations: a growing and shrinking symbol intuitively signals importance, intensity, or urgency. As suggested by Harrower (2007) and Harrower and Fabrikant (2008), rhythmic temporal change can heighten symbol salience when its meaning aligns with the underlying data. This idea traces back to DiBiase et al. (1992), who emphasized the importance of matching animation behavior to conceptual models in scientific visualization. A particularly interesting outcome of this study is that **pulsation proved helpful even when it was not encoding fatalities at all**—it originally represented magnitude, yet still enhanced users' ability to read size, the variable encoding fatalities. In other words, **the motion drew attention toward the symbol as a whole**, making users more aware of its visual properties—even those unrelated to the animated cue.

Comparative results suggest that static graduated symbols (variant D) and motion designs such as vibration or shaking (variants A and B) are less effective when the visual encoding lacks semantic clarity or introduces perceptual noise as in online study. Unlike pulsation—which provides a simple, rhythmically repeated cue—vibration and shaking offer motion without meaning. They attract attention but do not tell the user why they should be looking there. This aligns with concerns raised by Lai and Yeh (2004), who showed that poorly calibrated animation can overwhelm the viewer's perceptual system, increasing visual clutter and reducing the accuracy of spatial judgments. In this study, **vibrating symbols may have competed with the size encoding** rather than supporting it, producing a noisier and more cognitively demanding search pattern.

What makes the results even more intriguing is the behavior of pictorial motion (variant B). In the online survey, B performed moderately—stronger than static designs, but weaker than pulsation. Yet in the controlled environment of the eye-tracking laboratory, pictorial motion suddenly became one of the top performers. This shift suggests that **the effectiveness of motion is not only about how it moves, but also about where and how carefully users observe it**. In a stable, distraction-free setting with large, high-quality displays, pictorial movement becomes easier to distinguish and interpret. This finding echoes Battersby and Goldsberry (2010) and Griffin et al. (2006), who argued that animation style interacts with viewing conditions, screen resolution, and user attention. Under the right conditions, pictorial motion offers a more grounded, physically intuitive cue—one that users may interpret more effectively when the visual environment supports it.

When gender-based differences are examined more closely, the fatalities task reveals an interesting and somewhat contradictory pattern across the two research settings. In the online environment, men outperformed women in variant C (pulsation), and the difference was statistically significant. This finding stands out because it counters much of the existing literature, which often reports either gender parity or a slight female advantage in tasks requiring careful visual discrimination and pattern recognition (Castro-Alonso et al., 2019; Montello et al., 1999). One plausible explanation is that pulsation—highly salient and attention-grabbing—may have interacted with the noisier, **less-controlled viewing conditions typical of online studies, favoring broader visual-search styles that men are more likely to employ** (Voyer et al., 1995).

However, the pattern reversed in the eye-tracking laboratory. Under stable lighting, consistent screen quality, and reduced distractions, female participants outperformed males in several dynamic variants, particularly variant C. This suggests that **when the environment supports focused attention, women's typically more systematic and consistent scanning strategies become advantageous**—especially with animated symbols that require integrating multiple visual cues over time. This interpretation echoes findings by Tversky et al. (2002) and Griffin et al. (2006), who argued that visual search in animation-rich environments is shaped not by gender itself, but by cognitive strategy. In structured contexts, strategies emphasizing stability and repeated checking tend to yield better performance.

Male participants displayed a relative advantage in variant B (pictorial motion), though this effect weakened once near-correct answers were included. This reinforces the view that symbol type interacts with perceptual style: motion that resembles real-world physical behavior may fit better with more exploratory, spatially broad strategies. Meanwhile, the more rhythmic, attention-locking motion of pulsation favored the systematic visual patterns often associated with female participants in the lab.

Overall, these results highlight that gender effects in animated map reading are not fixed but highly context-dependent. **The combination of symbol design, viewing conditions, and cognitive strategy—not gender alone—shaped performance differences**. This reinforces the need for user-sensitive design and evaluation when implementing dynamic symbols in cartographic products.

When we look at how confident participants felt about their answers in the fatalities task, an interesting pattern emerges—one that shows confidence and accuracy do not always travel together. For example, in variant A (vibration), many people said they were "sure" about their answer, but their actual accuracy was quite low. In other words, they thought they were doing well, but they weren't. This is a classic pattern in psychology: when a task feels visually busy or stimulating, people often assume they understand it better than they do (Kruger & Dunning, 1999; Lichtenstein et al. 1982). Variant B (pictorial motion) showed a similar mismatch, reinforcing the idea that certain animated styles can create a false sense of clarity.

**Variant C (pulsation), however, told a different story**. Here, people who felt confident were usually correct, and people who felt unsure were more cautious for a good reason. The pulsating symbol seemed to give users clearer, more reliable cues about which earthquakes had the highest fatalities. In other words, users weren't just performing better—they also had a more realistic sense of how well they were performing. This matches research suggesting that when visual cues are clear and logically tied to the data, people make better judgments about their own understanding (Koriat, 2012).

The eye-tracking experiment revealed an even more surprising twist. In variant C, several participants who said they were unsure actually achieved very high accuracy. This type of "underconfidence" fits what psychologists call the hard-easy effect: sometimes, when a task looks complicated—like with a pulsing map symbol—people underestimate their own success even if they're doing well. Eye-tracking data supported this interpretation: **variant C users made fewer fixations and didn't look around the map as much, yet they still reached some of the best accuracy scores**. This suggests they were guided very efficiently by the pulsation, even if they didn't fully trust themselves.

Variant A, on the other hand, showed the opposite pattern. Participants scanned the map a lot—high fixation counts—but still performed poorly. This combination (lots of effort, little reward) suggests visual overload: the symbol moved so much that it kept pulling attention without helping users understand what needed to be found. This behavior perfectly matches findings by Çöltekin et al. (2010), who warn that too much animation can distract rather than inform.

Variant C also stood out in another important way: it produced the widest fixation distances, meaning participants' eyes jumped across larger portions of the map. This suggests that **pulsating symbols didn't just attract attention—they encouraged users to explore more of the spatial layout**. This pattern mirrors Lloyd's (2005a,

2005b) theory of spatial cognition, which argues that highly salient elements prompt people to scan the map more broadly, helping them efficiently spot extreme values or unusual cases. By contrast, participants using variants B and D showed much tighter, more localized eye movements. Variant D in particular combined few fixations with short eye jumps, a pattern often associated with quick, heuristic "skimming." While such behavior may work for simple lookups, it can cause users to miss important details and may lead to worse performance, as the familiarity of static symbols and overconfidence hinder accuracy in more demanding comparison tasks—especially those involving overlapping temporal or spatial patterns.

Effectiveness in size-based visual search improved dramatically whenever the symbol animation matched the meaning of the data being shown. **In this case, pulsating symbols stood out: they were the top performers in the online study and remained highly competitive in the lab**. This supports long-standing theories on motion salience and dynamic congruency (Harrower, 2007; Griffin et al., 2006), which argue that animation works best when its behavior "feels right" for what it represents. **Gender differences, while noticeable, depended, again, more on context**. Factors such as individual scanning habits, task difficulty, and the stability of the testing environment played a more important role than gender itself. **Confidence was an equally unreliable predictor of success**—participants often felt sure even when they were wrong—unless the animation provided strong visual cues that helped them judge their own performance realistically.

Eye-tracking data strengthened these conclusions. When animation was thoughtfully designed, participants scanned the map more efficiently, with fewer unnecessary revisits and clearer attention patterns. Pulsation, in particular, guided the eye smoothly without overwhelming the user. In contrast, less intuitive animations like vibration produced scattered, unfocused scanning. Overall, the results highlight a key principle: motion can significantly enhance cartographic communication when its perceptual qualities align with the thematic variable it represents. Importantly, motion did not disrupt participants' ability to read other visual variables such as color or size. Instead, it worked alongside them, suggesting that **dynamic point symbols can be confidently applied to many other point-based phenomena—ranging from volcanic eruptions and air quality measurements to meteorological events**—without sacrificing clarity or interpretability.

### 6.2.3 How Well People Identified the Most Common Earthquake Depth

Color-based search for earthquake depth turned out to be the most universally successful task in the entire study. Whether symbols were static, vibrating, pulsating, or moving pictorially, participants performed remarkably well. This reinforces a central idea in perceptual science: color is one of the most powerful preattentive features we have (Wolfe, 2020; Nothdurft, 1993). Our visual system is naturally tuned to detect color differences rapidly and accurately, even when other visual variables—like motion or size—compete for attention.

In the online study, pulsation-based variant C produced the highest accuracy of all symbol types, clearly outperforming both static and other dynamic designs. When the broader ±1 class tolerance was applied, accuracy rose above 99%, essentially reaching ceiling-level performance. This outcome shows that **pulsation does not interfere with color-based encoding; instead, it appears to amplify it**. The rhythmic expansion and contraction of the symbol draws attention without obscuring color, taking advantage of the visual system's natural sensitivity to motion (Wolfe, 2021; Nothdurft, 1993). Earlier concerns that animation might disrupt color perception (Brychtová & Çöltekin, 2016) were not supported here—on the contrary, this study demonstrates that well-designed motion can strengthen, rather than weaken, the clarity of color-coded information.

The laboratory experiment confirmed this pattern under controlled conditions. All symbol types performed extremely well: variant A reached 92.6% full correctness, variant B 83.3%, variant C 90.0%, and even the static variant D achieved 80.0%. When responses within one class were counted as correct, all variants approached perfect accuracy. These results—free from factors like poor screen quality or environmental distractions (Çöltekin et al., 2010)—suggest that **color encoding remains highly effective regardless of the presence or absence of motion**, especially when tasks are viewed on high-quality displays in distraction-free settings.

**Taken together, these findings show that color is not only a powerful preattentive cue but also a highly robust one. It maintains its clarity across different symbol behaviors, task conditions, and testing environments** (Healey, 1996; Ware, 2012). Pulsation, in particular, seems to enhance rather than hinder performance, boosting visual salience without introducing confusion. This aligns with earlier research suggesting that subtle, rhythmic motion can make visual targets more noticeable when paired with already strong visual attributes like color (Cybulski & Wielebski, 2019).

The remarkably stable performance observed across all symbol types also extended to gender, reinforcing just how robust and universally accessible color-based encoding is. Unlike motion- or size-based tasks—where subtle gender trends occasionally appeared—the depth task showed no statistically significant gender differences in either the online or laboratory study. In other words, regardless of whether symbols were static, vibrating, pulsating, or moving pictorially, **men and women interpreted the color scale with essentially the same level of effectiveness**.

Some small patterns did emerge—for instance, men performed slightly better in certain dynamic conditions online (variants A and B), while women showed marginally higher accuracy with the static variant D—but these differences were consistently nonsignificant across chi-square and Fisher's exact tests. Once the more inclusive scoring method (±1 class deviation) was applied in the laboratory setting, even these small discrepancies disappeared entirely. Both groups reached near-perfect accuracy across all variants.

This stability is important. It suggests that **color, as a perceptual channel, offers equal clarity and cognitive accessibility across gender lines, regardless of animation type or task complexity**. It also shows that adding motion—such as pulsation—does not introduce unequal cognitive load or bias toward one user group. In this sense, the results affirm the universal reliability of color-based depth encoding (Healey, 1996; Ware, 2012) and highlight its potential as a design choice that remains fair, intuitive, and effective for all map readers.

Confidence judgments in the depth task revealed a pattern that was both intuitive and reassuring: **participants who felt confident were usually the ones who**

**answered correctly**. This alignment between confidence and performance was especially strong in tasks that relied on color—a visual channel that is easy to interpret and quickly picked up even under time pressure. In the online study, for example, 80.6% of "very sure" and 91.1% of "sure" participants in variant C answered correctly, and the laboratory data showed similarly high correspondence across all map variants. In other words, when people felt they knew the answer, they usually did.

What makes the findings even more interesting is what happened at the other end of the confidence scale. Even participants who said they were unsure often performed exceptionally well—sometimes matching or even surpassing the high-confidence group. In variant C online, for instance, **100% of participants who described themselves as "unsure" still identified the correct depth class**. This suggests that the pulsating symbols, when combined with a clear color scale, created a strong enough visual cue that users could rely on their intuitive perception even when they lacked conscious certainty. As Koriat (2012) notes, such "intuition-based accuracy" emerges when perceptual cues are highly salient and require minimal cognitive interpretation.

Taken together, these results paint a very positive picture. **Color-based encoding—especially when paired with gentle pulsation—helped users not only find the correct answers but also judge their own performance more accurately**. High-confidence users tended to be right, and low-confidence users did surprisingly well, reflecting a design that supports both metacognitive clarity and strong perceptual guidance. This tight match between confidence and correctness underscores the cognitive strength of using color-related tasks.

Eye-tracking results strongly support the high effectiveness observed in the depth task. Even though participants showed different scanning behaviors across the map variants, accuracy stayed consistently high—especially in the pulsation-based variant C. When interpreting depth, variant C produced fewer fixations, wider fixation jumps, and faster eye movements than the other variants. This pattern suggests that pulsation encouraged users to scan the map quickly and broadly, yet still allowed them to spot the correct color category with ease.

Crucially, the eye-tracking data show that these scanning patterns did not reduce accuracy. In fact, the decoupling of eye-movement intensity from correctness indicates that broad, rapid scanning—triggered by the pulsating effect—was actually efficient rather than distracting. Meanwhile, the static variant D generated the fewest fixations and the shortest fixation durations, but still did not reach the same level of correctness. This demonstrates an important point: **looking less is not the same as performing better**. Instead, the combination of color and pulsation seems to guide attention in a way that supports fast, accurate searching without causing cognitive overload, as seen in the size-based task (Ge et al., 2025a, 2025b; Tasse et al., 2016).

Additional support comes from legend dwell time. Users spent noticeably less time looking at the legend in variant C than in variants A or B. This suggests that **pulsation helped make the color categories immediately clear, allowing participants to interpret depth "at a glance" instead of repeatedly checking the reference scale**. In other words, the design encouraged efficient, intuitive understanding rather than deliberate decoding.

Taken together, the behavioral and eye-tracking results point to a simple, powerful conclusion: color is an exceptionally reliable and cognitively efficient visual variable, even when combined with animation. **Across every setting—online, in the lab, among confident users, among unsure users—color allowed participants to identify depth categories with very high accuracy**. The pulsating variant, in particular, stood out by producing near-perfect results while still supporting fast and intuitive search behavior. The lack of gender differences and the close match between confidence and performance reinforce how universally accessible and easy-to-interpret color encoding is in this type of task.

Just as important, the findings reveal what does not work well. Earlier concerns suggested that motion might interfere with color perception (Brychtová & Çöltekin, 2016), but our results show the opposite: **when motion is subtle and meaningful—such as pulsation—it enhances rather than disrupts color-based search**. In other words, motion becomes problematic only when it adds noise (e.g., vibration) or pulls attention away from the relevant feature (Montello et al. 2018). But when motion supports the visual hierarchy instead of competing with it, the pairing of color and animation becomes a powerful design strategy. This aligns strongly with theories of preattentive processing (Wolfe, 2020; Nothdurft, 1993) and shows that color's strengths translate well to other geographic phenomena, whether they involve pollution levels, disease rates, weather intensities, or other value-based classifications. For these types of tasks, color—especially when enhanced with subtle dynamic cues—remains one of the most dependable tools in thematic cartography.

### 6.2.4 How Well People Interpreted the Earthquake Timeline

Unlike the earlier tasks, which relied on searching for spatial patterns using motion, size, or color, the timeline task asked participants to identify the year with the most earthquakes. This task is fundamentally different: instead of scanning spatial clusters, users must mentally combine spatial distribution with temporal aggregation, linking what they see on the map to the timeline. Because of this, animation can interact with the task in more complex and sometimes unexpected ways. In the online survey, pulsation (variant C) performed best, with 62.7% of participants giving the correct year. In a loosely controlled online environment—where distractions are common and screen quality varies—pulsation likely acted as a visual "beacon," making dense clusters of events from the dominant year more noticeable. In this context, pulsation appears to compensate for environmental uncertainty by helping participants notice temporal peaks even if they were not actively examining the timeline. This supports findings that rhythmic motion can heighten preattentive capture and help users detect prominent clusters under noisy viewing conditions (Wolfe, 2021; Nothdurft, 1993).

The other variants—A (vibration), B (pictorial motion), and D (static)—performed much worse online, with most users giving answers that were off by 3 or more year categories. This suggests that without a strong visual attractor like pulsation, many participants struggled to connect what they saw on the map with the timeline while working in a less-controlled environment (Kapaj et al., 2023; Segen et al., 2021).

The laboratory experiment revealed a very different pattern. Under controlled conditions—consistent monitor size, stable lighting, full attention—static symbols (variant D) and vibration (variant A) produced the highest accuracy (90% and 92.6%), while pulsation dropped to 70%. The statistical difference between variants C and D ($p = 0.022$) indicates that pulsation, which was helpful in the noisy online environment, could actually become a distraction when participants were fully focused. **In this structured setting, the task required careful temporal reasoning rather than simple visual capture**. Strong pulsation may have drawn attention back to the map instead of the timeline, subtly interfering with users' ability to extract year-based information. This contrast reinforces a key insight that temporal search relies far more on cognitive reasoning than the previous tasks, and therefore responds differently to animation.

Motion that attracts attention to spatial clusters does not necessarily help users interpret a timeline—and may even hinder them when deep focus is required. This principle echoes broader findings in perceptual research: motion support is highly context-dependent (Bartram et al., 2003; Hillstrom & Yantis, 1994).

Although pulsation can be very helpful in messy or distracting settings—because it quickly draws attention to important areas—it does not always offer the same advantage in calm, controlled environments. In fact, when users are fully focused, pulsation may slightly disrupt the careful reasoning needed for a timeline-based task (Tao et al., 2024). By contrast, the steady performance of vibration (variant A) in both online and laboratory settings shows that the type of motion really matters. A subtle, rhythmic animation can guide users without overwhelming them, helping them interpret temporal patterns without adding extra mental effort. This highlights an important design lesson: not all animations support map reading in the same way, and strong, attention-grabbing effects can sometimes be more effective than gentle motion (Wu et al., 2022).

Gender-based patterns in the timeline task add another layer to the story of how people interpret temporal information on animated maps. In the online survey—where viewing conditions vary widely—men performed slightly better than women across all variants. For example, in variant A, men reached 38% accuracy while women reached 23.7%. However, none of these differences were statistically significant, meaning the apparent gaps could be due to chance rather than a meaningful effect. In other words, gender did not reliably predict performance in the less-controlled online environment.

The laboratory results painted a different picture. In this structured, distraction-free space, women often equaled or surpassed men. The contrast was most visible for pulsation (variant C): women reached 87.5% accuracy, while men reached 58.3%. Although the sample size prevented this difference from reaching statistical significance, the direction of the results suggests that women may benefit more from dynamic symbol designs when conditions support focused, systematic visual search. A similar trend appeared in variant B, where women slightly outperformed men (91.7% vs. 81.8%).

These patterns likely reflect differences in visual strategies rather than ability. Prior research suggests that women may adopt more deliberate and structured scanning techniques, especially when handling dynamic or information-rich displays (Tlauka et al., 2005). The controlled laboratory environment amplifies the advantages of such strategies, whereas the noisy online environment may dampen them. What matters most is that **neither study revealed strong or consistent gender-based disadvantages**.

The relationship between how confident participants felt and how accurately they performed looked very different in the two study settings. In the **online survey, confidence was a fairly good indicator of performance**. For example, in variant C, 100% of participants who said they were "very sure" answered correctly, while only 40.9% of those who felt "very unsure" succeeded. Similar gaps appeared in variants B and D. These patterns suggest that, in an unstructured online environment—where distractions, screen differences, and other uncontrolled factors can influence perception—confidence works as a rough but useful signal of task mastery.

In the laboratory setting, however, this relationship became much less predictable. In several map variants, participants with low confidence performed just as well as, or even better than, those who felt highly confident. For instance, in variant A, all participants who reported being *unsure* still correctly identified the target year. The same occurred in variant D, where 100% of *unsure* participants also answered correctly, despite expressing low certainty. This reversal suggests that the controlled lab environment—quiet, distraction-free, and supported by high-quality equipment—helps even uncertain participants perform at a high level, likely because focused conditions and clear visual displays reduce cognitive strain (Schmidt-Weigand, 2005).

Taken together, these findings show that confidence aligns more closely with correctness in online settings, but becomes far less reliable in a structured laboratory environment. In the lab, correct answers sometimes came from participants who doubted their performance, partly because correct guessing can artificially boost success rates among low-confidence groups (Vuorre & Metcalfe, 2021). At the same time, the results highlight an important insight: **well-designed dynamic features—especially pulsation—can support intuitive understanding** (Koriat, 2012), helping users recognize patterns even when they are not consciously confident. These benefits are likely most valuable in situations where external distractions or perceptual uncertainty might otherwise interfere with visual attention.

Eye-tracking data reveal how participants distributed their attention during the year-based search task, and the number of fixations is a key indicator of how much visual effort each symbol type demanded. The results show clear differences between dynamic and static designs.

**Variant C** (pulsation) produced the fewest fixations of all map variants. At first glance, fewer fixations might suggest efficient processing—but in this task, it likely reflected something different. The pulsation drew attention quickly and broadly across the map, encouraging participants to make large visual jumps rather than carefully checking the timeline. This rapid, **wide scanning made the map feel intuitive but did not always guide users toward the correct year**, especially in the controlled laboratory setting where precision mattered more. This interpretation aligns with findings by Dong et al. (2018a), who showed that highly salient visual cues can cause users to lock onto striking areas early while overlooking subtler but important information.

In contrast, variant A (vibration) produced the highest number of fixations, indicating a slower, more deliberate search pattern. Users looked at more locations, made more checks, and showed a more effortful scanning strategy. Although this meant a higher visual load, it ultimately supported better accuracy in the lab, suggesting that the **extra visual effort helped participants carefully evaluate the temporal cues**

**on the map** (Dong et al., 2018a). In other words, a high fixation count here reflected useful analytical work, not confusion.

Fixation duration patterns offer another window into how participants processed the timeline task. Across the dynamic variants—pulsation (C), vibration (A), and pictorial motion (B)—median fixation durations were remarkably stable, suggesting that the presence of motion did not increase the cognitive effort required for each individual look. In contrast, the static **variant (D) produced slightly shorter fixation durations, indicating that participants needed less processing time per fixation when symbols were not moving.** This combination of short fixations and high correctness in the laboratory setting suggests that a calm, stable display may support more systematic and efficient examination of temporal information. In other words, when the task requires precise reading of the timeline, static symbols can reduce cognitive load enough to make careful year identification both easier and more accurate.

Saccadic measures—the speed and distance of eye movements between fixations—help clarify how each symbol type guided visual search. Variant C (pulsation) produced the highest saccadic velocities, meaning participants' eyes jumped quickly and across wider areas of the map. This pattern reflects how strongly pulsating symbols drew attention, often pulling the gaze toward visually intense regions even when they were not the correct temporal target. In contrast, the static variant D showed slower and shorter saccades, suggesting a calmer, more controlled scanning style. Combined with its high accuracy in the lab, these slower movements point to a simpler explanation: **people are naturally more familiar with static map symbols, so they require less cognitive effort to interpret them**. This familiarity allowed participants to navigate the timeline more effectively, without being pulled off-track by motion-driven salience.

Finally, the time participants spent looking at the legend also reflected these attentional differences. **Users working with variant C (pulsation) consulted the legend least often**, suggesting that the strong visual salience of pulsating symbols made the task feel more intuitive and required less deliberate checking of temporal categories. In contrast, variants A and B encouraged longer legend dwell times, indicating that vibration and pictorial motion prompted participants to double-check symbol meanings more frequently. However, the reduced reliance on the legend in variant C is a double-edged sword. While it points to faster, more intuitive recognition, it also means that some participants may have skipped necessary verification, which helps explain why pulsation performed worse in the controlled laboratory setting. Without careful checking, the eye is easily pulled toward bright or active symbol clusters—even when they are not the correct year. Taken together, the eye-tracking evidence shows a consistent pattern: **pulsation supports broad and attention-grabbing scanning but can unintentionally mislead users in tasks that require precise temporal reasoning**. In contrast, more stable symbol designs slow the viewer down just enough to encourage accurate, deliberate interpretation.

The time-based visual-search task revealed subtle but important differences in how motion interacts with the task environment. In the online setting, pulsation (variant C) stood out as the most helpful design. Its rhythmic size changes provided a strong visual cue that made certain years more noticeable, raising accuracy well above the levels achieved with static or other dynamic variants. This outcome supports earlier findings showing that changes in size are among the most powerful visual variables in cartography, helping users quickly spot important patterns even in less-controlled viewing conditions (Garlandini & Fabrikant, 2009).

In the laboratory setting, however, the story changed. Here, the static map (variant D) and the vibration map (variant A) produced the highest accuracy levels. Under conditions with high screen quality, no distractions, and focused viewing, these more stable designs allowed participants to examine the timeline more carefully and reason about the correct year with greater precision. The dip in pulsation's performance under these optimal conditions shows that motion-enhanced symbols do not behave the same way in every environment. Instead, their usefulness depends on how motion interacts with attention, displays clarity, and the level of cognitive control available to the user.

In sum, these findings highlight that the effectiveness of motion-based map designs depends strongly on the context in which they are used. Dynamic cues like pulsation can be extremely helpful in noisy, unstructured settings, where they guide attention and support quick, intuitive decisions. However, in precise tasks—such as identifying the correct year—motion must be used carefully to avoid distracting users. A broader pattern also emerged across both studies: **participants who felt more confident often worked faster but were not necessarily more accurate, and these fast-but-less-accurate responses were more common among men. Women, on the other hand, tended to be more accurate but less confident, and generally took more time to verify their answers**. This shows that motion design interacts not only with task demands and environment, but also with users' cognitive styles. Well-crafted dynamic symbols can therefore support different user groups effectively—but only when motion is thoughtfully matched to the type of information being communicated.

## 6.3 HOW FAST PARTICIPANTS WORKED: COMPARING RESPONSE TIMES ACROSS BOTH STUDIES

Response time patterns across the four map-reading tasks show that efficiency depends not only on how symbols are designed but also on the environment in which they are used. In the online study—where people work on different screens, in different lighting, and with varying levels of distraction—static symbols were generally the quickest to interpret, likely because they are familiar and visually stable. In the controlled eye-tracking laboratory, however, this advantage often disappeared. With high-quality displays and fewer distractions, well-designed dynamic symbols—especially pulsation—became just as fast as static versions. This demonstrates that motion cues can boost efficiency, but only when users can clearly perceive the animation.

Not all motion types helped equally. Vibration consistently slowed participants down in both studies, likely because its jittery behavior created visual noise and made the symbols harder to interpret. By contrast, pulsation offered a clear, rhythmic signal that enhanced symbol visibility without overwhelming the viewer, especially when paired with straightforward attributes like color or size. Gender differences in efficiency were generally minimal. When variations occurred, they reflected different strategies rather than differences in ability—for example, women sometimes took more time to verify their answers, while men more often responded quickly with higher confidence. Importantly, these differences did not affect overall accuracy. Overall, the efficiency results reinforce a key message: motion can make maps faster and easier to read, but only when the animation is clear, meaningful, and suited to the viewing environment.

### 6.3.1 How Quickly People Found the Strongest Earthquakes

Response time data across both online and eye-tracking laboratory studies revealed important contextual dependencies in how motion-based visual-search tasks are performed. Across all map variants, the static symbol design (variant D) consistently achieved the shortest response times, with median times ranging from 46 to 73 seconds, supporting prior findings that static symbols, due to their visual familiarity and minimal interpretive demand, optimize scanning speed (Cybulski & Wielebski, 2019). This result reinforces the long-standing principle that static cartographic representations enable faster decision-making, particularly in cognitively constrained settings; however, they are less effective for recognizing trends or patterns (Calvo et al., 2022; Cybulski, 2022b; Griffin et al., 2006).

Among the dynamic designs, efficiency differed sharply between environments. In the online setting, variant B (pictorial motion) emerged as the fastest of the animated types, with a median time of 92 seconds, while variant C (pulsation) was considerably slower at 159 seconds. One likely explanation is that pulsation—although highly effective for size-based tasks like identifying fatalities—was less intuitively aligned with the magnitude task. Its rhythmic expansion and contraction may have drawn attention toward symbol size rather than motion intensity, creating a mismatch between the animation and the attribute being searched.

In the laboratory environment—where distractions were minimized and the visual display was stable—variant C (pulsation) became the most efficient of the dynamic designs. Participants completed the magnitude task in a median of 78 seconds, making pulsation noticeably faster than both pictorial motion and vibration. This shift suggests that when fine-grained perceptual cues are easy to see, the rhythmic "breathing" of pulsation becomes a clear and helpful signal. In controlled settings, users can fully perceive this steady rhythm, allowing pulsation to guide attention smoothly and efficiently (Lobben, 2008).

By contrast, variant A (vibration) consistently produced the slowest response times—150 seconds online and 135 seconds in the lab. The jittery, irregular movement appears to have introduced visual noise rather than clarity, increasing cognitive load without improving search performance. These results echo Harrower's (2007) observations and align with recent findings by Cybulski (2024), which warn that poorly matched or overly chaotic motion can hinder interpretation instead of helping it.

Across both the online survey and the laboratory experiment, there were no statistically significant gender differences in response time—either when looking at all participants or only those who answered correctly. However, an interesting pattern did emerge: women often achieved higher accuracy even though their response times were similar to, or only slightly longer than, men's. In the online study, men and women worked at almost identical speeds, and small accuracy advantages for women (about 1%–5% in variants A and C) did not come with longer processing time. In some cases—such as variant B—women were even a bit faster, although the difference was not statistically significant.

In the laboratory, this pattern became even clearer. Women frequently achieved higher correctness rates while maintaining the same response times as men. For example, in variant C, men and women finished the task in almost exactly the same time, yet women produced over twice as many correct answers. This finding echoes earlier research (Fang et al., 2025; Jiao, 2023) showing that women may use more consistent and efficient scanning strategies when interpreting animated or dynamic map symbols.

Overall, while **men and women performed tasks in roughly the same amount of time**, women often demonstrated higher accuracy. This suggests that gender differences, when they appear, may be rooted not in speed but in how attention is allocated and how visual information is processed, even when timing alone does not reveal these differences.

Across both studies, participants' confidence levels did not reliably predict how quickly they completed the tasks. In the online survey, a surprising pattern emerged: participants who felt more confident ("very sure" or "sure") often took longer to finish, especially in variants A and B. Rather than reflecting actual mastery, this higher confidence seemed to signal more time spent double-checking or overthinking, which aligns with the well-known overconfidence bias in spatial decision-making (Stankov et al., 2012; Lundeberg et al., 1994). A clear example comes from variant A. Here, "very sure" participants had longer median response times than those who felt "neutral" or even "unsure," despite performing no better—or sometimes worse. This mirrors findings by Pallier et al. (2002) and Koriat (2012), who showed that confidence and competence frequently diverge, particularly in visually complex or ambiguous tasks such as reading animated maps.

In the controlled laboratory environment, this mismatch between confidence and efficiency persisted. At times, participants who felt less confident actually completed tasks faster, suggesting that feelings of certainty were shaped more by symbol complexity, task familiarity, or perceived difficulty than by real performance. As Griffin and Tversky (1992) explained, people often use the feeling of task fluency as a shortcut for confidence—even when that feeling is misleading. Together, these results highlight a key point: how confident participants feel has little to do with how efficiently they work when interacting with dynamic cartographic displays. Confidence is influenced by intuition, perceived clarity, or symbol salience, whereas efficiency depends on actual visual processing and decision-making. This reinforces long-standing guidance in cartography (Slocum et al., 2000, 2004): **user behavior must be assessed with objective measures rather than relying on subjective impressions alone**.

Eye-tracking results provide important insight into why some symbol types allowed participants to work more efficiently than others. **Variants A and B—both using continuous motion—produced the highest number of fixations and the longest fixation durations**. This pattern indicates that users spent more time examining the display and that the symbols required greater cognitive effort to interpret.

The visual system was clearly working harder to decode these more complex motion patterns. Variant C, in contrast, showed a very different profile. Participants made fewer fixations overall, yet these fixations were spread across wider distances and had moderate durations. This suggests a strategic balance in visual behavior: broad scanning to locate relevant areas, followed by focused attention once a target is identified. This combination—wide search with efficient processing—likely explains why pulsation supported faster performance in the controlled lab environment. The pulsing motion increased salience without overwhelming the visual system, enabling quick and confident decisions.

Static symbols in variant D produced the shortest fixation durations and the tightest, most localized gaze patterns. While this compact scanning led to fast responses, it also reflected shallow processing, which corresponded with lower accuracy. Participants appeared to skim rather than thoroughly inspect the map—an approach that works for simple tasks but breaks down when the display is dense or when temporal reasoning is required. These findings align with Wolfe's (2021) conclusion that motion-based cues enhance preattentive processing only when they harmonize with user expectations and the structural logic of the map. The statistically significant differences between variants (particularly between D and A/B/C, $p < 0.005$) confirm that **motion fundamentally changes how people search visually**—and that the nature of the motion is just as important as the presence of motion itself.

Taken together, these results show that context plays a major role in how effectively people use motion cues. In the controlled laboratory environment, participants were able to take advantage of subtle, rhythmic animations like pulsation, which supported precise and efficient search. In contrast, the online environment—where screen quality and distractions varied widely—favored simpler forms of animation, such as pictorial motion, which are easier to decode even under imperfect viewing conditions. Across both settings, however, one pattern remained clear: erratic motion (such as vibration) consistently slowed participants down. From a human factors perspective, gender did not directly affect response speed, but it did relate to different accuracy patterns, suggesting distinct processing strategies rather than differences in ability. Confidence, meanwhile, was a poor indicator of actual speed or performance, revealing a common gap between what users believe they understand and how efficiently they actually work—a well-documented metacognitive disconnect. Eye-tracking evidence makes this even clearer: efficiency is not only about finishing quickly, but about how attention is deployed. The most effective visual strategy combined broad exploratory scanning with focused, deliberate fixations, a pattern most strongly associated with the pulsation variant.

Ultimately, motion-based designs in cartography should be calibrated to both context and user characteristics, using animations that are salient but cognitively economical. These lessons extend far beyond earthquakes. Because the tested symbols represent a general model of how people interpret point-based spatiotemporal data, the same principles apply to many other phenomena—volcanic eruptions, meteorological events, pollution sensors, wildlife movement, or any spatial data involving change over time. In all such cases, carefully crafted motion can enhance comprehension while preserving the clarity of other visual variables.

### 6.3.2 How Quickly People Found the Deadliest Earthquakes

The analysis of response times in the "deadliest earthquakes" task reveals clear patterns in how symbol design and testing environment shape efficiency. In the online survey, participants worked fastest with static symbols (variant D), completing the task in a median of 45 seconds. This finding is consistent with earlier studies showing that familiar, stable symbols reduce cognitive effort and allow for quick decision-making (Goldsberry & Battersby, 2009; Battersby & Goldsberry, 2010). Among the dynamic designs, **pulsation (variant C) stood out with a strong median time of 89 seconds—especially notable given its high accuracy**. This suggests that pulsation not only draws attention but also supports efficient judgments, even in less-controlled conditions (Ge et al., 2025a, 2025b; Peng et al., 2017). At the opposite end of the spectrum, vibration (variant A) resulted in the slowest responses (median = 148 seconds). This aligns with earlier warnings that highly jittery or ambiguous motion can disrupt visual processing and slow users down by creating unnecessary noise (Harrower, 2007; Lai & Yeh, 2004).

The picture changed in the laboratory eye-tracking study, where environmental conditions were optimized. **Under controlled lighting, consistent screen quality, and reduced distraction, variant C (pulsation) became the fastest dynamic design**, with a median response time of 51 seconds—nearly matching the 47-second performance of static symbols (variant D). This shift demonstrates that when perceptual cues can be clearly seen and interpreted, well-designed motion can equal or even exceed the efficiency of static approaches (Cybulski & Medyńska-Gulij, 2018). Taken together, these findings highlight a key insight: efficiency in animated cartography is strongly shaped by context. In everyday, real-world online settings, screen variability and distractions make simple, static designs more efficient.

Efficiency results showed no meaningful differences between men and women in how quickly they completed the task, whether in the online survey or the laboratory eye-tracking study. In variant C, for example, male and female participants recorded almost identical median response times, and similar patterns appeared across all other symbol types. This finding contrasts with earlier research suggesting that men and women might differ in spatial abilities or visual processing strategies (Voyer et al., 1995; Montello et al., 1999). Instead, our **results support a strategy-equivalent interpretation**: regardless of gender, participants devoted similar amounts of time to complete the task and achieved comparable levels of efficiency. What this implies is that symbol motion—not gender—is the key factor shaping efficiency in this visual-search context. Still, **equal timing does not necessarily mean identical thinking**. Men and women may reach these similar response times through different cognitive strategies—for example, broader vs. more focused scanning.

Confidence patterns reveal an interesting and sometimes counterintuitive relationship with response time. In the online survey, participants who said they were "very sure" or "sure" often took longer to finish the task—especially when using the vibration (A) or pictorial motion (B) variants. This suggests that **high confidence did not always come from quick insight**. Instead, confident participants may have spent extra time double-checking their answers or interpreting the dynamic symbols. This pattern echoes **earlier work showing that people often base confidence on how**

**fluent or effortful a task feels**, rather than on actual accuracy (Koriat, 2012; Griffin & Tversky, 1992). In dynamic symbol conditions—particularly those involving pulsating size (variant C)—confidence tended to align better with correctness, meaning participants who felt confident were often right. Here, the motion was clear, semantically meaningful, and easier to interpret, which likely helped users judge their own performance more accurately.

The laboratory results add an important twist: even participants who **reported low confidence while using variant C still performed well**. This reflects the classic hard-easy effect (Gigerenzer et al., 1991)—people often underestimate how well they are doing on tasks that are actually manageable. In the controlled lab environment, where distractions are minimized and display quality is high, participants could succeed even when they doubted themselves. Taken together, these findings suggest that good symbol design—especially when motion aligns with the meaning of the data—improves not only speed and accuracy but also how well people judge their own performance. In other words, clear animation helps users not just answer correctly, but also know they're answering correctly.

Eye-tracking results help explain why some symbol designs felt easier or faster to use than others. Even though variant C (pulsation) was dynamic, it actually produced the fewest fixations and the widest fixation distances. In simple terms, participants looked around the map less often but covered more ground with each eye movement. This pattern suggests efficient, purposeful scanning: users quickly spotted where important information was likely to be, instead of repeatedly checking the same areas. This fits well with Wolfe's (2021) theory of preattentive processing, which proposes that certain types of motion—when they match what the data represents—can guide the eye naturally without increasing mental effort.

Other symbol types behaved differently. Variant B (pictorial motion) led to more fixations that were closer together, indicating that participants spent more time scanning small areas in detail. While this approach can support accuracy, it usually slows down task completion. **Variant D (static) produced short fixation durations, meaning participants didn't spend long on each item, but their eye movements stayed narrowly concentrated**. This limited, "tight" scanning can be quick, but it may also reduce the chance of noticing important size differences—especially in tasks that require comparing multiple symbols. Statistical tests (Mann–Whitney) confirmed that the eye-movement patterns for variant C were significantly different from those of other designs, especially A and D ($p < 0.005$). Taken together, these findings show that pulsation encourages a broad, efficient visual sweep of the map, helping users complete tasks faster without sacrificing accuracy.

In sum, the size-based visual-search tasks showed that well-designed dynamic symbols—especially pulsating size encodings—can support both fast and accurate interpretation in controlled conditions. Static symbols remained reliably efficient, but when motion was both perceptually and semantically aligned with the data, animated symbols offered comparable or even superior efficiency. Gender differences in efficiency were negligible, while the match between confidence and actual performance depended on how clear and intuitive the symbols were, underlining the value of supporting users' self-awareness in cartographic interfaces. Eye-tracking results added an important layer: broader, strategically guided fixations—most clearly seen in variant C—were linked with the most efficient spatial decision-making. Taken together, these findings reinforce current cartographic theory that animation, when carefully aligned with data meaning and perceptual design principles, can enhance both cognitive performance and user satisfaction (Harrower, 2007; Cybulski & Krassanakis, 2023; Griffin et al., 2006).

### 6.3.3 How Quickly People Identified Earthquake Depth

The efficiency of color-based visual search in the earthquake depth task depended strongly on both the symbol design and the testing environment, revealing several telling patterns in how quickly participants worked. In the online study, static symbols (variant D) produced the fastest median response time at 24 seconds, showing that familiar, nonmoving symbols encouraged rapid decisions. However, these quick responses did not always lead to accurate answers. Many participants appeared to rely on fast, surface-level judgments—prioritizing speed over careful inspection—which ultimately reduced their effectiveness. This disconnect highlights an important point: speed alone is not a reliable indicator of good performance in color-based search.

Dynamic symbol designs introduced a wider range of processing times. The pulsation variant (C), although slower than the static design in the online environment, achieved a more favorable balance between response time and accuracy. Participants spent slightly longer analyzing the map, but reached consistently high correctness, suggesting that moderate attentional engagement supported better decision-making. In contrast, variant A (vibration) produced the longest response times and revealed a clear divide between successful and unsuccessful participants. Those who answered incorrectly tended to respond faster, indicating premature decisions driven by less intuitive or visually disruptive motion. This pattern suggests that vibration may encourage impulsive scanning behaviors—where users believe they understand the map more quickly than they actually do—resulting in lower accuracy despite rapid responses.

The controlled laboratory environment compressed response times across all map variants, showing how much stable conditions matter in visual-search tasks (Clifford & Jerit, 2014; Riva et al., 2004). In the lab, **the pulsation variant (C) not only kept its high accuracy but also became the fastest solution**, even outperforming the static variant D. The difference in response time between variants C and B was statistically significant, confirming a clear efficiency advantage for pulsation under optimal viewing conditions. Unlike in the online study, the speed advantage of static symbols disappeared in the lab. This suggests that when distractions are minimized and screens are clear and consistent, dynamic cues like pulsation can be processed quickly and fluently, rather than slowing users down.

These patterns fit well with theories from vision science, which argue that **motion and color together can activate preattentive visual mechanisms**, helping users detect targets rapidly when the cues are perceptually well aligned (Wolfe, 2021; Nothdurft, 1993). The laboratory findings are also consistent with Cybulski and Wielebski (2019) and Ge et al. (2025a, 2025b), who showed that pulsation and outline pulsation improve detection efficiency when users work in clear, low-distraction settings. The fact that variant C remained both fast and accurate across different environments suggests that pulsating color symbols are a strong candidate for real-world mapping applications, especially where users need to work quickly and reliably—for example, in monitoring hazards or environmental change.

Gender-based comparisons of response times revealed a few interesting differences, even though accuracy stayed similar for both men and women. In the online study, men and women worked at almost the same speed across all map variants. For example, in variant A, their median response times were nearly identical, suggesting that both groups processed the task at a similar pace when working in an uncontrolled, everyday environment.

In the laboratory setting, however, small but statistically significant differences appeared for some symbol types. In variants A and B, men completed tasks faster than women, even though both groups achieved perfect or near-perfect accuracy. This pattern is consistent with earlier research (Tlauka et al., 2005), which suggests that women may adopt more careful or deliberate search strategies when maps become visually complex. Other studies (Keskin et al., 2018) propose an alternative explanation: men and women may simply focus on different aspects of spatial information, leading to slightly different processing speeds without affecting correctness. Importantly, despite these timing differences, accuracy remained the same, showing that color-based symbol interpretation works equally well for all users.

No gender differences in response time were found for variants C and D in the lab, which reinforces an important point: pulsation (variant C) seems to be a very fluent and easy-to-process type of dynamic symbol. Once users are familiar with the map and the symbols are clearly visible, pulsation can be interpreted equally quickly by everyone, regardless of cognitive style or strategy. Taken together, these results suggest that while men and women may use different approaches to scanning a map, the cognitive accessibility of color—especially when enhanced by pulsation—remains consistently efficient across diverse user groups (Keskin et al., 2018).

The relationship between response time and confidence adds another layer to understanding how efficiently people work. In the online survey, some surprising patterns appeared—especially in variant A. Participants who said they were "very unsure" actually finished the task faster than those who felt "very sure," and this difference was statistically significant. This pattern suggests that unsure participants may have stopped searching too early, possibly due to frustration or decision fatigue. This matches what cognitive psychology calls the confidence–effort paradox, where lower confidence sometimes leads to less effort and faster—but not better—decisions (Koriat, 2012).

In the laboratory environment, confidence related to response time in more expected way. Highly confident participants usually worked faster than those with lower confidence, particularly in variants A and B. For example, in variant A, "very sure" participants completed the task in about 44 seconds, while neutral participants took around 20 seconds—an oddly faster result that may simply indicate shallow, surface-level processing rather than careful analysis. Overall, the lab results show that in controlled environments, confidence tends to match the amount of visual effort and attention participants invest.

Across both settings, the pulsation variant (C) stood out for its consistency. Participants at all confidence levels completed tasks quickly, and in the lab, both "very sure" and "unsure" groups finished in roughly the same short time (around 36 seconds). This suggests that pulsation supports automatic, easy-to-recognize cues that speed up interpretation whether or not users feel confident. Taken together, these findings show that color—especially when combined with pulsation—supports fast and effective visual search by helping users rely on both intuitive and deliberate processing with very little extra time cost (Dong et al., 2012, 2014a, 2014b).

Eye-tracking results help explain why some symbol types were more efficient than others. For variant C (pulsation), participants made fewer fixations, moved their eyes over larger distances, and showed the fastest saccadic movements. In simple terms, pulsation encouraged people to scan the map quickly and broadly, helping them spot important colors and patterns without getting stuck on individual points. This kind of wide, rapid scanning is often linked to more efficient visual search when the symbol's motion naturally draws attention (Dong et al., 2018a).

By comparison, variant D (static symbols) produced lower fixation counts and shorter fixation durations—metrics typically associated with ease of processing and visual familiarity (Dong et al., 2018b). However, this did not always result in faster responses or better accuracy in the lab. This mismatch shows that familiar static symbols are not always the most efficient choice, especially when the task requires looking across larger or more complex spatial patterns. In these situations, the added salience from motion—such as pulsation—can guide attention more effectively (Lloyd, 2005b; Midtbø & Baroš, 2005).

Legend use provides another clue. Participants using pulsation (variant C) spent far less time checking the legend—about 8 seconds, compared with over 13 seconds in variants A and B. This means they understood the symbols more quickly and didn't need to repeatedly confirm what the colors meant. Previous research (Gołębiowska et al., 2017) suggests that reduced legend dependency is a strong indicator of intuitive, low-effort interpretation.

Overall, the efficiency results from both the online and laboratory studies show that color-based encoding becomes even more powerful when paired with pulsation. In the online survey, static symbols seemed fastest, mainly because people relied on quick, familiar strategies in an uncontrolled environment. But once the task moved into the laboratory—where distractions were minimized—this speed advantage disappeared. Under these stable conditions, pulsation (variant C) became the most efficient design, delivering both fast response times and high accuracy.

Pulsation helped users scan widely and decide quickly, which was confirmed by the eye-tracking data: their eyes moved efficiently, made fewer unnecessary fixations, and rarely returned to the legend. Gender differences in response time reflected different search styles rather than differences in ability, since accuracy stayed the same. Confidence results also showed that pulsation supports both fast and confident decision-making, and even users who felt unsure still performed well. This suggests that pulsation encourages intuitive recognition, especially when certainty is low.

These findings fit well with theories on preattentive processing, which argue that color and motion work together to speed up visual search (Wolfe, 2021; Nothdurft, 1993). They also support earlier research by Cybulski and Wielebski (2019) and Ge et al. (2025a, 2025b) showing that pulsation improves detection efficiency when visual conditions are clear. From a design perspective, the message is straightforward: using pulsation alongside color makes color-based mapping both faster and more reliable, making it a strong choice for many real-world mapping applications.

### 6.3.4 How Quickly People Interpreted the Earthquake Timeline

The comparison of response times in the year-based search task shows how strongly motion design, the testing environment, and the nature of the task influence visual efficiency. In the online survey, static symbols (variant D) were the fastest to interpret, with a median response time of 52 seconds. Pictorial motion (variant B) came next, while pulsation (variant C) required more time to process. The slowest responses appeared with vibration (variant A)—especially among participants who answered correctly. This suggests that confusing or hard-to-interpret animations force users to slow down and spend more time analyzing the map (Lai & Yeh, 2004; Meyer et al., 2010).

These results agree with what cognitive load theory tells us: when people face motion types that feel unfamiliar or visually chaotic—like vibration—they need extra mental effort to figure out what the symbols mean, which increases response time (Wolfe, 2020, 2021). In contrast, static symbols are easy and familiar, so people can respond quickly, even if their interpretation is sometimes shallow or rushed (Piątek & Gołębiowska, 2022). The fact that incorrect answers in variant A were completed much faster highlights a classic speed–accuracy trade-off in online environments—people may decide too quickly when the animation is confusing or distracting.

Laboratory results showed almost the opposite pattern from the online survey. In the controlled lab environment, pulsation (variant C) became the fastest and most efficient design, clearly outperforming both vibration (A) and pictorial motion (B). Even the static variant (D), which was the quickest online—lost its advantage once distractions, screen differences, and environmental noise were removed. This shift highlights how strongly efficiency depends on viewing conditions. When the environment supports focused attention and high-quality perception, dynamic cues like pulsation can use their preattentive salience to speed up decision-making (Clifford & Jerit, 2014; Cybulski & Wielebski, 2019; Ge et al., 2025a, 2025b).

Across both settings, pictorial motion (B) consistently landed in the middle—neither as efficient as pulsation nor as slow as vibration. It was easy to follow, but did not fully benefit from motion-driven salience. Meanwhile, vibration (A) was the slowest in both environments, confirming that overly chaotic or unclear motion patterns increase cognitive load and slow down users (Cybulski, 2024; Harrower, 2007; Lloyd, 2005a).

Overall, these results show that symbol motion must be matched to the viewing context (Wolfe, 2020). When motion is clear and meaningful, it can greatly improve search efficiency. When it is confusing or noisy, it can backfire—especially if users are distracted (Hohnsbein & Mateeff, 1998; Meyer et al., 2010; Pratt et al., 2010). Pulsation, in particular, proved to be flexible and effective in both settings, making it a strong choice for real-world applications where users may be working under very different conditions.

Gender-based analysis of efficiency revealed only small, mostly nonsignificant differences. In the online survey, men and women completed the task at almost the same speed in every map variant. For example, in variant C, men needed 107 seconds and women 100 seconds—essentially the same. This suggests that in everyday, unstructured settings, both groups process temporal information at similar speeds (Keskin et al., 2018).

The laboratory results showed a bit more variation. In variant B (pictorial motion), women completed the task significantly faster than men, even though both groups were highly accurate. This may reflect slight differences in search strategy or visual fluency when interpreting pictorial movement. In other variants, including A and D, men and women worked at very similar speeds. For pulsation (variant C), women were only slightly slower than men, and the difference was not statistically meaningful. Overall, these findings show that while small strategy-related differences may appear—especially with certain types of motion in controlled settings—men and women are generally equally efficient in temporal-visual search. This fits with earlier studies showing that gender differences in spatial tasks depend more on the type of task and symbol design than on ability itself (Montello et al., 1999; Voyer et al., 1995). Importantly, motion-enhanced designs such as pulsation remain efficient and easy to use for everyone when the viewing conditions support focused attention.

The link between confidence and speed looked very different in the two study environments. In the online survey, people who felt less confident often finished faster. For example, in variant A, "very sure" participants needed almost 5 minutes (294 seconds), while "very unsure" participants answered in less than 2 minutes (117 seconds). This pattern suggests that low-confidence users sometimes rush through the task, while highly confident users spend more time double-checking their answers. These differences were not always statistically significant, but they point to a clear trend: online, confidence did not reliably predict efficiency.

In the laboratory, this pattern was still present but much weaker. In variant C, confident participants took 87 seconds, while unsure participants finished slightly faster. The same happened in variant D, where response times differed only by a few seconds. Despite these small differences, accuracy remained high for all confidence levels in the lab. This shows that in a supportive environment—quiet room, high-quality screen, no distractions—even participants who felt unsure could still work efficiently and give correct answers.

These results fit well with Koriat's (2012) self-consistency model, which explains that confidence depends not only on correctness but also on how easy or difficult the task feels. In a stable, well-designed setting, even low-confidence users can perform well because the visual conditions help them stay focused. For map design, this suggests a clear lesson: symbols should be easy to read and free from unnecessary distractions. When visual clarity is high (Gar-on et al., 2001), users do not need to rely on confidence to guide their decisions—making the map more accessible and effective for everyone.

Eye-tracking data adds an important layer to understanding why some symbol types were more efficient than others. In the laboratory setting—where participants tended to have stronger geographic and cartographic backgrounds, the pulsation design (variant C) encouraged a very distinctive scanning pattern. Participants made the fewest fixations, meaning their eyes jumped across the map in broader, more global movements instead of stopping frequently to inspect individual elements. Combined with the fast response times recorded in the lab, this suggests that pulsation helped users quickly take in the overall pattern and make decisions efficiently when their attentional resources were stable and the display quality was high (Keskin et al., 2020; Kiefer et al., 2016; Popelka et al., 2019).

Vibration (variant A) produced the opposite effect. It triggered the highest number of fixations, showing that participants felt the need to inspect the map more often and more carefully. This kind of dense, effortful scanning is typical when the visual cue is noisy or ambiguous. Even though the laboratory group had greater expertise, they still spent more time and attention interpreting vibration, resulting in slower responses. This supports earlier findings that unclear motion forces users to compensate with extra scanning rather than improving performance (Cybulski et al., 2023). Fixation duration did not differ dramatically between variants, although static symbols (D) produced slightly shorter fixations, consistent with the ease of processing familiar, motionless symbols.

Saccadic behavior—the speed and distance of eye movements, offers further insight. Pulsation generated the fastest saccades, meaning participants moved their eyes quickly across large portions of the map. This fits with their low fixation count and supports the idea of rapid, global scanning. However, high saccadic velocity can be a double-edged sword: while it speeds up search, it also increases the chance of "overshooting" important details, especially in tasks requiring precise timeline reasoning (Opach et al., 2014). Static symbols, in contrast, produced slower, more deliberate eye movements—reflecting a calmer, more controlled search strategy that can help in detail-oriented tasks but does not always speed up overall performance.

Legend dwell times tie these patterns together. With pulsation, participants needed very little time with the legend, indicating that the animated color-coded symbols were immediately recognizable without constant reference checking. In vibration and pictorial motion (A and B), users spent more time looking at the legend—likely because these motion types were harder to interpret and required repeated verification. This confirms a broader pattern: efficient visual search depends not just on speed, but on how clearly the visual design guides attention. When symbol behavior is intuitive—as with pulsation—users can work quickly and confidently with minimal extra checking. When the animation is ambiguous—as in vibration—users compensate with heavier scanning and more legend consultation, ultimately slowing the entire process (Peterson, 1999).

The efficiency results for the year-based search task show that how fast people work depends strongly on both the motion design and the environment in which the map is used. Online, where distractions are common, static symbols were the quickest because they are familiar and easy to interpret—though this speed sometimes reduced accuracy. In the laboratory, however, pulsation became the fastest and most effective option, using its strong visual salience to support rapid and accurate decisions. Eye-tracking data help explain this shift: pulsation encouraged broad, fast scanning and required very little time spent checking the legend, while vibration forced participants into slow, effortful, closely focused scanning that hurt efficiency. Static symbols led to calm and controlled eye movements, but lost their speed advantage once dynamic cues became easier to perceive under ideal lab conditions. Overall, the results show that efficiency is closely tied to context, and that pulsation—when carefully designed—can adapt well to different environments and support quick, accurate temporal interpretation.

# 7 Practical Lessons for Using Animated Symbols on Maps

To summarize the study as a whole, it is important to look at both sides of the results: the objective measures (such as accuracy, time, and eye-tracking data) and the subjective elements (such as user confidence, preferences, and feelings). When we bring these together, a consistent picture emerges. Dynamic point symbols generally perform at a similar level of effectiveness as traditional, "classic" static map symbols. In other words, using animation is not harmful by default. In some cases, dynamic symbols may even reduce performance slightly, but there are also clear situations where a specific type of animation can be more effective and more efficient than the static designs we are familiar with. This means that motion-based symbols are a valid option—and sometimes an even better one—when thoughtfully designed and appropriately applied.

A second important pattern concerns the relationship between confidence, speed, and performance. In general, participants who felt very confident tended to respond more quickly than those who reported neutral or lower confidence. Their answers came more freely and decisively. However, the less confident groups often achieved higher accuracy, suggesting that taking a little more time may have helped them think more carefully before responding. This does not mean the pattern was identical in every task, but across the study, it was noticeable: fast, confident answers were not always the most accurate, while slower, less confident answers sometimes led to better results. This highlights that confidence and speed do not necessarily reflect true understanding—and that thoughtful processing can pay off, even when users feel unsure.

## 7.1 ACCURACY, SPEED, AND EYE MOVEMENTS: THE BIG PICTURE

Overall, the results show that dynamic point symbols—especially pulsating ones—can boost both accuracy and speed in map-based tasks. However, this advantage only appears when the motion clearly relates to what the data represents and supports the specific goal of the task. In other words, animation is most effective not because it moves, but because the *right* kind of movement helps users notice what matters, search more efficiently, and make decisions with less effort.

- For the magnitude task, animated symbols worked about as well as classic size-based static symbols. This means motion can be just as effective as familiar static designs, but it isn't automatically better. However,

DOI: 10.1201/9781003728535-7

in the lab—where people could focus more—pictorial motion performed especially well. Importantly, this success came not from changing the size or another variable, but from the motion acting as a symbolic representation of the earthquake itself. In other words, even when the movement wasn't tied to the data value (like size or color), a motion that looks and feels like the real phenomenon was just as accurate in communicating magnitude as the traditional visual variables people already know.

- In the tasks based on symbol size (showing the number of fatalities), pulsating symbols led to strong performance—especially online, where they often outperformed the other designs. Even though pulsation was originally meant to express magnitude, its rhythmic movement helped users notice and compare sizes more easily. Importantly, the motion did not hide or interfere with the size information. Instead, it made the symbols more visible and supported better results in size-based tasks. Pictorial motion helped to some extent, but vibration tended to distract users, which lowered both accuracy and speed. Overall, this suggests that when motion directly supports the visual variable being interpreted—such as pulsation reinforcing size—it can make that variable easier to compare, count, and detect, leading to clearer and more confident decisions.
- In the tasks where color was used to show earthquake depth, participants performed extremely well in both the online and laboratory studies, confirming that color is a strong and reliable visual cue that works accurately even when users are distracted or have different levels of experience. When color was paired with pulsation, the symbols became more noticeable while accuracy remained just as high, showing that motion did not compete with or confuse the color information but simply helped it stand out. However, these strong results were also linked to the simplicity of the task—finding the most common color is a relatively easy visual judgment—so motion did not significantly change performance. It neither improved nor worsened results, suggesting that in simple map-reading tasks, such as spotting the most frequent color, adding motion is essentially neutral and can be used confidently without risking confusion or reduced accuracy.
- In the task where participants had to identify the correct year of the earthquakes, pulsation helped users in the online study, where distractions and split attention were more likely. The moving symbols made certain years stand out and guided users toward the right answer. However, in the laboratory—where people could focus more closely—static and vibration-based symbols worked better. In this setting, the attention-grabbing nature of pulsation sometimes got in the way of the careful, step-by-step reasoning needed to compare years accurately. This shows that motion isn't always helpful by default. Instead, it needs to be matched to the kind of thinking the task requires so that it supports interpretation rather than distracts from it.
- Eye-tracking supported these findings by showing that different symbol types led people to look at the map in different ways. Pulsation encouraged users to scan widely and efficiently—they made fewer fixations and relied less on the legend, suggesting that the symbols were easy to interpret

at a glance. Vibration, however, caused users to fixate more often, look in a narrower area, and check the legend more frequently—signs that the task felt harder and required more mental effort. Most importantly, success wasn't determined by how many fixations someone made, but by how well their attention was spread across the map. Users who explored the map in a broad, balanced way tended to be both more accurate and more efficient.

- Overall, gender differences in performance were minimal, with only a few instances reaching statistical significance. In the laboratory setting, there were some indications that women performed slightly better with structured, dynamic designs—especially with pictorial and pulsating symbols—which may reflect different approaches to visual attention or scanning strategies. However, both men and women achieved strong results across tasks, showing that well-designed animated symbols are broadly accessible and can be used effectively by all users, regardless of gender.
- Analysis of confidence levels showed that how sure participants felt did not always match how well—or how accurately—they performed. In the online study, highly confident users often answered quickly but made more mistakes, especially when motion made the task harder to interpret. By contrast, in the laboratory, even participants who felt unsure frequently answered correctly, likely because the controlled setting helped them focus and process information more carefully. Across the study, a clear pattern emerged: very confident participants tended to respond faster, while those who felt less confident often took more time—and sometimes achieved higher accuracy. In other words, fast and confident did not automatically mean correct, and slower, more cautious responses sometimes led to better outcomes. This pattern was reflected at times in gender differences as well: women tended to use a more careful strategy, taking slightly longer and feeling less certain, but making fewer mistakes, while men were often faster and more confident, yet occasionally less accurate.
- The online surveys made it possible to gather data from various users and identify broad trends in how people interpreted the symbols. However, because participants completed the tasks in uncontrolled environments, their responses could be influenced by factors like distractions, lighting, or screen quality. These external conditions may have indirectly affected how quickly or accurately they answered.
- Laboratory eye-tracking studies offered a detailed view of how people visually explored the maps, helping to explain the broader patterns found in the online survey. In the controlled lab setting—with no distractions and high-quality displays—participants were able to take better advantage of subtle motion effects like pulsation. This shows that the effectiveness of animated symbols is closely tied to the viewing environment and how clearly the map is presented.

Dynamic symbols can work very well for mapping earthquakes. The study shows that animations—especially pulsation and pictorial motion—can improve performance in some tasks, but they are not always the best choice in every situation.

When the movement fits the purpose of the task and makes sense visually, animated symbols can help users find answers more accurately and more quickly. Pulsation was especially effective in the lab, where its clear rhythm associated with the specific variable of size helped people make confident decisions with less effort. However, the success of motion depends on several factors: how complex the task is, the conditions in which the map is viewed, and how users process visual information. This means that choosing the right type of motion—and using it in the right context—is crucial. The results also show that good design should be based on how people pay attention and perceive visual cues, not just on how fast or accurate they are. In short, animation can be a powerful tool, but it needs thoughtful, science-based design to work effectively.

In this study, earthquakes were used as a clear example of a phenomenon that changes across space and time. They were shown on the map using three visual variables: color for depth, size for fatalities, and motion for magnitude. By treating motion as a visual variable alongside familiar ones like size and color, the study demonstrated that animation can be part of a general cartographic toolkit—not just a special effect. Most importantly, this approach is not limited to earthquakes. The same method can be applied to many other geographic processes, such as volcanic eruptions, floods, pollution events, migration, or any situation where events happen at different locations and times. The real contribution of this research goes beyond proving that dynamic symbols can work—it provides practical, evidence-based design principles grounded in how people see, think, and process visual information. By explaining why certain animated designs help users and when they should be used, the study builds a bridge between cartography and cognitive science, making its findings useful for a wide range of spatial visualization challenges.

## 7.2 TIPS FOR DESIGNING EFFECTIVE ANIMATED MAP SYMBOLS

The results of this study—covering four different tasks, two testing environments, and several types of animated symbols—provide a solid basis for practical design tips. Although the research focused on earthquakes, the lessons apply to many other maps that show things changing over space and time. Examples include volcanic eruptions, floods, wildfires, disease spread, migration, or even social and economic patterns like unemployment. All of these involve events happening in different places and at different moments, so they require symbols that can show both where something occurs and how it changes over time. The key takeaway is that animation can be a useful addition to traditional visual tools like size and color—but only when the motion makes sense for what it represents or matches other visual variables. When the movement feels natural and matches the meaning of the data, it can help users understand patterns more clearly and quickly. When motion is unclear, confusing, or unrelated to the information, it can become distracting instead. For example, using a vibrating animation on circles of different sizes may make it harder for users to read the map accurately, whereas a smooth pulsating motion can highlight size differences and make interpretation easier than with no animation at all.

These design tips are grounded in three areas of evidence from the study:

- Effectiveness (how accurate users were),
- Efficiency (how fast they worked), and
- Eye-tracking (how they visually explored the map).

Together, these findings show that well-designed motion can support understanding without overwhelming users. The approach tested here is therefore not just a method for visualizing earthquakes—it offers a transferable design toolkit that other cartographers can use for many types of dynamic geographic information. By combining real user data with principles from visual perception and cartographic theory, this study helps build a clearer picture of how animated symbols can support spatial thinking and improve communication in both physical and human geography.

### 7.2.1 Design Motion That Matches the Meaning of the Data

A central finding of this study is that motion can be just as effective as traditional static symbols when it is meaningfully linked to the data it represents. In the magnitude task, dynamic motion performed on par with classical size-based symbols, showing that animation itself does not reduce accuracy and may provide a viable alternative for cartographic design. Pulsation, in particular, proved useful in tasks where its rhythmic character supported the interpretation of other visual variables, such as symbol size (fatalities) and color (depth). In these cases, participants achieved high accuracy and showed efficient gaze patterns, suggesting that motion can draw attention without causing confusion when it reflects the meaning of the data. Eye-tracking data confirmed this: pulsation encouraged broader and more efficient scanning across the map, helping users explore the full spatial pattern rather than getting "stuck" in narrow areas of focus, as often happened with static symbols. This suggests that motion can guide attention in a productive way, helping users see the bigger picture more quickly. Because of this, the motion types tested in this study are not only useful for earthquake maps—they can also be applied to other geographic phenomena that change across space and time.

### 7.2.2 Choose the Right Type of Motion for the Task

Another major takeaway from the study is that not all motion works the same way. The effectiveness of an animation depends on what the user is trying to find and how well the motion fits that goal. Pulsation was the strongest overall performer. It worked especially well in tasks based on color (depth) and size (fatalities) because its steady "in-and-out" rhythm made important symbols stand out without causing confusion. In the lab, where users could focus more easily, pulsation led to both high accuracy and efficient searching. Pictorial motion—movement inside the symbol that resembled shaking—was the best choice for magnitude tasks. Because its motion looked like real earthquake activity, users could intuitively link what they saw with what the data meant. Importantly, pictorial motion did not conflict with other visual variables like size or color, because it carried its own distinct meaning. This shows that motion works best when it visually echoes the real-world phenomenon it represents. Vibration, however, was the weakest motion type. It led to more fixations, longer search times, and signs of cognitive overload. Its fast, jittery movement made it harder for users to stay focused, which reduced accuracy and slowed down performance across most tasks. Overall, these results show that motion shouldn't be used the same way in every situation. The type of animation needs to match the task and the meaning of the data. When the motion makes sense and supports interpretation, it can be a powerful cartographic tool. When it doesn't, it can easily become distracting and counterproductive.

### 7.2.3 Motion Performance Depends on Viewing Conditions

This study showed that the environment in which people view a map has a major impact on how well motion-based symbols work. In the large online survey—where participants used different screens, worked in unknown environments, and were likely surrounded by distractions—pulsation often helped users perform better in more difficult tasks, such as identifying the correct year. In this setting, the motion seemed to help draw attention and compensate for noise and uncertainty. However, the laboratory told a different story. In the controlled, quiet setting—with high-quality screens and focused participants—static designs sometimes worked better than animated ones, especially for tasks that required careful, step-by-step reasoning about time. With fewer distractions, users did not need motion to guide their attention, and in some cases, animation added extra visual load instead of helping. Together, these findings show that online and lab studies differ not only in the kind of data they provide—broad behavioral patterns versus detailed cognitive insights—but also in how people actually perform. Motion seems most helpful when attention is fragile, such as in noisy or distracting environments. Under ideal viewing conditions, it may be less necessary—or even counterproductive—if it adds complexity without supporting the task.

### 7.2.4 Motion-Based Symbols Aid Users with Different Backgrounds

A key finding of this study is that well-designed animated symbols worked effectively for various types of users—regardless of gender, map experience, or confidence level. Pulsating symbols, in particular, supported accurate and efficient answers even among people who felt unsure about their performance. This suggests that motion can help level the playing field by guiding attention in a way that reduces the impact of individual differences in skill or experience. This makes dynamic symbols especially valuable in situations where maps are used by the general public, not just experts. In real-world settings, users vary widely—some are confident, some are uncertain, some are experienced with maps, and others are not. The fact that pulsation still helped users perform well shows that motion-based designs can support clear understanding even when users have different backgrounds and abilities. Importantly, this lesson is not limited to earthquakes. Many geographic phenomena also change over space and time and must be communicated to mixed audiences. Because motion-enhanced symbols proved effective across diverse users in this study, they have strong potential as a general tool for explaining dynamic processes in both natural and human environments.

## 7.3 WHAT THIS STUDY DIDN'T COVER—AND WHAT SHOULD COME NEXT

Although this study offers strong evidence about how animated point symbols work in earthquake maps, it's important to recognize its limits—and how they can guide future research. Like any empirical project, the study had practical boundaries. Time, equipment, and participant access limited both the number of people involved and the variety of tasks that could be tested. The research used two complementary methods—a large online survey and a focused laboratory eye-tracking experiment—but repeating these studies with different user groups, devices, or regions would help confirm how broadly the results apply.

A key limitation lies in the restricted design space that was tested. Only a small subset of possible motion types and parameter settings was examined: each animation had a fixed rhythm, speed, and amplitude, and all symbols were simple point markers placed on a relatively clean base map. In reality, cartographers work with a much richer palette—different animation speeds, easing functions, motion paths, symbol shapes, background complexity, and combinations with interactivity (such as data filtering and brushing). The tasks in this study were also deliberately focused on fairly "classical" operations: searching, counting, and identifying the most frequent values. More complex tasks—such as spotting emerging trends, comparing time periods, detecting anomalies, or making forecasts—might interact with motion in different ways. Future work should therefore explore a wider motion design space, include more diverse task types, and consider interactive controls, accessibility issues (e.g., motion sensitivity, color vision deficiencies), and long-term use. Only by systematically extending this experimental framework can we fully understand when animated symbols support users—and when they risk getting in the way.

Acknowledging these limits is not a weakness—it points to the next steps. To keep advancing dynamic cartography in a meaningful way, future studies should build gradually, repeat key findings, and explore new contexts. By doing so, researchers can ensure that animated map symbols continue to develop in ways that are both scientifically grounded and useful for real-world communication.

### 7.3.1 SAMPLE SIZE AND PARTICIPANT DIVERSITY

A core limitation of research involving human participants is the number of people who can realistically be included in a study. Large samples are ideal, but laboratory-based cartographic research must operate within practical time, staffing, and financial constraints. As a result, the lab component of this study involved a relatively small number of participants. To compensate for this, a dual-method approach was used. The controlled eye-tracking experiment was paired with a large-scale online survey that gathered responses from more than 1,000 participants representing a broad range of ages and educational backgrounds. Together, these two datasets provided both depth (from controlled lab conditions) and breadth (from diverse online participants), helping to strengthen the overall reliability of the findings.

However, each method has its own limitations. The laboratory group consisted of students from the Faculty of Geographical and Geological Sciences, who likely had greater experience with maps and spatial thinking than the general public. This introduces a bias toward more skilled users. The online questionnaire, while broader and more inclusive, lacked control over participants' environments, screen quality, attention levels, and devices. These uncontrolled factors could influence visual perception and response accuracy, and must therefore be taken into account when interpreting the results.

Many previous studies suggest focusing on specific user groups, but a major challenge for the field now is not just studying isolated groups—it is integrating findings across them. What is missing is a consolidated, cross-group understanding of how different users interact with maps. Instead of scattered case studies, cartography needs a clearer, unified picture of user behavior.

In this regard, emerging tools such as machine learning and artificial intelligence offer promising opportunities. These methods can analyze very large datasets, identify shared patterns within and across user groups, and detect individual differences in how people scan or interpret maps. Such approaches could eventually support adaptive cartographic systems—maps that respond to users' needs, skills, and attention patterns—instead of relying on one-size-fits-all designs.

### 7.3.2 Task Type and Domain Specificity

This study focused on four tasks using an earthquake map: identifying magnitude, determining depth, estimating fatalities, and finding the year with the most events. Although these tasks were framed around earthquakes, they were built on universal cartographic principles—specifically, the use of motion, size, and color as visual variables. Because these variables are widely used in geographic visualization, the methods and conclusions from this study can be applied beyond seismology to many other physical and socioeconomic phenomena. This makes the findings relevant for topics such as natural hazards, environmental change, disease spread, migration, or other processes that unfold across space and time.

However, the study's scope was intentionally narrow. Only a limited set of visual variables and task types was examined. More complex, multistep tasks—or additional visual encodings—could have offered a richer picture of user interaction, but expanding the task set would have significantly increased the length and difficulty of both the online survey and the laboratory experiment. Eye-tracking sessions are already time-intensive, and extending them further risks participant fatigue, which can distort visual behavior and reduce data quality. Likewise, more complicated tasks in the online survey could have discouraged participation, led to incomplete responses, or increased misunderstanding among users with varied backgrounds.

For these reasons, the study focused on clear, manageable tasks that balanced scientific rigor with practical feasibility. Still, this creates an important direction for future work. Research should move toward examining more advanced, multilayered map interactions—such as decision-making across multiple data layers, dynamic filtering, or real-time monitoring. These complex scenarios remain understudied in cartography, yet they increasingly reflect how maps are used in modern applications. Expanding research in this area would help fill a critical gap in understanding how people interpret and act on more sophisticated cartographic products.

### 7.3.3 Symbol Density and Scene Complexity

In this study, the number of map symbols was intentionally controlled so that all map variants could be fairly compared. However, real-world maps are often much more crowded. They may contain hundreds or thousands of points, multiple data layers, labels, and changing information. Under these conditions, animated symbols might behave very differently from how they did in this experiment. The findings—especially the visual overload observed with vibration-based motion—suggest that dense displays could quickly overwhelm users. This raises an important question for future research: How well do animated symbols scale when the map becomes crowded? Increasing symbol density may affect attention, slow down visual search, and increase cognitive fatigue, especially when users have limited time to interpret the map.

Although psychology and vision science offer extensive research on visual overload, only a small portion of this knowledge has been applied directly to cartography. Cartographic experiments face unique constraints: the number of variables that can be tested is limited, and maps must be standardized to keep conditions comparable across participants. This simplification is a typical limitation of experimental map studies, but it is also necessary to produce reliable and controlled results.

A promising direction for future work is the use of machine learning and artificial intelligence to identify when maps are likely to become visually difficult. These tools could automatically detect situations such as overly dense symbol placement or symbols that are too similar to distinguish and then suggest design adjustments based on established cartographic research. This points toward the development of adaptive cartography—maps that can adjust to the user's needs or help designers optimize symbol density during the map-making process. Such approaches could help maintain readability and reduce visual strain, especially in complex, data-rich environments.

### 7.3.4 Motion Type Scope

This study examined three motion types—pulsation, vibration, and pictorial motion—applied consistently across all map symbols. Even within this small set, the results showed clear differences in how people searched for information and interpreted the maps. However, these three designs represent only a narrow portion of what animated cartography could explore. Many other motion forms—such as objects moving along paths, changes in acceleration, flickering or blinking, rotation, or complex irregular rhythms—remain largely untested in cartographic research. Each of these could potentially convey different kinds of information, such as direction of movement, flow intensity, change over time, or even uncertainty.

For this reason, future studies should move beyond these basic motion types and investigate a wider range of animation possibilities. This might include combining multiple motion parameters—for example, altering both speed and direction—or testing nonlinear and irregular motion patterns. Such research would help identify perceptual thresholds: points at which motion becomes too weak to notice or too strong and therefore distracting. It would also clarify how different motion types influence cognitive load and visual attention.

Expanding this motion "vocabulary" would give cartographers a richer set of tools for representing complex geographic processes, such as traffic and migration flows, climate dynamics, wildfire spread, or other natural and human-driven phenomena. Ultimately, a broader understanding of motion design would allow animations to be more precisely matched to the meaning of the data, improving both clarity and usability in dynamic maps.

### 7.3.5 Device and Environmental Variability

A key limitation of this study is the difference between the two testing environments. The laboratory experiment was carried out under highly controlled conditions—participants used the same equipment, worked in similar lighting, and were free from outside distractions. This makes the results reliable and easier to repeat in future studies. However, it also means the setting does not fully reflect how people typically use maps in everyday life.

By contrast, the online study involved very little control. Participants may have viewed the maps on laptops, desktops, tablets, or phones, in bright or dim environments, while multitasking, tired, or distracted. These uncontrolled factors are common in online research and impossible to eliminate, but they also mirror real-world map use much more closely. People rarely study maps under perfect conditions.

Because of this gap, future research should move toward more realistic usage scenarios—testing multilayered maps on different devices, in varied environments, and during real activities. For example, studies could explore how animated maps support pedestrians navigating a city, cyclists dealing with changing traffic, or field workers monitoring hazards outdoors. Such approaches would reveal how design choices perform when users must act quickly, divide their attention, or interact on the move.

It is also important to note that research on mobile cartography is still relatively limited, even though mobile devices are now the most common way people access maps. Today, smartphones dominate map use, but in the near future, smartwatches, smartbands, or even smart rings may become important interfaces for spatial information. These devices introduce new constraints—small screens, brief interactions, and constant movement—which may change how motion-based symbols are perceived. Future studies on animated cartographic methods should therefore consider these technologies, as they pose both challenges and exciting opportunities for dynamic map design and user interaction.

### 7.3.6 Integration of Real-Time Eye-Tracking Feedback

Eye-tracking offers powerful insight into how people look at maps—where they focus, what they ignore, and how they search for information. However, in current research and practice, eye-tracking is almost always used after a study is finished.

It helps us evaluate a design, but it does not influence the map while the user is actually working with it.

A promising future direction is the development of real-time, adaptive map systems that respond directly to a user's gaze. In such systems, the map could automatically adjust symbol visibility, motion intensity, or highlight important areas based on where the user is looking and how long their eyes stay in one place. If the system detects confusion—for example, long searching in one area or repeatedly returning to the legend—it could increase symbol salience or guide attention more clearly.

This idea has strong practical value. In real-world contexts—such as pedestrian navigation, land surveying, or emergency response—users often need to locate critical information quickly and under pressure. A gaze-responsive map could assist them in real-time if they miss important symbols or struggle to interpret the display. By reducing cognitive effort and improving clarity, such adaptive interfaces could support more personalized and efficient map use in both everyday situations and professional fieldwork.

Technologically, this direction is becoming increasingly realistic. Remote eye-tracking—using webcams, mobile cameras, or facial tracking algorithms—is already possible. However, a major gap remains: we still lack clear guidelines for how a map should adapt based on specific gaze behaviors. For example, what should a map do if users scan too narrowly, overlook key areas, or repeatedly misread symbols?

Future research will need to define these rules, developing a framework that links gaze patterns to meaningful map adjustments. Doing so would move cartography toward a new generation of smart, adaptive, and cognitively aware mapping systems—maps that don't just display information, but actively help users understand it.

# Appendix 1

A catalogue of earthquake maps scrutinized by the author, encompassing an assessment of the information provided, mapping methods utilized, and interactive functionalities offered.

1. https://earth3dmap.com/earthquake-live-map/

| Information provided | Mapping method | Interactive functions |
|---|---|---|
| Magnitude; Depth; Date; Coords | Graduated color symbols | Selection; Zoom; Pan |

2. https://erccportal.jrc.ec.europa.eu/ECHO-Products/Maps#/maps?pageIndex=1&mt=Daily%20Map

| Information provided | Mapping method | Interactive functions |
|---|---|---|
| Magnitude; Depth; Date; Fatalities; People affected; Geographic location | Graduated color symbols | – |

3. https://earthquake.usgs.gov/earthquakes/map/?extent=21.0845,-130.16602&extent=52.3756,-59.85352

| Information provided | Mapping method | Interactive functions |
|---|---|---|
| Magnitude; Depth; Date; Coords; Faults | Proportional color symbols | Selection; Zoom; Pan; Brushing |

4. https://seismic-explorer.concord.org/

| Information provided | Mapping method | Interactive functions |
|---|---|---|
| Magnitude; Depth; Geographic location; Date | Graduated color symbols | Selection; Zoom; Pan; Brushing; Animation |

5. https://nagix.github.io/world-eq-locator/

| Information provided | Mapping method | Interactive functions |
|---|---|---|
| Magnitude; Depth; Geographic location; Date | Color point symbols | Selecting; Zoom; Pan |

6. https://www.youtube.com/@PacificTWC

| Information provided | Mapping method | Interactive functions |
|---|---|---|
| Magnitude; Depth; Geographic location; Year | Proportional color symbols | Animation |

7. https://www.bgs.ac.uk/discovering-geology/earth-hazards/earthquakes/where-do-earthquakes-occur/

| Information provided | Mapping method | Interactive functions |
|---|---|---|
| Magnitude; Geographic location | Point symbols | Selecting; Zoom; Pan |

8. https://www.nsf.gov/news/mmg/mmg_disp.jsp?med_id=64691

| Information provided | Mapping method | Interactive functions |
|---|---|---|
| Depth; Geographic location | Color point symbols | – |

9. https://earthquakescanada.nrcan.gc.ca/historic-historique/caneqmap-en.php

| Information provided | Mapping method | Interactive functions |
|---|---|---|
| Magnitude; Geographic location | Graduated point symbols | – |

10. https://ds.iris.edu/seismon/index.phtml

| Information provided | Mapping method | Interactive functions |
|---|---|---|
| Magnitude; Depth; Date; Coords | Graduated color symbols | Selecting |

11. https://projects.oregonlive.com/maps/earthquakes/

| Information provided | Mapping method | Interactive functions |
|---|---|---|
| Magnitude; Depth; Date; Coords | Proportional point symbols | Selecting; Zoom; Pan |

12. https://arcgis-content.maps.arcgis.com/apps/mapviewer/index.html?webmap=c85b477ca58a4b968fc3ce1878b0b564

| Information provided | Mapping method | Interactive functions |
|---|---|---|
| Magnitude; Depth; Date; Geographic location | Graduated color symbols | Selecting; Zoom; Pan; Brushing |

13. https://www.christchurchquakemap.co.nz/

| Information provided | Mapping method | Interactive functions |
|---|---|---|
| Magnitude; Depth; Date; Geographic location | Color point symbols | Selecting; Zoom; Pan; Animation |

14. https://www.japanquakemap.com/

| Information provided | Mapping method | Interactive functions |
|---|---|---|
| Magnitude; Depth; Date; Geographic location | Color point symbols | Selecting; Zoom; Pan; Animation |

15. https://www.ngdc.noaa.gov/hazard/data/publications/significant-earthquakes-poster.pdf

| Information provided | Mapping method | Interactive functions |
|---|---|---|
| Magnitude; Geographic location; Fatalities | Graduated color symbols | – |

16. https://earthquakes.volcanodiscovery.com/

| Information provided | Mapping method | Interactive functions |
|---|---|---|
| Magnitude; Depth; Date; Coords | Graduated point symbols | Selecting; Zoom; Pan |

17. https://glowy-earthquakes.glitch.me/?adumkts=social&aduc=social&adum=external&aduca=social_technical&adusf=linkedin&adut=3d76a431-1519-4725-b169-939f5120340f

| Information provided | Mapping method | Interactive functions |
|---|---|---|
| Magnitude; Depth; Hypocenter location; Date | Proportional color symbols | Selecting; Zoom; Pan |

18. https://www.arcgis.com/apps/Minimalist/index.html?appid=62915c919a5842348edd1dd62feab263

| Information provided | Mapping method | Interactive functions |
|---|---|---|
| Magnitude; Geographic location | Graduated point symbols | Zoom; Pan |

19. https://www.visualcapitalist.com/wp-content/uploads/2023/02/Mapping-The-Worlds-Major-Earthquakes-from-1956%E2%80%922022.html

| Information provided | Mapping method | Interactive functions |
|---|---|---|
| Magnitude; Geographic location | Color point symbols | – |

20. https://ralucanicola.github.io/JSAPI_demos/last-earthquakes/

| Information provided | Mapping method | Interactive functions |
|---|---|---|
| Magnitude; Depth; Date; Coords; | Graduated point symbols; | Selecting; Zoom; Pan; |

21. https://ttweathercenter.com/category/recent-earthquakes/

| Information provided | Mapping method | Interactive functions |
|---|---|---|
| Magnitude; Depth; Date; Coords; Faults | Point symbols | – |

22. https://www.frontiersin.org/articles/10.3389/feart.2019.00042/full

| Information provided | Mapping method | Interactive functions |
|---|---|---|
| Magnitude; Geographic location; Fatalities; Plate boundaries | Graduated color symbols | – |

23. https://seismo.berkeley.edu/seismo.real.time.map.html

| Information provided | Mapping method | Interactive functions |
|---|---|---|
| Magnitude; Depth; Date; Geographic location | Proportional color symbols | Selecting; Zoom; Pan; Brushing |

24. https://www.aljazeera.com/news/2023/2/6/turkey-syria-earthquake-death-toll-and-devastation-live-tracker

| Information provided | Mapping method | Interactive functions |
|---|---|---|
| Magnitude; Geographic location; Fatalities | Proportional point symbols | – |

25. https://www.bgs.ac.uk/news/the-kahraman-maras-earthquake-sequence-turkey-syria/

| Information provided | Mapping method | Interactive functions |
|---|---|---|
| Magnitude; Geographic location | Proportional point symbols | – |

26. https://peteresis.github.io/earthquake_mapping_demo_site/

| Information provided | Mapping method | Interactive functions |
|---|---|---|
| Magnitude; Geographic location | Proportional color symbols | Selecting; Zoom; Pan; Layers |

27. https://mediaspace.esri.com/media/t/1_yib094sg

| Information provided | Mapping method | Interactive functions |
|---|---|---|
| Magnitude; Depth; Geographic location; Year | Color point symbols | Animation |

28. https://www.businessinsider.com/map-earthquakes-magnitude-animated-1900-2016-8?IR=T

| Information provided | Mapping method | Interactive functions |
|---|---|---|
| Magnitude; Geographic location; Year | Color point symbols | Animation |

29. http://eqseis.geosc.psu.edu/cammon/Japan2011EQ/

| Information provided | Mapping method | Interactive functions |
|---|---|---|
| Magnitude; Depth; Geographic location | Proportional color symbols | – |

30. https://stefcroo.medium.com/exploring-groningen-earthquake-data-with-python-4dde2586b7ca

| Information provided | Mapping method | Interactive functions |
|---|---|---|
| Geographic location; Date | Heatmap | – |

# References

Acevedo, W., Masuoka, P. (1997). Time-series animation techniques for visualization urban growth. *Computers & Geosciences*, 23(4), 423–435. https://doi.org/10.1016/S0098-3004(97)00006-X

Amante, C., Eakins, B. W. (2009). ETOPO1 1 arc-minute global relief model: Procedures, data sources and analysis. *NOAA Technical Memorandum NESDIS NGDC-24*. National Geophysical Data Center, NOAA. https://doi.org/10.7289/V5C8276M

Amorim, F. R., Schmidt, M. A. R. (2021). Classification of dynamic cartographic symbols applied to Augmented Reality (AR) systems. *Proceedings of the International Cartographic Association*, 4, 8. https://doi.org/10.5194/ica-proc-4-8-2021

Andrienko, G., Andrienko, N., Demsar, U., Dransch, D., Dykes, J., Fabrikant, S. I., Jern, M., Kraak, M.-J., Schumann, H., Tominski, C. (2010). Space, time and visual analytics. *International Journal of Geographic Information Science*, 24(10), 1577–1600. https://doi.org/10.1080/13658816.2010.508043

Andrienko, N., Andrienko, G., Gatalsky, P. (2003). Exploratory spatio-temporal visualization: An analytical review. *Journal of Visual Languages & Computing*, 14(6), 503–541. https://doi.org/10.1016/S1045-926X(03)00046-6

Andrienko, N., Andrienko, G., Pelekis, N., Spaccapietra, S. (2008). Basic concepts of movement data. In: Giannotti, F., Pedreschi, D. (eds) *Mobility, Data Mining and Privacy – Geographic Knowledge Discovery*. Springer, Berlin, pp. 15–38. https://doi.org/10.1007/978-3-540-75177-9_2

Archambault, D., Purchase, H. C. (2016). Can animation support the visualisation of dynamic graphs? *Information Sciences*, 330, 495–509. https://doi.org/10.1016/j.ins.2015.04.017

Archambault, D., Purchase, H., Pinaud, B. (2011). Animation, small multiples, and the effect of mental map preservation in dynamic graphs. *IEEE Transactions on Visualization and Computer Graphics*, 17(4), 539–552. https://doi.org/10.1109/TVCG.2010.78

Baranowski, M., Bielecka, E., Dukaczewski, D. (2009). Methods of portraying spatial data used in official geoinformation services in Poland – A comparative study. In: Gartner, G., Ortag, G. (eds) *Cartography in Central and Eastern Europe*. Lecture Notes in Geoinformation and Cartography, pp. 41–62. Springer, Berlin, Heidelberg. https://doi.org/10.1007/978-3-642-03294-3_3

Bartram, L. (1997). Perceptual and interpretative properties of motion for information visualization. *NPIV ,97: Proceedings of the 1997 Workshop on New Paradigms in Information Visualization and Manipulation*, Las Vegas Nevada USA, November 10–14, pp. 3–7. https://doi.org/10.1145/275519.275520

Bartram, L., Ware, C., Calvert, T. (2003). Moticons: Detection, distraction and task. *International Journal of Human-Computer Studies*, 58(5), 515–545. https://doi.org/10.1016/S1071-5819(03)00021-1

Battersby, S. E., Goldsberry, K. P. (2010). Consideration in design of transition behaviors for dynamic thematic maps. *Cartographic Perspectives*, 65, 16–32. https://doi.org/10.14714/CP65.127

Beaumont, P., Longley, P. A., Maguire, D. J. (2005). Geographic information portals: A UK perspective. *Computers, Environment and Urban Systems*, 29(1), 49–69. https://doi.org/10.1016/j.compenvurbsys.2004.05.010

Behroozi, M., Lui, A., Moore, I., Ford, D., Parnin, C. (2018). Dazed: Measuring the cognitive load of solving technical interview problems at the whiteboard. *Proceedings of the 40th International Conference on Software Engineering: New Ideas and Emerging Results*, Gothenburg, Sweden, 27th May–3rd June, 93–96. https://doi.org/10.1145/3183399.3183415

Beitlova, M., Popelka, S., Konopka, M., Macku, K. (2023). Verification of cartographic communication models using detection of map reading strategies based on eye movement recording. *The Cartographic Journal*, 60(3), 5–24. https://doi.org/10.1080/00087041.2022.2150362

Bertin, J. (1981). *Graphics and Graphic Information-Processing*. Walter de Gruyter, Berlin, New York.

Biland, J., Çöltekin, A. (2017). An empirical assessment of the impact of the light direction on the relief inversion effect in shaded relief maps: NNW is better than NW. *Cartography and Geographic Information Science*, 44(4), 358–372. https://doi.org/10.1080/15230406.2016.1185647

Bláha, J. D. (2009). Various Ways of Assessment of Cartographic Works. In: Gartner, G., Ortag, F. (eds) *Cartography in Central and Easter Europe*. Lecture Notes in Geoinformation and Cartography, pp. 211–229. Springer, Berlin, Heidelberg. https://doi.org/10.1007/978-3-642-03294-3_14

Blok, C. (1997). Dynamic visualization in a developing framework for the representation of geographic data. *Cybergeo: European Journal of Geography*, https://doi.org/10.4000/cybergeo.509

Board, C. (1978). Map reading tasks appropriate in experimental studies in cartographic communication. *Cartographica*, 15(1), 1–12. https://doi.org/10.3138/AG15-V252-3726-W346

Board, C. (1981). Cartographic communication. *Cartographica*, 18(2), 42–78. https://doi.org/10.3138/8R07-2125-L843-0767

Bormann, P., Dewej, J. W. (2014). The new IASPEI standards for determining magnitudes from digital data and their relation to classical magnitudes. In: Bormann, P. (ed) *New Manual for Seismological Observatory Practice 2 (NMSOP-2)*, pp. 1–44. Deutsches GeoForschungZentrum GFZ, Potsdam, Germany. https://doi.org/10.2312/GFZ.NMSOP-2_IS_3.3

Bradley, M. M., Miccoli, L., Escrig, M. A., Lang, P. J. (2008). The pupil as a measure of emotional arousal and autonomic activation. *Psychophysiology*, 45(4), 602–607. https://doi.org/10.1111/j.1469-8986.2008.00654.x

Braile, L. W. (2009). Seismic monitoring. In: Young, R., Norby, L. (eds) *Geological Monitoring*, pp. 229–244. The Geological Society of America, Boulder, CO. https://doi.org/10.1130/2009.monitoring(10)

Brewer, C. A., Hatchard, G. W., Harrower, M. A. (2003). ColorBrewer in print: A catalog of color schemes for maps. *Cartography and Geographic Information Science*, 30(1), 5–32. https://doi.org/10.1559/152304003100010929

Brown, L. E., Halpert, B. A., Goodale, M. A. (2005). Peripheral vision for perception and action. *Experimental Brain Research*, 165(1), 97–106. https://doi.org/10.1007/s00221-005-2285-y

Bryant, E. (2005). *Natural Hazards*, 2nd ed. Cambridge University Press, New York.

Brychtová, A., Çöltekin, A. (2016). An empirical user study for measuring the influence of colour distance and font size in map reading using eye tracking. *The Cartographic Journal*, 53(3), 202–212. https://doi.org/10.1179/1743277414Y.0000000103

Brychtová, A., Çöltekin, A. (2017). The effect of spatial distance on the discriminability of colors in maps. *Cartography and Geographic Information Science*, 44(3), 229–245. https://doi.org/10.1080/15230406.2016.1140074

Buziek, G. (2000). Legend designs for noninteractive cartographic animations. *Computers & Geosciences*, 26(1), 21–28. https://doi.org/10.1016/S0098-3004(99)00029-1

Bylinskii, Z., Borkin, M.A., Kim, N. W., Pfister, H., Oliva, A. (2017). Eye fixation metrics for large scale evaluation and comparison of information visualizations. In: Burch, M., Chuang, L., Fisher, B., Schmidt, A., Weiskopf, D. (eds) *Eye Tracking and Visualization*. ETVIS 2015. Mathematics and Visualization, pp. 235–255. Springer, Cham. https://doi.org/10.1007/978-3-319-47024-5_14

Calvo, L., Cucchietti, F., Pérez-Montoro, M. (2022). Measuring the effectiveness of static maps to communicate changes over time. *IEEE Transactions on Visualization and Computer Graphics*, 29(10), 4243–4255. https://doi.org/10.1109/TVCG.2022.3188940

Campbell, C. S., Egbert, S. L. (1990). Animated cartography/thirty years of scratching the surface. *Cartographica*, 27(2), 24–46. https://doi.org/10.3138/V321-5367-W742-1587

Caquard, S. (2009). Foreshadowing contemporary digital cartography: A historical review of cinematic maps in films. *The Cartographic Journal*, 46(1), 46–55. https://doi.org/10.1179/000870409X415589

Caramanna, G., Ciotoli, G., Nisio, S. (2008). A review of natural sinkhole phenomenon in Italian plain areas. *Natural Hazards*, 45, 145–172. https://doi.org/10.1007/s11069-007-9165-7

Carpenter, R. H. S. (1977). *Movements of the Eyes*. Pion, London.

Cartwright, W., Peterson, M. P., Gartner, G. (2007). *Multimedia Cartography*. Springer-Verlag, Berlin, Heidelberg.

Cassin, B., Solomon, S. (1990). *Dictionary of Eye Terminology*. Triad Publishing Company, Gainesville, FL.

Casti, E. (2005). Towards a theory of interpretation: Cartographic semiosis. *Cartographica*, 40(3), 1–16. https://doi.org/10.3138/M4M1-R663-32V2-W151

Castro-Alonso, J. C., Wong, M., Adesope, O. O., Ayres, P., Paas, F. (2019). Gender imbalance in instructional dynamic vs. static visualizations: A meta-analysis. *Educational Psychology Review*, 31, 361–387. https://doi.org/10.1007/s10648-019-09469-1

Cavanagh, P. (1991). What's up in top-down processing? In: Gorea, A. (ed) *Representations of Vision: Trends and Tacit Assumptions in Vision Research*, pp. 295–304, Cambridge University Press, Cambridge.

Cawthon, N., Moere, A. V. (2007). The effect of aesthetic on the usability of data visualization. *Proceeding of the 11th International Conference Information Visualization*, 4–6th July, Switzerland, Zurich. https://doi.org/10.1109/IV.2007.147

Cheng, B., Lin, E., Wunderlich, A., Gramann, K., Fabrikant, S. I. (2023). Using spontaneous eye blink-related brain activity to investigate cognitive load during mobile map-assisted navigation. *Frontiers in Neuroscience*, 17, 1024583, https://doi.org/10.3389/fnins.2023.1024583

Chevalier, F., Dragicevic, P., Franconeri, S. (2014). The not-so-staggering effect of staggered animated transitions on visual tracking. *IEEE Transactions on Visualization and Computer Graphics*, 20(12), 2241–2250. https://doi.org/10.1109/TVCG.2014.2346424

Chun, M. M., Wolfe, J. M. (2005). Visual attention. In: Goldstein, E. B. (ed) *Blackwell Handbook of Sensation and Perception*, pp. 272–310. Blackwell Publishing Ltd, Malden, MA. https://doi.org/10.1002/9780470753477.ch9

Cinnamon, J., Rinner, C., Cusimano, M. D., Marshall, S., Bekele, T., Hernandez, T., Glazier, R. H., Chipman, M. L. (2009). Evaluating web-based static, animated and interactive maps for injury prevention. *Geospatial Health*, 4(1), 3–16. https://doi.org/10.4081/gh.2009.206

Clifford, S., Jerit, J. (2014). Is there a cost to convenience? An experimental comparison of data quality in laboratory and online studies. *Journal of Experimental Political Science*, 1(2), 120–131. https://doi.org/10.1017/xps.2014.5

Çöltekin, A., Fabrikant, S. I., Lacayo, M. (2010). Exploring the efficiency of users' visual analytics strategies based on sequence analysis of eye movement recordings. *International Journal of Geographical Information Science*, 24(10), 1559–1575. https://doi.org/10.1080/13658816.2010.511718

Çöltekin, A., Francelet, R., Richter, K.-F., Thoresen, J., Fabrikant, S. I. (2017). The effects of visual realism, spatial abilities, and competition on performance in map-based route learning in men. *Cartography and Geographic Information Science*, 45(4), 339–353. https://doi.org/10.1080/15230406.2017.1344569

Couclelis, H. (1998). Worlds of information: The geographic metaphor in the visualization of complex information. *Cartography and Geographic Information Systems*, 25(4), 209–220. https://doi.org/10.1559/152304098782383034

Cua, G., Wald, D. J., Allen, T. I., Garcia, D., Worden, C. B., Gerstenberger, M., Lin, K., Marano, K. (2010). "Best Practices" for Using Macroseismic Intensity and Ground Motion-Intensity Conversion Equations for Hazard and Loss Models in GEM1. GEM Technical Report 2010-4, GEM Foundation, Pavia, Italy.

Cunty, C., Mathian, H. (2017). Les pratiques de cartographie animée pour représenter le changement. *Mappe Monde*, 120, 1–21. https://doi.org/10.4000/mappemonde.2602

Cybulski, P. (2014a). Rotating point symbols on animated maps for the presentation of quantitative data. *KN – Journal of Cartography and Geographic Information*, 64, 198–203. https://link.springer.com/article/10.1007/BF03544165

Cybulski, P. (2014b). Prezentacja dynamiki stężenia fosforu całkowitego w jeziorach reperowych w Polsce w latach 2000-2008 za pomocą sygnatur animowanych [Presenting the dynamics of the concentration of the total phosphorus in the benchmark lakes in poland in the years 2000-2008 using animated signatures]. *Badania Fizjograficzne, Seria A*, 65, 85–94. https://doi.org/10.14746/bgf.2014.5.7

Cybulski, P. (2016). Design rules and practices for animated maps online. *Journal of Spatial Science*, 61(2), 461–471. https://doi.org/10.1080/14498596.2016.1147394

Cybulski, P. (2020). Spatial distance and cartographic background complexity in graduated point symbol map-reading task. *Cartography and Geographic Information Science*, 47(3), 244–260. https://doi.org/10.1080/15230406.2019.1702102

Cybulski, P. (2021). Effectiveness of memorizing an animated route—Comparing satellite and road map differences in the eye-tracking study. *ISPRS International Journal of Geo-Information*, 10(3), 159. https://doi.org/10.3390/ijgi10030159

Cybulski, P. (2022a). Preattentive features of cartographic symbols in animated mapping. *Abstracts of the International Cartographic Association*, 5, 97. https://doi.org/10.5194/ica-abs-5-97-2022

Cybulski, P. (2022b). An empirical study on the effects of temporal trends in spatial patterns on animated choropleth maps. *ISPRS International Journal of Geo-Information*, 11(5), 273. https://doi.org/10.3390/ijgi11050273

Cybulski, P. (2024). Animating cartographic meaning: unveiling the impact of pictorial symbol motion speed in preattentive processing. *ISPRS International Journal of Geo-Information*, 13(4), 118. https://doi.org/10.3390/ijgi13040118

Cybulski, P., Krassanakis, V. (2021). The role of the magnitude of change in detecting fixed enumeration units on dynamic choropleth maps. *The Cartographic Journal*, 58(3), 251–267. https://doi.org/10.1080/00087041.2020.1842146

Cybulski, P., Krassanakis, V. (2023). Motion velocity as a preattentive feature in cartographic symbolization. *Journal of Eye Movement Research*, 16(4), 1–14. https://doi.org/10.16910/jemr.16.4.1

Cybulski, P., Medyńska-Gulij, B. (2018). Cartographic redundancy in reducing change blindness in detecting extreme values in spatio-temporal maps. *ISPRS International Journal of Geo-Information*, 7(1), 8. https://doi.org/10.3390/ijgi7010008

Cybulski, P., Medyńska-Gulij, B., Horbiński, T. (2023). Users' visual experience during temporal navigation in forecast weather maps on mobile devices. *Journal of Geovisualization and Spatial Analysis*, 7, 32. https://doi.org/10.1007/s41651-023-00160-2

Cybulski, P., Wielebski, Ł. (2019). Effectiveness of dynamic point symbols in quantitative mapping. *The Cartographic Journal*, 56(2), 146–160. https://doi.org/10.1080/00087041.2018.1507183

Dawidowicz, A., Sońta, M. (2019). Comparative analysis of selected European national geoportals. *Acta Scientiarum Polonorum Administratio Locorum*, 13(2). https://czasopisma.uwm.edu.pl/index.php/aspal/article/view/4208

De Cock, L., Van de Weghe, N., Ooms, K., Saenen, I., Van Kets, N., Van Wallendael, G., Lambert, P., De Maeyer, P. (2022). Linking the cognitive load induced by route instruction types and building configuration during indoor route guidance, a usability study in VR. *International Journal of Geographical Information Science*, 36(10), 1978–2008. https://doi.org/10.1080/13658816.2022.2032080

DiBiase, D., MacEachren, A. M., Krygier, J. B., Reeves, C. (1992). Animation and the role of map design in scientific visualization. *Cartography and Geographic Information Systems*, 19(4), 201–214. https://doi.org/10.1559/152304092783721295

Dillman, D. A., Bowker, D. K. (2001). The web questionnaire challenge to survey methodologists. In: Reips, U. D., Bosnjak, M. (eds) *Dimensions of Internet Science*, pp. 53–72. Pabst Science Publishers, Lengerich.

Dobson, M. W. (1979). The influence of map information on fixation localization. *The American Cartographer*, 6(1), 51–65. https://doi.org/10.1559/152304079784022772

Dong, W., Jiang, Y., Zheng, L., Liu, B., Meng, L. (2018a). Assessing map-reading skills using eye tracking and bayesian structural equation modeling. *Sustainability*, 10(9), 3050. https://doi.org/10.3390/su10093050

Dong, W., Liao, H., Roth, R. E., Wang, S. (2014b). Eye tracking to explore the potential of enhanced imagery basemaps in web mapping. *The Cartographic Journal*, 51(4), 313–329. https://doi.org/10.1179/1743277413Y.0000000071

Dong, W., Liao, H., Xu, Liu, Z., Zhang, S. B. (2014a). Using eye tracking to evaluate the usability of animated maps. *Science China Earth Scienes*, 57, 512–522. https://doi.org/10.1007/s11430-013-4685-3

Dong, W., Ran, J., Wang, J. (2012). Effectiveness and efficiency of map symbols for dynamic geographic information visualization. *Cartography and Geographic Information Science*, 39(2), 98–106. https://doi.org/10.1559/1523040639298

Dong, W., Wu, Y., Qin, T., Bian, X., Zhao, Y., He, Y., Xu, Y., Yu, C. (2021). What is the difference between augmented reality and 2D navigation electronic maps in pedestrian wayfinding? *Cartography and Geographic Information Science*, 48(3), 225–240. https://doi.org/10.1080/15230406.2021.1871646

Dong, W., Zheng, L., Meng, L. (2018b). Using eye tracking to explore differences in map-based spatial ability between geographers and non-geographers. *ISPRS International Journal of Geo-Information*, 7(9), 337. https://doi.org/10.3390/ijgi7090337

Dorling, D. (1992). Stretching space and splicing time: From cartographic animation to interactive visualization. *Cartography and Geographic Information Systems*, 19(4), 215–227. https://doi.org/10.1559/152304092783721259

Dorling, D., Openshaw, S. (1992). Using computer animation to visualize space-time patterns. *Environment and Planning B: Urban Analytics and City Science*, 19(6), 639–650. https://journals.sagepub.com/doi/10.1068/b190639

Douglas, J. (2003). Earthquake ground motion estimation using strong-motion records: A review of equations for the estimation of peak ground acceleration and response spectral ordinates. *Earth-Science Reviews*, 61(1-2), 43–104. https://doi.org/10.1016/S0012-8252(02)00112-5

Duchowski, A. T. (2017). *Eye Tracking Methodology*, 3rd ed. Springer Cham. https://doi.org/10.1007/978-3-319-57883-5

Dukaczewski, D. (2014). Designing simple and complex animated maps for users from different age groups, employing the appropriate selection of static and dynamic visual and sound variables. In: Bandrova, T., Konecny, M., Zlatanova, S. (eds) *Thematic Cartography for Society. Lecture Notes in Geoinformation and Cartography*, pp. 47–60. Springer, Cham. https://doi.org/10.1007/978-3-319-08180-9_5

Dukaczewski, D., Bielecka, E. (2009). Analiza porównawcza krajowych geoportali w Europie. *Roczniki Geomatyki, Tom VII*, 36(6), 35–60.

Duncan, J., Humphreys, G. W. (1989). Visual search and stimulus similarity. *Psychological Review*, 96(3), 433–458. https://doi.org/10.1037/0033-295X.96.3.433

Dykes, J. A. (1997). Exploring spatial data representation with dynamic graphics. *Computers & Geosciences*, 23(4), 345–370. https://doi.org/10.1016/S0098-3004(97)00009-5

Dykes, J. A. (2002). Cartographic visualization. *Journal of the Royal Statistical Society: Series D*, 47(3), 485–497. https://doi.org/10.1111/1467-9884.00149

Dykes, J. A., MacEachren, A. M., Kraak, M.-J. (2005). Advancing geovisualizaiton. In: Dykes, J. A., MacEachren, A. M., Kraak, M.-J. (eds) *Exploring Geovisualization*, pp. 693–703. Elsevier, Amsterdam.

Edler, D., Keil, J., Tuller, M.-C., Bestgen, A.-K., Dickmann, F. (2020). Searching for the 'Right' legend: The impact of legend position on legend decoding in a cartographic memory task. *The Cartographic Journal*, 57(1). 6–17. https://doi.org/10.1080/00087041.2018.1533293

Elzakker, C. P. J. M., Wealands, K. (2007). Use and users of multimedia cartography. In: Cartwright, W., Peterson, M. P., Gartner, G. (eds) *Multimedia Cartography*, pp. 487–504. Springer, Berlin, Heidelberg. https://doi.org/10.1007/978-3-540-36651-5_34

Enns, J. T., Lleras, A. (2008). What's next? New evidence for prediction in human vision. *Trends in Cognitive Sciences*, 12(9), 327–333. https://doi.org/10.1016/j.tics.2008.06.001

Fabrikant, S. I. (2005). Towards an understanding of geovisualizaation with dynamic displays: Issues and Prospects. In: *American Association for Artificial Intelligence (AAAI) 2005 Spring Symposium Series: Reasoning with Metal and External Diagrams: Computational Modeling and Spatial Assistance*, pp. 6–11, Stanford University, Stanford, CA.

Fabrikant, S., Goldsberry, K. (2005). Thematic relevance and perceptual salience of dynamic geovisualisation displays. *Proceedings of the 22nd International Cartographic Conference*, A Coruña, Spain, 9–16 July.

Fairbairn, D., Hepburn, J. (2023). Eye-tracking in map use, map user and map usability research: What are we looking for? *International Journal of Cartography*, 9(2), 231–254. https://doi.org/10.1080/23729333.2023.2189064

Faliszewska, A. (2012). Badania czytelności kartograficznych anamorfoz powierzchni. *Polski Przegląd Kartograficzny*, 44(3), 225–238.

Fang, H., Guo, H., Song, Z., Yang, N., Wang, R., Guo, F. (2025). The influence of the relationship between landmark symbol types, annotations, and colors on search performance in mobile maps based on eye tracking. *ISPRS International Journal of Geo-Information*, 14(3), 129. https://doi.org/10.3390/ijgi14030129

Fidotta, G. (2014). Animated maps and the power of the trace. *NECSUS. European Journal of Media Studies*, 3(1), 267–298. https://dx. doi.org/10.25969/mediarep/15117

Fish, C. S. (2021). Elements of vivid cartography. *The Cartographic Journal*, 58(2), 150–166. https://doi.org/10.1080/00087041.2020.1800160

Fish, C. S., Raposo, P., Guidero, E. M., Roth, R. E. (2023). Research approaches in cartography: A preliminary review. *Abstracts of the International Cartographic Association*, 6, 63. https://dx. doi.org/10.5194/ica-abs-6-63-2023

Fish, C., Goldsberry, K. P., Battersby, S. (2011). Change blindness in animated choropleth maps: An empirical study. *Cartography and Geographic Information Science*, 38(4), 350–362. https://doi.org/10.1559/15230406384350

Fisher, P., Dykes, J., Wood, J. (1993). Map design and visualization. *The Cartographic Journal*, 30(2), 136–142. https://doi.org/10.1179/000870493787859960

Fisher, S., Lowe, R. K., Schwan, S. (2008). Effects of presentation speed of a dynamic visualization on the understanding of mechanical systems. *Applied Cognitive Psychology*, 22(8), 1126–1141. https://doi.org/10.1002/acp.1426

Flannery, J. J. (1971). The relative effectiveness of some common graduated point symbols in the presentation of quantitative data. *Cartographica*, 8(2), 96–109. https://doi.org/10.3138/J647-1776-745H-3667

Foulger, G. R., Wilson, M. P., Gluyas, J. G., Julian, B. R., Davies, R. J. (2018). Global review of human-induced earthquakes. *Earth-Science Reviews*, 178, 438–514. https://doi.org/10.1016/j.earscirev.2017.07.008

Fowler, C. M. R. (2005). *The Solid Earth: An Introduction to Global Geophysics*. Cambridge University Press, Cambridge.

Franconeri, S. L., Padilla, L. M., Shah, P., Zacks, J. M., Hullman, J. (2021). The science of visual data communication: What works. *Psychological Science in the Public Interest*, 22(3), 110–161. https://doi.org/10.1177/15291006211051956

Franke, C., Schweikart, J. (2017). Mental representation of landmarks on maps: Investigating cartographic visualization methods with eye tracking technology. *Spatial Cognition & Computation*, 17(1-2), 20–38. https://doi.org/10.1080/13875868.2016.1219912

Garde, G., Larumbe-Bergera, A., Bossavit, B., Porta S., Cabeza, R., Villanueva, A. (2021) Low-cost eye tracking calibration: A knowledge-based study. *Sensors*, 21(15), 5109. https://doi.org/10.3390/s21155109

Garlandini, S., Fabrikant, S. I. (2009). Evaluating the effectiveness and efficiency of visual variables for geographic information visualization. In: Hornsby, K. S., Claramunt, C., Denis, M., Ligozat, G. (eds) *Spatial Information Theory. COSIT 2009*. Lecture Notes in Computer Science, vol 5756, pp. 195–211. Springer, Berlin, Heidelberg. https://doi.org/10.1007/978-3-642-03832-7_12

Gar-on, Y., Shuangming, Z., Jianya, G., Xinyan, Z. (2001). Design and implementation of dynamic symbols in dynamic GIS. *Geo-spatial Information Science*, 4, 32–36. https://doi.org/10.1007/BF02826575

Gates, A. E., Ritchie, D. (2006). *Encyclopedia of Earthquakes and Volcanoes*, 3rd ed. Infobase Publishing, New York.

Ge, W., Zhang, J., Jiang, S., Shi, X., Zhou, Y. (2025b). Effects of dynamic visual coding of point symbols in map-based information visualization design: an eye-tracking study. In: Mori, H., Asahi, Y. (eds) *Human Interface and the Management of Information*, vol. 15773. Springer, Cham. https://doi.org/10.1007/978-3-031-93819-1_3

Ge, W., Zhang, J., Shi, X., Tang, W., Qian, L. (2025a). Effect of dynamic point symbol visual coding on user search performance in map-based visualizations. *ISPRS International Journal of Geo-Information*, 14(8), 305. https://doi.org/10.3390/ijgi14080305

Gersmehl, P. J. (1990). Choosing tools: Nine metaphors of four-dimensional cartography. *Cartographic Perspectives*, 5, 3–16. https://doi.org/10.14714/CP05.1122

Gigerenzer, G., Hoffrage, U., Kleinbölting, H. (1991). Probabilistic mental models: A Brunswikian theory of confidence. *Psychological Review*, 98(4), 506–528. https://doi.org/10.1037/0033-295X.98.4.506

Gleicher, M. (2018). Considerations for visualizing comparison. *IEEE Transactions on Visualization and Computer Graphics*, 24(1), 413–423. https://doi.org/10.1109/TVCG.2017.2744199

Goldsberry, K. P., Battersby, S. E. (2009). Issues of change detection in animated choropleth maps. *Cartographica*, 44(3), 201–215. https://doi.org/10.3138/carto.44.3.201

Gołębiowska, I., Opach, T., Rød, J. K. (2017). For your eyes only? Evaluating a coordinated and multiple views tool with a map, a parallel coordinated plot and a table using an eye-tracking approach. *International Journal of Geographical Information Science*, 31(2), 237–252. https://doi.org/10.1080/13658816.2016.1191636

Gołębiowska, I., Opach, T., Rød, J. K. (2020). Breaking the eyes: How do users get started with a coordinated and multiple view geovisualization tool? *The Cartographic Journal*, 57(3), 235–248. https://doi.org/10.1080/00087041.2019.1660513

Graniczny, M., Mizerski, W. (2017). *Geozagrożenia*. PWN, Warszawa.

Gregory, R. L. (1990). *Eye and Brain: The Psychology of Seeing*. Princeton University Press, Princeton, NJ.

Griffin, A. L., MacEchren, A. M., Hardisty, F., Steiner, E., Li, B. (2006). A comparison of animated maps with static small-multiple maps for visually identifying space-time clusters. *Annals of the Association of American Geographers*, 96(4), 740–753. https://doi.org/10.1111/j.1467-8306.2006.00514.x

Griffin, A. L., Robinson, A. C., Roth, R. E. (2017). Envisioning the future of cartographic research. *International Journal of Cartography*, 3(sup1), 1–8. https://doi.org/10.1080/23729333.2017.1316466

Griffin, D., Tversky, A. (1992). The weighing of evidence and the determinants of confidence. *Cognitive Psychology*, 24(3), 411–435. https://doi.org/10.1016/0010-0285(92)90013-R

Guorui, Z., Zhiyong, X., Xiaofang, W. (2006). Design of dynamic map symbol based on multi-transform assembly. *Journal of Geomatics and Information Science of Wuhan University*, 31(6), 548–554.

Gutenberg, B. (1945). Magnitude determination for deep-focus earthquakes. *Bulletin of the Seismological Society of America*, 35(3), 117–130. https://doi.org/10.1785/BSSA0350030117

Gutiérrez, F., Cooper, A. H., Johnson, K. S. (2008). Identification, prediction, and mitigation of sinkhole hazards in evaporite karst areas. *Environmental Geology*, 53, 1007–1022. https://doi.org/10.1007/s00254-007-0728-4

Harrower, M. (2004). A look at the history and future of animated maps. *Cartographica*, 39(3), 33–42. https://doi.org/10.3138/7MN7-5132-1MW6-4V62

Harrower, M. (2007). The cognitive limits of animated maps. *Cartographica*, 42(2), 349–357. https://doi.org/10.3138/carto.42.4.349

Harrower, M., Brewer, C. A. (2003). ColorBrewer.org: an online tool for selecting colour schemes for map. *The Cartographic Journal*, 40(1), 27–37. https://doi.org/10.1179/000870403235002042

Harrower, M., Fabrikant, S. I. (2008). The role of map animation in geographic visualization. In: Dodge, M. (eds) *Geographic Visualization: Concepts, Tools and Applications*, pp. 49–65. Wiley, Chichester, UK. https://doi.org/10.5167/uzh-8979

Healey, C. G. (1996). Choosing effective colors for data visualization. *Proceedings of Seventh Annual IEEE Visualization ,96*, San Francisco, USA, 27th October–1st November. https://doi.org/10.1109/VISUAL.1996.568118

Healey, C. G., Booth, K. S., Enns, J. T. (1995). Visualizing real-time multivariate data using preattentive processing. *ACM Transactions on Modeling and Computer Simulation*, 5(3), 190–221. https://doi.org/10.1145/217853.217855

Hegarty, M. (2004). Dynamic visualizations and learning: Getting to the difficult questions. *Learning and Instruction*, 14(3), 343–351. https://doi.org/10.1016/j.learninstruc.2004.06.007

Hegarty, M., Kriz, S. (2008). Effects of knowledge and spatial ability on learning from animation. In: Lowe, R., Schnotz, W. (eds), *Learning with Animation: Research Implications for Design*, pp. 1–27. Cambridge UP, New York.

Hegarty, M., Montello, D. R., Richardson, A. E., Ishikawa, T., Lovelace, K. (2006). Spatial abilities at different scales: Individual differences in aptitude-test performance and spatial-layout learning. *Intelligence*, 34(2), 151–176. https://doi.org/10.1016/j.intell.2005.09.005

Hegarty, M., Richardson, A. E., Montello, D. R., Lovelace, K., Subbiah, I. (2002). Development of a self-report measure of environmental spatial ability. *Intelligence*, 30(5), 425–447. https://doi.org/10.1016/S0160-2896(02)00116-2

Hennig, B. D. (2014). Gridded cartograms as a method for visualizing earthquake risk at the global scale. *Journal of Maps*, 10(2), 186–194. https://doi.org/10.1080/17445647.2013.806229

Hillstrom, A. P., Yantis, S. (1994). Visual motion and attentional capture. *Perception & Psychophysics*, 55, 399–411. https://doi.org/10.3758/BF03205298

Hohnsbein, J., Mateeff, S. (1998). The time it takes to detect changes in the speed and direction of visual motion. *Vision Research* 38(17), 2569–2573. https://doi.org/10.1016/S0042-6989(98)00014-5

Holmqvist, K., Nyström, M., Andersson, R., Dewhurst, R., Jarodzka, H., Van de Weijer, J. (2011). *Eye Tracking: A Comprehensive Guide to Methods and Measures*. Oxford University Press, Oxford.

Horbiński, T., Cybulski, P. (2018). Similarities of global web mapping services functionality in the context of responsive web design. *Geodesy and Cartography*, 67(2), 159–177. https://doi.org/10.1007/978-3-642-03294-3_3

Horbiński, T., Zagata, K. (2022). Interpretation of map symbols in the context of gamers' age and experience. *ISPRS International Journal of Geo-Information*, 11(2), 150. https://doi.org/10.3390/ijgi11020150

Jenks, G. F. (1967). The data model concept in statistical mapping. *International Yearbook of Cartography*, 7, 186–190.

Jenks, G. F., Coulson, M. (1963). Class intervals for statistical maps. *International Yearbook of Cartography*, 3, 119–134.

Jiang, B. (1996). Cartographic visualizations: Analytical and communication tools. *Cartography*, 25(2), 1–11. https://doi.org/10.1080/00690805.1996.9714027

Jiang, O., Matuk, C., Franconeri, S. (2023). How to ensure that animated data visualizations respect visual capacity limits. *Journal of Vision* 23(4), 4770. https://doi.org/10.1167/jov.23.9.4770

Jiao, Z. (2023). Evaluating techniques and design gaps between static and animated flow maps. Master Thesis, https://cartographymaster.eu/wp-content/theses/2023_Jiao_Thesis.pdf

Jobard, B., Lefer, W. (1997). The motion map: Efficient computation of steady flow animations. *Proceedings Visualization'97*, Phoenix, USA, 24 October. https://doi.org/10.1109/VISUAL.1997.663899

Joshi, A., Kale, S., Chandel, S., Pal, D. K. (2015). Likert scale: Explored and explained. *British Journal of Applied Science & Technology*, 7(4), 396–403. https://doi.org/10.9734/BJAST/2015/14975

Julész, B. (1981). A theory of preattentive texture discrimination based on first-order statistics of textons. *Biological Cybernetics* 41, 131–138. https://doi.org/10.1007/BF00335367

Kanamori, H. (1977). The energy release in great earthquakes. *Journal of Geophysical Research*, 82(20), 2981–2987. https://doi.org/10.1029/JB082i020p02981

Kapaj, A., Lanini-Maggi, S., Hilton, C., Cheng, B., Fabrikant, S. I. (2023). How does the design of landmarks on a mobile map influence wayfinding experts' spatial learning during a real-world navigation task? *Cartography and Geographic Information Science*, 50(2), 197–213. https://doi.org/10.1080/15230406.2023.2183525

Karl, D. (1992). Cartographic animation: Potential and research issues. *Cartographic Perspectives*, 13, 3–9. https://doi.org/10.14714/CP13.999

Kasprowski, P., Herężlak, K., Stasch, M. (2014). Guidelines for the eye tracking calibration using points of regards. In: Piętka, E., Kawa, J., Wieclawek, W. (eds) *Information Technologies in Biomedicine*, vol 4, pp. 225–236. Springer, Cham. https://doi.org/10.1007/978-3-319-06596-0_21

Keil, J., Dickmann, F., Kuchinke, L. (2024). Effects of landmark position and design in VGI-based maps on visual attention and cognitive processing. In: Burghardt, D., Demidova, E., Keim, D. A. (eds) *Volunteered Geographic Information*, pp. 161–183. Springer, Cham. https://doi.org/10.1007/978-3-031-35374-1_8

Keskin, M., Ooms, K., Dogru, A. O., De Maeyer, P. (2018). Digital sketch maps and eye tracking statistics as instruments to obtain insights into spatial cognition. *Journal of Eye Movement Research*, 11(3), 4. https://doi.org/10.16910/jemr.11.3.4

Keskin, M., Ooms, K., Dogru, A. O., De Maeyer, P. (2019). EEG & eye tracking user experiments for spatial memory task on maps ISPRS *International Journal of Geo-Information*, 8(12), 546. https://doi.org/10.3390/ijgi8120546

Keskin, M., Ooms, K., Dogru, A. O., De Maeyer, P. (2020). Exploring the cognitive load of expert and novice map users using EEG and eye tracking. *ISPRS International Journal of Geo-Information* 9(7), 429. https://doi.org/10.3390/ijgi9070429

Khan, N., Rahman, A. U. (2018). Rethinking the mini-map: A navigational aid to support spatial learning in urban game environment. *International Journal of Human-Computer Interaction*, 34(12), 1135–1147. https://doi.org/10.1080/10447318.2017.1418804

Kiefer, P., Giannopoulos, I., Duchowski, A., Raubal, M. (2016). Measuring cognitive load for map tasks through pupil diameter. In: Miller, J., O'Sullivan, D., Wiegand, N. (eds) *Geographic Information Science. GIScience 2016*. Lecture Notes in Computer Science, vol. 9927, pp. 323–337, Springer, Cham. https://doi.org/10.1007/978-3-319-45738-3_21

Kiefer, P., Giannopoulos, I., Raubal, M. (2013). Using eye movements to recognize activities on cartographic maps. *SIGSPATIAL'13: Proceedings of the 21st ACM SIGSPATIAL International Conference on Advances in Geographic Information Systems*, Orlando Florida, November 5–8, pp. 488–491. https://doi.org/10.1145/2525314.2525467

Kiefer, P., Giannopoulos, I., Raubal, M., Duchowski, A. (2017). Eye tracking for spatial research: Cognition, computation, challenges. *Spatial Cognition & Computation*, 17(1-2), 1–19. https://doi.org/10.1080/13875868.2016.1254634

Kiik, A., Nyström, M., Harrie, L. (2017). Cartographic design matters – A comparison of thematic polygon design. *The Cartographic Journal*, 54(1), 24–35. https://doi.org/10.1080/00087041.2016.1147191

Kinchla, R. A., Wolfe, J. M. (1979). The order of visual processing: "Top-down," "bottom-up," or "middle-out". *Perception & Psychophysics*, 25, 225–231. https://doi.org/10.3758/BF03202991

Koriat, A. (2012). The self-consistency model of subjective confidence. *Psychological Review*, 119(1), 80–113. https://doi.org/10.1037/a0025648

Korycka-Skorupa, J. (2000). From data to cartographic presentation forms. *Miscellanea Geographica*, 9(1), 253–262. https://doi.org/10.2478/mgrsd-2000-090130

Koussoulakou, A., Kraak, M.-J. (1992). Spatio-temporal maps and cartographic communication. *The Cartographic Journal*, 29(2), 101–108. https://doi.org/10.1179/000870492787859745

Kraak, M.-J. (1998). The cartographic visualization process: From presentation to exploration. *The Cartographic Journal*, 35(1), https://doi.org/10.1179/000870498787074100

Kraak, M.-J. (2007). Cartography and the use of animation. In: Cartwright, W., Peterson, M. P., Gartner, G. (eds) *Multimedia Cartography*, pp. 317–326. Springer-Verlag, Berlin, Heidelberg. https://doi.org/10.1007/978-3-540-36651-5_22

Kraak, M.-J. (2025). Maps and changes in time representation. *International Journal of Cartography*, 11(2), 256–263. https://doi.org/10.1080/23729333.2025.2464404

Kraak, M.-J., Edsall, R., MacEachren, A. M. (1997). Cartographic animation and legends for temporal maps: Exploration and or interaction. *Proceedings of the 18th International Cartographic Conference*, Stockholm, Sweden, 23–27 June, pp. 253–261.

Krassanakis, V. (2013). Exploring the map reading process with eye movement analysis. *Proceeding of 1st International Workshop on Eye Tracking for Spatial Research*, pp. 2–7, Scarborough, UK, 2nd September.

Krassanakis, V., Cybulski, P. (2021). Eye tracking research in cartography: Looking into the future. *ISPRS International Journal of Geo-Information*, 10(6), 411. https://doi.org/10.3390/ijgi10060411

Krassanakis, V., Filippakopoulou, V., Nakos, B. (2016). Detection of moving point symbols on cartographic backgrounds. *Journal of Eye Movement Research*, 9(2). https://doi.org/10.16910/jemr.9.2.2

Kriz, S., Hegarty, M. (2007). Top-down and bottom-up influences on learning from animations. *International Journal of Human-Computer Studies*, 65(11), 911–930. https://doi.org/10.1016/j.ijhcs.2007.06.005

Kruger, J., Dunning, D. (1999). Unskilled and unaware of it: How difficulties in recognizing one's own incompetence lead to inflated self-assessments. *Journal of Personality and Social Psychology*, 77(6), 1121–1134. https://doi.org/10.1037/0022-3514.77.6.1121

Kwon, O. S., Tadin, D., Knill, D. C. (2015). Unifying account of visual motion and position perception. *Proceedings of the National Academy of Sciences*, 112(8), 8142–8147. https://doi.org/10.1073/pnas.1500361112

Lai, P.-C., Yeh, A. G.-O. (2004). Assessing the effectiveness of dynamic symbols in cartographic communication. *The Cartographic Journal*, 41(3), 229–244. https://doi.org/10.1179/000870404X13300

Land, M. F. (2009). Vision, eye movement, and natural behavior. *Visual Neuroscience*, 26(1), 51–62. https://doi.org/10.1017/S0952523808080899

Lepetiuk, V., Ostroukh, V. (2021). Methodological approaches to creation of educational electronic cartographic guides. *The Cartographic Journal*, 58(2), 196–205. https://doi.org/10.1080/00087041.2021.1930668

Liao, H., Dong, W., Peng, C., Liu, H. (2017). Exploring differences of visual attention in pedestrian navigation when using 2D maps and 3D geo-browsers. *Cartography and Geographic Information Science*, 44(6), 474–490. https://doi.org/10.1080/15230406.2016.1174886

Lichtenstein, S., Fischhoff, B., Phillips, L. D. (1982). Calibration of probabilities: The state of the art to 1980. In: Kahneman, D., Slovic, P., Tversky, A. (eds) *Judgment under Uncertainty: Heuristics and Biases*, pp. 306–334. Cambridge University Press. https://doi.org/10.1017/CBO9780511809477.023

Liu, X., Zhou, Y., Zhang, Y. (2025). The effects of landmark salience on drivers' spatial cognition and takeover performance in autonomous driving scenarios. *Behavioral Sciences*, 15(7), 966. https://doi.org/10.3390/bs15070966

Liu, Z., Heer, J. (2014). The effects of interactive latency on exploratory visual analysis. *IEEE Transactions on Visualization and Computer Graphics*, 20(12), 2122–2131. https://doi.org/10.1109/TVCG.2014.2346452

Lloyd, R. (1997). Visual search processes used in map reading. *Cartographica*, 34(1), https://doi.org/10.3138/F342-1217-827R-0777

Lloyd, R. E. (2005a). Attention on maps. *Cartographic Perspectives*, 52, 28–57. https://doi.org/10.14714/CP52.377

Lloyd, R. E. (2005b). Assessment of simulated cognitive maps: The influence of prior knowledge from cartographic maps. *Cartography and Geographic Information Science*, 32(3), 161–180. https://doi.org/10.1559/1523040054738963

Lobben, A. (2003). Classification and application of cartographic animation. *The Professional Geographer*, 55(3), 318–328. https://doi.org/10.1111/0033-0124.5503016

Lobben, A. (2008). Influence of data properties on animated maps. *Annals of the Association of American Geographers*, 98(3), 583–603. https://doi.org/10.1080/00045600802046577

Lowe, R. (2019). Designing animated information graphics: Challenges and opportunities. *Information Design Journal*, 25(3), 300–306. https://doi.org/10.1075/idj.25.3.07low

Lowe, R. K., Boucheix, J.-M. (2016). Principled animation design improves comprehension of complex dynamics. *Learning and Instruction*, 45(1), 72–84. https://doi.org/10.1016/j.learninstruc.2016.06.005

Łucjan, K. (2016). Perception of the contents of animated maps. *Polish Cartographical Review*, 48(4), 149–160. https://doi.org/10.1515/pcr-2016-0015

Łucjan, K., Wojtanowicz, P. (2018). Continuous or discontinuous? Empirical study on animated maps. *Polish Cartographical Review*, 50(3), 127–140. https://doi.org/10.2478/pcr-2018-0008

Lundeberg, M. A., Fox, P. W., Punćcohaŕ, J. (1994). Highly confident but wrong: Gender differences and similarities in confidence judgments. *Journal of Educational Psychology*, 86(1), 114–121. https://doi.org/10.1037/0022-0663.86.1.114

MacEachren, A. M. (1992). Visualizing uncertain information. *Cartographic Perspectives*, 13, 10–19. https://doi.org/10.14714/CP13.1000

MacEachren, A. M. (1994). Time as a cartographic variable. In: Hearnshaw, H., Unwin D. J. (eds) *Visualization in Geographical Information Systems*, pp. 115–130. John Wiley and Sons, Chichester, UK.

MacEachren, A. M. (1995). *How Maps Work? Representation, Visualization and Design*. The Guilford Press, New-York, London.

MacEachren, A. M., DiBiase, D. (1991). Animated maps of aggregate data: conceptual problems. *Cartography and Geographic Information Systems*, 18(4), 221–229. https://doi.org/10.1559/152304091783786790

MacEachren, A. M., Kraak, M.-J. (1997). Exploratory cartographic visualization: Advancing the agenda. *Computers & Geosciences*, 24(4), 335–343. https://doi.org/10.1016/S0098-3004(97)00018-6

MacEachren, A. M., Kraak, M.-J. (2001). Research challenges in geovisualization. *Cartography and Geographic Information Science*, 28(1), 3–12. https://doi.org/10.1559/152304001782173970

Maggi, S., Fabrikant, S. I., Imbert, J.-P., Hurter, C. (2016). How do display design and user characteristics matter in animations? An empirical study with air traffic control displays. *Cartographica*, 51(1), 25–37. https://doi.org/10.3138/cart.51.1.3176

Maggi, S.-L., Ruginski, I. T., Shipley, T. F., Hurter, C., Duchowski, A. T., Briesemeister, B. B., Lee, J., Fabrikant, S. I. (2021). Assessing how visual search entropy and engagement predict performance in a multiple-objects tracking air traffic control task. *Computers in Human Behavior Reports*, 4, 100127. https://doi.org/10.1016/j.chbr.2021.100127

Mann, C., Steward, F. (2000). *Internet* Communication and *Qualitative Research: A Handbook for Researching Online*. Sage Publications, Thousand Oaks, CA.

McManus, I. C. (1999). Eye-dominance, writing hand, and throwing hand. *Laterality*, 4(2), 173–192. https://doi.org/10.1080/713754334

Medyńska-Gulij, B., Forrest, D., Cybulski, P. (2021). Modern cartographic forms of expression: The renaissance of multimedia cartography. *ISPRS International Journal of Geo-Information*, 10(7), 484. https://doi.org/10.3390/ijgi10070484

Medyńska-Gulij, B., Forrest, D., Kersten, T. P. (2023). Cartography and geomedia in pragmatic dimension. *ISPRS International Journal of Geo-Information*, 12(8), 326. https://doi.org/10.3390/ijgi12080326

Meyer, K., Rasch, T., Schnotz, W. (2010). Effects of animation's speed of presentation on perceptual processing and learning. *Learning and Instruction*, 20(2), 136–145. https://doi.org/10.1016/j.learninstruc.2009.02.016

Michaelidou, E., Filippakopoulou, V., Nakos, B., Petropoulou, A. (2005). Designing point map symbols: The effect of preattentive attribute of shape. *Proceedings of the 22nd International Cartographic Conference*, La Coruna, Spain, 9–16 July.

Midtbø, T. (2003). Multivariable "seeing maps" through interactive animations. *Proceedings of the 21st International Cartographic Conference*, Durban, South Africa, 10–16 August.

Midtbø, T., Baroš, R. (2005). Interactive cartographic animations – Analysis functionality in a web environment. *Proceedings of the 22nd International Cartographic Conference*, La Coruna, Spain, 9–16 July.

Midtbø, T., Clarke, K., Fabrikant, S. I. (2007). Human interaction with animated maps: The portrayal of the passage of time. *Proceedings of the 11th Scandinavian Research Conference on Geographical Information Science*, Ås, Norway, 5th–7th September, pp. 45–60.

Miller, G. A. (1956). The magical number seven, plus or minus two: Some limits on our capacity for processing information. *Psychological Review*, 63(2), 81–97. https://doi.org/10.1037/h0043158

Milsom, J. (2003). *Field Geophysics*, 3rd ed. Wiley & Sons, Chichester.

Moellering, H. (1973). The computer animated film: A dynamic cartography. *Proceedings of the ACM Annual Conference*, Atlanta Georgia, August 27–29, pp. 64–69. https://doi.org/10.1145/800192.805682

Moellering, H. (1976). The potential uses of a computer animated film in the analysis of geographical patterns of traffic crashes. *Accident Analysis & Prevention*, 8(4), 215–227. https://doi.org/10.1016/0001-4575(76)90007-5

Monmonier, M. (1990). Strategies for the visualization of geographic time-series data. *Cartographica*, 27(1), 30–45. https://doi.org/10.3138/U558-H737-6577-8U31

Monmonier, M., Gluck, M. (1994). Focus groups for design improvement in dynamic cartography. *Cartography and Geographic Information Systems*, 21(1), 37–47. https://doi.org/10.1559/152304094782563948

Montello, D. R., Fabrikant, S. I., Davies, C. (2018). Cognitive perspectives on cartography and other geographic information visualizations. In: Montello, D. R. (ed) *Handbook of Behavioral and Cognitive Geography*, pp. 177–196. Edward Elgar Publishing Cheltenham, . https://doi.org/10.4337/9781784717544.00018

Montello, D. R., Lovelace, K. L., Golledge, R. G., Self, C. M. (1999). Sex-related differences and similarities in geographic and environmental spatial abilities. *Annals of the Association of American Geographers*, 89(3): 515–534. https://doi.org/10.1111/0004-5608.00160

Moon, S., Kim, E.-K., Hwang, C.-S. (2014). Effects of Spatial distribution on change detection in animated choropleth maps. *Journal of the Korean Society of Surveying, Geodesy, Photogrammetry and Cartography*, 32(6), 571–580. https://doi.org/10.7848/ksgpc.2014.32.6.571

Morrison, J. L. (1974). A theoretical framework for cartographic generalization with emphasis on the process of symbolization. *International Yearbook of Cartography*, 14, 114–127.

Multimäki, S., Hall, A., Ahonen-Rainio, P. (2015). Comparison of temporally classified and unclassified map animations. *Cartographic Perspecitves*, 82, 5–14. https://doi.org/10.14714/CP82.1296

Navakanesh, B., Shah, A. A., Prasanna, M. V. (2019). Earthquake education through the use of documentary movies. *Frontiers in Earth Science*, 7. https://doi.org/10.3389/feart.2019.00042

Neider, M. B., Zelinksy, G. J. (2006). Searching for camouflaged targets: Effect of target-background similarity on visual search. *Vision Research*, 46(14), 2217–2235. https://doi.org/10.1016/j.visres.2006.01.006

NOAA National Centers for Environmental Information. (2022). *ETOPO 2022 15 Arc-Second Global Relief Model*. NOAA National Centers for Environmental Information. https://doi.org/10.25921/fd45-gt74. Accessed [15.10.2024].

Nothdurft, H.-C. (1993). The role of features in preattentive vision: Comparison of orientation, motion and color cues. *Vision Research*, 33(14), 1937–1958. https://doi.org/10.1016/0042-6989(93)90020-w

Oberholzer, C., Hurni, L. (2000). Visualization of change in the interactive multimedia Atlas of Switzerland. *Computers & Geosciences*, 26(1), 37–43. https://doi.org/10.1016/S0098-3004(99)00031-X

Obuhov, L., Panidi, E. (2024). Cartographic animations, a supporting technology for education. *The International Archives of the Photogrammetry, Remote Sensing and Spatial Information Sciences*, XLVIII-5-2024, 69–74. https://doi.org/10.5194/isprs-archives-XLVIII-5-2024-69-2024

Ogao, P. J., Kraak, M.-J. (2002). Defining visualization operations for temporal cartographic animation design. *International Journal of Applied Earth Observation and Geoinformation*, 4(1), 23–31. https://doi.org/10.1016/S0303-2434(02)00005-3

Ooms, K., De Maeyer, P., Fack, V. (2014). Study of the attentive behavior of novice and expert map users using eye tracking. *Cartography and Geographic Information Science*, 41(1), 37–54. https://doi.org/10.1080/15230406.2013.860255

Ooms, K., De Maeyer, P., Fack, V., Van Assche, E., Witlox, F. (2012). Interpreting maps through the eyes of expert and novice users. *International Journal of Geographical Information Science*, 26(10), 1773–1788. https://doi.org/10.1080/13658816.2011.642801

Ooms, K., Dupont, L., Lapon, L., Popelka, S. (2015). Accuracy and precision of fixation locations recorded with the low-cost eye tribe tracker in different experimental set-ups. *Journal of Eye Movement Research*, 8(1), 1–24. https://doi.org/10.16910/jemr.8.1.5

Ooms, K., Krassanakis, V. (2018). Measuring the spatial noise of a low-cost eye tracker to enhance fixation detection. *Journal of Imaging*, 4(8), 96. https://doi.org/10.3390/jimaging4080096

Opach, T. (2006). Z problematyki wykorzystania animacji w kartografii. *Polski Przegląd Kartograficzny*, 38(4), 301–305.

Opach, T., Gołębiowska, I., Fabrikant, S. I. (2014). How do people view multi-component animated maps? *The Cartographic Journal*, 51(4), 330–342. https://doi.org/10.1179/1743277413Y.0000000049

Opach, T., Midtbø, T., Nossum, A. (2011). A new concept of multi-scenario, multi-component animated maps for the visualization of spatio-temporal landscape evolution. *Miscellanea Geographica*, 15(1), 215–229. https://doi.org/10.2478/v10288-012-0013-6

Opach, T., Nossum, A. (2011). Evaluating the usability of cartographic animations with eye-movement analysis. *Proceeding of the 25th International Cartographic Conference*, Paris, France, 3–8 July.

Ousler, G. W., Hangberg, K. W., Schindelar, M., Welch, D., Abelson, M. B. (2008). The ocular protection index. *Cornea*, 27(5), 509–513. https://doi.org/10.1097/ICO.0b013e31816583f6

Özalp, S., Kürçer, A., Özdemir, E., Duman, T. Y. (2016). The Bekten Fault: The palaeoseismic behaviour and kinematic characteristics of an intervening segment of the North Anatolian Fault Zone, Southern Marmara Region, Turkey. *Geodinamica Acta*, 28(4), 347–362. https://doi.org/10.1080/09853111.2016.1208524

Pagani, M., Garcia-Pelaez, J., Gee, R., Johnson, K., Poggi, V., Silva, V., Simionato, M., Styron, R., Viganò, D., Danciu, L., Monelli, D., Weatherill, G. (2020). The 2018 version of the Global Earthquake Model: Hazard component. *Earthquake Spectra*, 36(1), 226–251. https://doi.org/10.1177/8755293020931866

Pallier, G., Wilkinson, R., Danthiir, V., Kleitman, S., Knezevic, G., Stankov, L., Roberts, R. D. (2002). The role of individual differences in the accuracy of confidence judgments. *The Journal of General Psychology*, 129(3), 257–299. https://doi.org/10.1080/00221300209602099

Pearce, M. W. (2008). Framing the days: Place and narrative in cartography. *Cartography and Geographic Information Science*, 35(1), 17–32. https://doi.org/10.1559/152304008783475661

Peng, G., Yue, S., Li, Y., Song, Z., & Wen, Y. (2017). A procedural construction method for interactive map symbols used for disasters and emergency response. *ISPRS International Journal of Geo-Information*, 6(4), 95. https://doi.org/10.3390/ijgi6040095

Pequet, D. J. (1994). It's about time: A conceptual framework for the representation of temporal dynamics in geographic information systems. *Annals of the Association of American Geographers*, 84(3), 441–461. https://doi.org/10.1111/j.1467-8306.1994.tb01869.x

Peterson, M. P. (1993). Interactive cartographic animation. *Cartography and Geographic Information Systems*, 20(1), 40–44. https://doi.org/10.1559/152304093782616724

Peterson, M. P. (1999). Active legends for interactive cartographic animation. *International Journal of Geographical Information Science*, 13(4), 375–383. https://doi.org/10.1080/136588199241256

Piątek, R., Gołębiowska, I. (2022). Do navigation maps need a legend? Empirical assessment of the intuitiveness of point symbols on mobile maps. *Polish Cartographical Review*, 54(1), 23–34. https://doi.org/10.2478/pcr-2022-0003

Pieniążek, M., Zych, M. (2017). *Mapy statystyczne. Opracowanie i prezentacja danych.* Główny Urząd Statystyczny, Warszawa.

Pődör, A., Kiszely, M. (2014). Experimental investigation of visualization methods of earthquake catalogue maps. *Geodesy and Cartography*, 40(4), 156–162. https://doi.org/10.3846/20296991.2014.987451

Pomerantz, J. R. (2003). Wholes, holes, and basic features in vision. *Trends in Cognitive Sciences*, 7(11), 471–473. https://doi.org/10.1016/j.tics.2003.09.007

Popelka, S., Vondrakova, A., Hujnakova, P. (2019). Eye-tracking evalutaion of weather maps. *ISPRS International Journal of Geo-Information*, 8(6), 256. https://doi.org/10.3390/ijgi8060256

Pourgolmohammadgolshani, A. (2022). Development of a volcanic hazard visualisation tool for risk communication, Master Thesis, University of Canterbury. https:// doi.org/10.26021/13940

Prasannakumar, V., Shiny, R., Geetha, N., Vijith, H. (2011). Applicability of SRTM data for landform characterisation and geomorphometry: A comparison with contour-derived parameters. *International Journal of Digital Earth*, 4(5), 387–401. https://doi.org/10.1080/17538947.2010.514010

Pratt, J., Radulescu, P., Guo, R. M., Abrams, R. A. (2010). It's Alive! Animate motion captures visual attention. *Psychological Science*, 21(11), 1724–1730. https://doi.org/10.1177/0956797610387440

Qin, T., Fias, W., Van de Weghe, N., Huang, H. (2024). Recognition of map activities using eye tracking and EEG data. *International Journal of Geographical Information Science*, 38(3), 550–576. https://doi.org/10.1080/13658816.2024.2309188

Ratajski, L. (1971). Logiczno-semiotyczne zasady uporządkowania i standaryzacji znaków kartograficznych [Logical-semiotic principles of arrangement and standarization of cartographic signs]. *Polish Cartographical Review*, 3(4), 156–166.

Rautian, T. G., Khalturin, V. I., Fujita, K., Mackey, K. G., Kendall, A. D. (2007). Origins and methodology of the Russian energy K-class system and its relationship to magnitude scales. *Seismology Research Letters*, 78(6), 579–590. https://doi.org/10.1785/gssrl.78.6.579

Richter, C. F. (1935). An instrumental earthquake magnitude scale. *Bulletin of the Seismological Society of America*, 35(1), 1–32. https://doi.org/10.1785/BSSA0250010001

Riva, G., Teruzzi, T., Anolli, L. (2004). The use of the internet in psychological research: comparison of online and offline questionnaires. *CyberPsychology & Behavior*, 6(1), 73–80. https://doi.org/10.1089/109493103321167983

Robinson, A. C. (2011). Highlighting in geovisualization. *Cartography and Geographic Information Science*, 38(4), 373–383. https://doi.org/10.1559/15230406384373

Robinson, A. C. (2014). Reflections on 'ColorBrewer.org: An Online Tool for Selecting Colour Schemes for Maps'. In: Kent, A. (ed) *Landmarks in Mapping*. Routledge, London. https://doi.org/10.4324/9781351191234

Robinson, D. A. (1968). The oculomotor control system: A review. *Proceedings of the IEEE,* 56(6), 1032–1049. https://doi.org/10.1109/PROC.1968.6455

Rød, J. (2004). Cartographic signs and arbitrariness. *Cartographica*, 39(4), 27–36. https://doi.org/10.3138/4462-2125-1312-217T

Roth, R. E. (2012). Cartographic interaction primitives: Framework and synthesis. *The Cartographic Journal*, 49(4), 376–395. https://doi.org/10.1179/1743277412Y.0000000019

Roth, R. E. (2013a). An empirically-derived taxonomy of interaction primitives for interactive cartography and geovisualization. *IEEE Transactions on Visualization and Computer Graphics*, 19(12), 2356–2365. https://doi.org/10.1109/TVCG.2013.130

Roth, R. E. (2013b). Interactive maps: What we know and what we need to know. *Journal of Spatial Information Science*, 6, 59–115. https://doi.org/10.5311/JOSIS.2013.6.105

Roth, R. E. (2015). Interactivity and cartography: A contemporary perspective on user interface and user experience design from geospatial professionals. *Cartographica*, 50(2), 94–115. https://doi.org/10.3138/cart.50.2.2427

Roth, R. E., Çöltekin, A., Delazari, L., Filho, H. F., Griffin, A., Hall, A., ... van Elzakker, C. P. J. M. (2017). User studies in cartography: Opportunities for empirical research on interactive maps and visualizations. *International Journal of Cartography*, 3(sup1), 61–89. https://doi.org/10.1080/23729333.2017.1288534

Roth, R. E., Ross, K. S., MacEchren, A. M. (2015). User-centered design for interactive maps: A case study in crime analysis. *ISPRS International Journal of Geo-Information*, 4(1), 262–301. https://doi.org/10.3390/ijgi4010262

Rucci, M., Poletti, M. (2015). Control and functions of fixational eye movements. *Annual Reviews*, 1, 499–518. https://doi.org/10.1146/annurev-vision-082114-035742

Ruddro, R. A., Mohna, H. A. (2023). Visual communication in industrial safety systems: A review of Ui/Ux design for risk alerts and warnings. *American Journal of Scholarly Research and Innovation*, 3(2), 217–245. https://doi.org/10.63125/wbv4z521

Ruginski, I., Giudice, N., Creem-Regehr, S., Ishikawa, T. (2022). Designing mobile spatial navigation systems from the user's perspective: An interdisciplinary review. *Spatial Cognition & Computation*, 22(1-2), 1–29. https://doi.org/10.1080/13875868.2022.2053382

Sarın, P., Uluğtekin, N. (2019). Analyzing newspaper maps for earthquake news through cartographic approach. *ISPRS International Journal of Geo-Information*, 8(5), 235. https://doi.org/10.3390/ijgi8050235

Schiewe, J. (2019). Empirical studies on the visual perception of spatial patterns in choropleth maps. *KN – Journal of Cartography and Geographic Information*, 69, 217–228. https://doi.org/10.1007/s42489-019-00026-y

Schill, H. M., Cain, M. S., Josephs, E. L., Wolfe, J. M. (2020). Axis of rotation as a basic feature in visual search. *Attention, Perception & Psychophysics*, 82(1), 31–43. https://doi.org/10.3758/s13414-019-01834-0

Schiller, P. H., Tehovnik, E. J. (2001). Look and see: How the brain moves your eyes about. *Progress in Brain Research*, 134, 127–142. https://doi.org/10.1016/S0079-6123(01)34010-4

Schmidt-Weigand, F. (2005). Dynamic visualizations in multimedia learning: The influence of verbal explanations on visual attention, cognitive load and learning outcome. PhD Thesis, Justus-Liebig-Universität Gießen. https:// doi.org/10.22029/jlupub-15356

Schmitt, C., Klingenhoefer, S., Bremmer, F. (2018). Preattentive and predictive processing of visual motion. *Scientific Reports*, 8(1), 12399. https://doi.org/10.1038/s41598-018-30832-9

Segen, V., Avraamides, M., Slattery, T., Colombo, G., Wiener, J. M. (2021). Comparable performance on a spatial memory task in data collected in the lab and online. *PLoS One* 16(11), e0259367. https://doi.org/10.1371/journal.pone.0259367

Shebilske, W. L., Fisher, D. F. (1983). Extended discourse through the eyes: How and why. In: Groner, R., Menz, C., Fisher, D. F., Monty, R. A. (eds) *Eye Movements and Psychological Functions*, pp. 303–314. Lawrence Erlbaum, Hillsdale, NJ. https://doi.org/10.4324/9781003165538

Shortridge, B. G., Welch, R. B. (1982). The effect of stimulus redundancy on the discrimination of town size on maps. *The American Cartographer*, 9(1), 69–78. https://doi.org/10.1559/152304082783948277

Simon, D. J., Chabris, C. F. (1999). Gorillas in our midst: Sustained inattentional blindness for dynamic events. *Perception*, 28(9), 1059–1074. https://doi.org/10.1068/p281059

Skaramagkas, V., Giannakakis, G., Ktistakis, E., Manousos, D., Karatzanis, I., Tachos, N. S., Tripoliti, E., Marias, K., Fotiadis, D. I., Tsiknakis, M. (2021). Review of eye tracking metrics involved in emotional and cognitive processes. *IEEE Reviews in Biomedical Engineering*, 16, 260–277. https://doi.org/10.1109/RBME.2021.3066072

Slocum, T. A., Blok, C., Jiang, B., Koussoulakou, A., Montello, D. R., Fuhrmann, S., Hedley, N. R. (2001). Cognitive and usability issues in geovisualization. *Cartography and Geographic Information Science*, 28(1), 61–75. https://doi.org/10.1559/152304001782173998

Slocum, T. A., McMaster, R. B., Kessler, F. C., Howard, H. H. (2023). *Thematic Cartography and Geovisualization*, 4th ed. CRC Press, Boca Raton, FL. https://doi.org/10.1201/9781003150527

Slocum, T. A., Sluter, R., Kessler, F., Yoder, S. (2004). A qualitative evaluation of maptime, a program for exploring spatiotemporal point data. *Cartographica*, 39(3), https://doi.org/10.3138/92T3-T928-8105-88X7

Slocum, T. A., Yoder, S., Kessler, F., Sluter, R. (2000). MapTime: Software for exploring spatiotemporal data associated with point locations. *Cartographica*, 37(1), 15–32. https://doi.org/10.3138/T91X-1N21-5336-2R73

Stachoň, Z., Šašinka, Č., Konečný, M., Popelka, S., Lacko, D. (2020). An eye-tracking analysis of visual search task on cartographic stimuli. *Proceedings of the 8th International Conference on Cartography and GIS*, vol. 1, pp. 36–41. Bulgarian Cartographic Association, Nessebar, Bulgaria, 15–20 June.

Stankov, L., Crawford, J. D. (1997). Self-confidence and performance on tests of cognitive abilities. *Intelligence*, 25(2), 93–109. https://doi.org/10.1016/S0160-2896(97)90047-7

Stankov, L., Lee, J., Luo, W., Hogan, D. J. (2012). Confidence: A better predictor of academic achievement than self-efficacy, self-concept and anxiety? *Learning and Individual Differences*, 22(6), 747–758. https://doi.org/10.1016/j.lindif.2012.05.013

Steinke, T. R. (1987). Eye movement studies in cartography and related fields. *Cartographica*, 24(2), 40–73. https://doi.org/10.3138/J166-635U-7R56-X2L1

Štěrba, Z., Šašinka, C., Stachoň, Z. (2014). Usability testing of cartographic visualizations: principles and research design. *Proceedings of the 5th International Conference on Cartography and GIS*, Riviera, Bulgaria, June 15–21, 333–340.

Štěrba, Z., Šašinka, C., Stachoň, Z., Štampach, R., Morong, K. (2015). *Selected Issues of Experimental Testing in Cartography*. Masaryk University, Brno.

Stucchi, M. et al. (2013). The SHARE European Earthquake Catalogue (SHEEC) 1000-1899. *Journal of Seismology*, 17, 523–544. https://doi.org/10.1007/s10950-012-9335-2

Tao, D., Ren, X., Liu, K., Mao, Q., Cai, J., Wang, H. (2024). Effects of color scheme and visual fatigue on visual search performance and perceptions under vibration conditions. *Displays*, 82, 102667. https://doi.org/10.1016/j.displa.2024.102667

Tasse, D., Ankolekar, A., Hailpern, J. (2016). Getting users' attention in web apps in likable, minimally annoying ways. *Proceedings of the 2016 CHI Conference on Human Factors in Computing Systems*, San Jose California, May 7–12, pp. 3324–3334. https://doi.org/10.1145/2858036.2858174

Tatler, B. W., Wade, N. J. (2003). On nystagmus, saccades, and fixations. *Perception*, 32(2), 167–184. https://doi.org/10.1068/p3395

Thompson, L. F., Sufrace, E. A., Martin, D. L., Sanders, M. G. (2003). From paper to pixels: Moving personnel surveys to the Web. *Personnel Psychology*, 56(1), 197–227. https://doi.org/10.1111/j.1744-6570.2003.tb00149.x

Thrower, N. J. W. (1959). Animated cartography. *The Professional Geographer*, 11(6), 9–12. https://doi.org/10.1111/j.0033-0124.1959.116_9.x

Tinghua, A. (1998). Dynamic symbol and dynamic map. *Journal of Geomatics and Information Science of Wuhan University*, 23(1), 47–51.

Tlauka, M., Williams, M. D., Williamson, J. (2005). Gender differences in spatial knowledge acquired through simulated experience of a virtual shopping centre. *Journal of Environmental Psychology*, 25(1), 111–118. https://doi.org/10.1016/j.jenvp.2004.12.002

Tobler, W. R. (1970). A computer movie simulating urban growth in the Detroit region. *Economic Geography*, 46, 234–240. https://doi.org/10.2307/143141

Treisman, A. (1985). Preattentive processing in vision. *Computer Vision, Graphics, and Image Processing* 31(2), 156–177. https://doi.org/10.1016/S0734-189X(85)80004-9

Treisman, A., Gelade, D. (1980). A feature-integration theory of attention. *Cognitive Psychology* 12, 97–136. https://doi.org/10.1016/0010-0285(80)90005-5

Tse, P. U. (2008). Mapping visual attention with change blindness: New directions for a new method. *Cognitive Science*, 28(2), 241–258. https://doi.org/10.1207/s15516709cog2802_6

Tufte, E. R. (1983). *The Visual Display of Quantitative Information*. Graphics Press, Cheshire.

Tversky, B., Morrison, J. B., Betrancourt, M. (2002). Animation: Can it facilitate? *International Journal of Human-Computer Studies*, 57(4), 247–262. https://doi.org/10.1006/ijhc.2002.1017

Tynan, P. D., Sekuler, R. (1982). Motion processing in peripheral vision: Reaction time and perceived velocity. *Vision Research*, 22(1), 61–68. https://doi.org/10.1016/0042-6989(82)90167-5

Tyner, J. A. (1982). Persuasive cartography. *Journal of Geography*, 81(4), 140–144. https://doi.org/10.1080/00221348208980868

Vater, C. (2019). How selective attention affects the detection of motion changes with peripheral vision in MOT. *Heliyon*, 5(12), e02952. https://doi.org/10.1016/j.heliyon.2019.e02282

Vincent, K., Roth, R. E., Moore, S. A., Huang, Q., Lally, N., Sack, C. M., Nost, E., Rosenfeld, H. (2019). Improving spatial decision making using interactive maps: An empirical study on interface complexity and decision complexity in the North American hazardous waste trade. *Environment and Planning B: Urban Analytics and City Science*, 46(9), 1706–1723. https://doi.org/10.1177/2399808318764122

Vondrakova, A., Vozenilek, V. (2016). User issue in geovisualization. *Proceedings of the 16th International Multidisciplinary Scientific Geoconference (SGEM 2016), Informatics, Geoinformatics and Remote Sensing Conference Proceedings, SGEM 2016*, vol. III, Albena, Bulgaria.

Voyer, D., Voyer, S., Bryden, M. P. (1995). Magnitude of sex differences in spatial abilities: A meta-analysis and consideration of critical variables. *Psychological Bulletin*, 117(2), 250–270. https://doi.org/10.1037/0033-2909.117.2.250

Vuorre, M., Metcalfe, J. (2021). Measures of relative metacognitive accuracy are confounded with task performance in tasks that permit guessing. *Metacognition Learning* 17, 269–291 (2022). https://doi.org/10.1007/s11409-020-09257-1

Wang, Z. (2008). A technical note on seismic microzonation in the central United States. *Journal of Earth System Science*, 117(2), 749–756. https://doi.org/10.1007/s12040-008-0060-8

Ware, C. (2012). *Information Visualization*. 3rd ed. Elsevier. https://doi.org/10.1016/C2009-0-62432-6

Ware, C., Bobrow, R. (2004). Motion to support rapid interactive queries on node-link diagrams. *ACM Transactions on Applied Perception*, 1(1), 3–18. https://doi.org/10.1145/1008722.1008724

Ware, C., Bolan, D., Miller, R., Rogers, D. H., Ahrens, J. P. (2016). Animated versus static views of steady flow patterns. *SAP ,16: Proceedings of the ACM Symposium on Applied Perception*, Anaheim California, July 22–23, pp. 77–84. https://doi.org/10.1145/2931002.2931012

Wenclik, L., Touya, G. (2023). Where do people look at during multi-scale map tasks? *AGILE: GIScience Series*, 4, 51. https://doi.org/10.5194/agile-giss-4-51-2023

Wexler, M., Klam, F. (2001). Movement prediction and movement production. *Journal of Experimental Psychology: Human Perception and Performance*, 27(1), 48–64. https://doi.org/10.1037/0096-1523.27.1.48

White, R., McCausland, W. (2016). Volcano-tectonic earthquakes: A new tool for estimating intrusive volumes and forecasting eruptions. *Journal of Volcanology and Geothermal Research*, 309, 139–155. https://doi.org/10.1016/j.jvolgeores.2015.10.020

Wielebski, Ł., Medyńska-Gulij, B. (2019). Graphically supported evaluation of mapping techniques used in presenting spatial accessibility. *Cartography and Geographic Information Science*, 46(4), 311–333. https://doi.org/10.1080/15230406.2018.1479311

Wilkening, J., Fabrikant, S. I. (2011). How do decision time and realism affect map-based decision making? In: Hutchison, D., Kanade, T., Kittler, J., Kleinberg, J. M., Mattern, F., Mitchell, J. C., Naor, M., Nierstrasz, O., Pandu Rangan, C., Steffen, B., Sudan, M., Terzopoulos, D., Tygar, D., Vardi, M. Y., Weikum, G., Egenhofer, M., Giudice, N., Moratz, R., Worboys, M. (eds) *Lecture Notes in Computer Science. Spatial Information Theory*, vol. 6899, pp. 1–19. Springer, Berlin, Heidelberg. https://doi.org/10.1007/978-3-642-23196-4_1

Willmes, C., Weskamm, J., Baaser, U., Hinzen, K.-G., Bareth, G. (2008). SeismoGIS: A tool for the visualization of earthquake data. *Proceedings of the International Archives of the Photogrammetry, Remote Sensing and Spatial Information Sciences*, XXXVII(B8), 1239–1244.

Winn, B., Whitaker, D., Elliott, D. B., Philips, N. J. (1994). Factors affecting light-adapted pupil size in normal human subjects. *Investigative Ophthalmology & Visual Science*, 35(3), 1132–1137.

Wolfe, J. M. (1994). Guided Search 2.0: A revised model of visual search. *Psychonomic Bulletin and Review*, 1(2), 202–238. https://doi.org/10.3758/BF03200774

Wolfe, J. M. (2000). Visual attention. In: De Valois, K. K. (eds) *Seeing. Handbook of Perception and Cognition*, pp. 335–386. Academic Press Amsterdam, Netherlands, . https://doi.org/10.1016/B978-012443760-9/50010-6

Wolfe, J. M. (2020). Visual search: How do we find what we are looking for?. *Annual Review of Vision Science*, 15(6), 539–562. https://doi.org/10.1146/annurev-vision-091718-015048

Wolfe, J. M. (2021). Guided Search 6.0: An updated model of visual search. *Psychonomic Bulletin & Review*, 28, 1060–1092. https://doi.org/10.3758/s13423-020-01859-9

Wolfe, J. M., Horowitz, T. S. (2004). What attributes guide the deployment of visual attention and how do they do it? *Nature Reviews Neuroscience*, 5, 495–501. https://doi.org/10.1038/nrn1411

Woźniak, J., Rymarkiewicz, W., Łyszczak, D., Horbiński, T. (2024). Effectiveness study on the correct interpretation of point symbols using the online survey method on the example of the Witcher 3: Wild hunt. *KN – Journal of Cartography and Geographic Information*. https://doi.org/10.1007/s42489-023-00158-2

Wright, K. B. (2017). Researching internet-based populations: Advantages and disadvantages of online survey research, online questionnaire authoring software packages, and web survey services. *Journal of Computer-Mediated Communication*, 10(3), JCMC1034. https://doi.org/10.1111/j.1083-6101.2005.tb00259.x

Wu, C.-F., Gao, C., Lin, K.-C., & Chang, Y.-H. (2022). Evaluating impacts of bus route map design and dynamic real-time information presentation on bus route map search efficiency and cognitive load. *ISPRS International Journal of Geo-Information*, 11(6), 338. https://doi.org/10.3390/ijgi11060338

Wu, M., Lv, G., Zhang, K. (2020). A similarity-based approach for improving the efficiency of drawing spatiotemporal point features. *The Cartographic Journal*, 57(1), 57–69. https://doi.org/10.1080/00087041.2018.1500767

Xiaofang, W., Qingyun, D., Zhiyoung, X., Na, L. (2005). Research and design of dynamic symbols in GIS. *Proceeding of International Symposium on Spatio-temporal Modeling. Spatial Reasoning, Analysis, Data Mining and Data Fusion*, Beijing, China.

Yang, B., Li, H. (2021). A visual attention model based on eye tracking in 3D scene maps. *ISPRS International Journal of Geo-Information*, 10(1), 664. https://doi.org/10.3390/ijgi10100664

Zacks, J. M. (2004). Using movement and intentions to understand simple events. *Cognitive Science*, 28(6), 979–1008. https://doi.org/10.1207/s15516709cog2806_5

Zagata, K., Gulij, J., Halik, Ł., Medyńska-Gulij, B. (2021). Mini-map for gamers who walk and teleport in a virtual stronghold. *ISPRS International Journal of Geo-Information*, 10(2), 96. https://doi.org/10.3390/ijgi10020096

Zagata, K., Medyńska-Gulij, B. (2023). Mini-map design features as a navigation aid in the virtual geographical space based on video games. *ISPRS International Journal of Geo-Information*, 12(2), 58. https://doi.org/10.3390/ijgi12020058

Żyszkowska, W. (2000). *Semiotyczne aspekty wizualizacji kartograficznej*. Wydawnictwo Uniwersytetu Wrocławskiego, Wrocław.

Żyszkowska, W., Spallek, W., Borowicz, D. (2012). *Kartografia tematyczna*. Wydawnictwo Naukowe PWN, Warszawa.

# Index


## A

Accuracy, 9
Animation, 4, 26
  animated cartography, 23
  animation speed, 30
Area of Interest (AOI), 39

## B

Bottom-up processing, 25

## C

Calibration (eye tracking), 38
Cartography, 2, 23, 51
  cartographic communication, 6
  cartographic animation, 29
  cartographic visualization, 15
  cognitive cartography, 80
  dynamic cartography, 26
Change blindness, 43
Cognitive load, 40
Confidence (self-reported), 49

## D

Dwell time, 209
Dynamic point symbols, 2, 8

## E

Earthquake, 1, 15
  earthquake depth, 18, 19
  earthquake magnitude, 14
  earthquake map, 4
Effectiveness, 6
  map effectiveness, 6
Efficiency, 6
Eye movement, 9
Eye-tracking, 3, 8

## F

Fixation, 36
Flickering, 33

## G

Geographic Information Systems (GIS), 15

## I

Inattentional blindness, 25

## M

Map design, 5
Map variant, 76

## O

Online survey, 8

## P

Pictorial symbol, 33
  pictorial motion, 72
Preattentive processing, 30
Pulsation, 33

## R

Rotation, 26

## S

Saccades, 36
Seismic map, 1
Semiotic coherence, 33
Spatio-temporal representation, 7
Statistical significance, 102, 117

## U

Usability, 7

## V

Vibration, 33
Visual attention, 39
Visual search, 42
Visual variables, 24

For Product Safety Concerns and Information please contact our
EU representative GPSR@taylorandfrancis.com Taylor & Francis
Verlag GmbH, Kaufingerstraße 24, 80331 München, Germany